Cardiac and Vascular Biology

Volume 5

The book series gives an overview on all aspects of state-of-the-art research on the cardiovascular system in health and disease. Basic research aspects of medically relevant topics are covered and the latest advances and methods covering diverse disciplines as epigenetics, genetics, mechanobiology, platelet research or stem cell biology are featured. The book series is intended for researchers, experts and graduates, both basic and clinically oriented, that look for a carefully selected collection of high quality review articles on their respective field of expertise.

More information about this series at http://www.springer.com/series/13128

Andreas Zirlik • Christoph Bode
Meinrad Gawaz
Editors

Platelets, Haemostasis and Inflammation

Springer

Editors
Andreas Zirlik
Cardiology and Angiology I
University-Heart Center Freiburg
Freiburg, Germany

Christoph Bode
Cardiology and Angiology I
University-Heart Center Freiburg
Freiburg, Germany

Meinrad Gawaz
Department of Cardiology
University Hospital Tübingen
Tübingen, Germany

ISSN 2509-7830 ISSN 2509-7849 (electronic)
Cardiac and Vascular Biology
ISBN 978-3-319-66223-7 ISBN 978-3-319-66224-4 (eBook)
https://doi.org/10.1007/978-3-319-66224-4

Library of Congress Control Number: 2017960851

Printed on acid-free paper

This Springer imprint is published by Springer Nature
The registered company is Springer International Publishing AG
The registered company address is: Gewerbestrasse 11, 6330 Cham, Switzerland

Preface

Over the last two decades, inflammation has emerged as the key underlying pathology of a variety of diseases including but not limited to rheumatoid disorders, cancer, and cardiovascular disease. Inflammation within the vessel wall clearly promotes the nascence of atherosclerotic lesions and consequently the rise of clinical complications such as myocardial infarction and stroke. A plethora of basic and clinical evidence clearly links this inflammatory process with disease burden. We appreciate today that atherosclerotic lesions containing a multitude of inflammatory cells tend to be much weaker in composition and consistence, rendering them more prone to rupture and subsequent clinical sequelae. While the role of classic inflammatory cells and immunologic cell types has been extensively characterized throughout the last decade, it only recently became evident that nontraditional inflammatory cell types such as the platelet take a center stage in initiation, promotion, and ultimately complication of vascular inflammation. This book focuses on the platelet as a versatile cell type unraveling its role as a mediator between hemostasis and inflammation. Finally, we propose several platelet-targeting and alternate anti-inflammatory therapies as novel and promising therapeutic approaches to ultimately combat the high residual risk of cardiovascular disease in our world.

We thank our internationally renowned faculty for their outstanding contribution and wish you, our readers, joy and enlightenment with our book.

Freiburg, Germany Andreas Zirlik
Freiburg, Germany Christoph Bode
Tübingen, Germany Meinrad Gawaz

The original version of this book was revised. The correction is available at https://doi.org/10.1007/978-3-319-66224-4_18

Contents

Cardiac Imaging of Platelets and Inflammation

1

Constantin von zur Mühlen and Robin P. Choudhury

Abstract

Platelets and inflammation play a pivotal role in a wide range of cardiac pathophysiologies, such as coronary vessel atherosclerosis, ischemia/reperfusion injury, or myocarditis. Imaging of early stages of these diseases would be helpful. Molecular imaging is a promising approach for characterizing biological processes and especially atherosclerosis, which presents numerous mechanistically important targets. Inflammation and thrombus formation as key events are reflected by a wide range of potential targets, e.g., inflammatory adhesion molecules, inflammatory cells and proteases, or fibrin and platelets. Molecular imaging of these processes is possible by applying single imaging techniques, such as MRI, or the combination of different imaging modalities, such as PET and CT. In this chapter, we describe current concepts, challenges, and the future potential of molecular imaging in the context of platelets and inflammation involved in atherosclerosis.

Contents

C. von zur Mühlen (✉)
Faculty of Medicine, Heart Center Freiburg University, University of Freiburg, Freiburg, Germany
e-mail: constantin.vonzurmuehlen@universitaets-herzzentrum.de

R.P. Choudhury
Radcliffe Department of Medicine, Acute Vascular Imaging Centre, University of Oxford, Oxford, UK

A. Zirlik et al. (eds.), *Platelets, Haemostasis and Inflammation*,
Cardiac and Vascular Biology 5, https://doi.org/10.1007/978-3-319-66224-4_1

1.1 Introduction

Platelets and inflammation play a pivotal role in a wide range of cardiac pathophysiologies, such as coronary vessel atherosclerosis, ischemia/reperfusion injury, or myocarditis [1–3]. Imaging of early stages of these diseases would be helpful to manage the patient, and imaging of established disease could help to guide or optimize treatment. Various imaging techniques are available, either already clinically established or at an experimental level. In this chapter, different imaging approaches to target platelets and/or inflammation will be described, with a focus on vascular inflammation and thrombosis in atherosclerosis. We describe techniques applied in animal studies, but also in humans, and the challenging path from "bench to bedside."

1.2 Understanding Coronary Atherosclerosis: Still a Long Way to Go

Years ago the idea of atherosclerosis development was very simple. Depositions of fatty tissue, so-called "fatty streaks," already develop during early childhood [4]. Over time and with the exposure toward certain risk factors, they progress toward atherosclerotic plaques, progressively resulting in luminal narrowing and symptoms in the patient. At some point, plaque rupture occurs, and a rapid superimposed thrombosis results in immediate vessel occlusion and therefore myocardial infarction or stroke [1, 5]. However, we have learned that it is not such a linear progression of disease and that smaller and nonobstructive plaques can rupture abruptly and cause vascular occlusion [6]. Such plaques are often missed by conventional imaging techniques available in routine clinical practice. A coronary angiogram, which is routinely performed in patients with symptoms suggestive of coronary artery disease (CAD), only shows the luminal filling with contrast agent but cannot characterize the occult vascular inflammation of the vascular wall that does not result in significant luminal narrowing. In 2011, the "PROSPECT" study was published, which tried to characterize nonocclusive lesion in patients with an acute coronary syndrome (ACS) [7]. Patients with an ACS and therefore subtotal/total occlusion of a coronary vessel were treated by percutaneous coronary intervention (PCI) and stent placement at the so-called "culprit" lesion. Nonocclusive, non-culprit lesions were characterized by gray-scale and radiofrequency intravascular ultrasonographic imaging (IVUS) after PCI, and median follow-up period was 3.4 years. 20.4% of patients came back with new major adverse cardiac events (death from cardiac causes, cardiac arrest, myocardial infarction, or rehospitalization due to unstable or progressive angina). However, only 12.9% of these new events were related to the initially treated culprit lesion; the other 11.6% were related to non-culprit lesions, which were angiographically mild at baseline. These potentially "vulnerable" non-culprit lesions were characterized by a plaque burden of 70% or greater, or a minimal luminal area of 4.0 mm^2 or less, or were classified as thin-cap fibroatheromas (TCFA) by virtual histology in IVUS. However, also the combination of these different nonocclusive plaque characteristics did not result in a reliable prediction of MACE: when combining TCFA, plaque burden, and MLA, only 18.2% of patients with MACE had these characteristics present in the initial coronary angiogram.

Although these data might provide us with some prognostic information on IVUS-VH, we need other techniques to image more selectively plaque components and characteristics that might allow us to more precisely predict the fate of a coronary plaque. Especially targeting cells involved in certain stages of vascular inflammation or thrombosis by molecular imaging is an interesting and promising strategy.

1.3 Molecular Imaging: Definition, Goals, and Imaging Techniques

Molecular imaging can be defined as visualization, characterization, and noninvasive measurement of biological processes at the molecular and cellular levels in humans and other living systems [8]. This could help to accelerate and refine diagnosis, provide insights that reveal disease diversity, and monitor the effects of therapies. Molecular imaging contrast agents usually consist of two components: an antibody or peptide-mimetic targeting a certain cell or cellular receptor, conjugated toward a signal-giving carrier element. It now depends on which imaging will be used: for magnetic resonance imaging (MRI), paramagnetic chalets such as gadolinium (Gd) or superparamagnetic iron oxide particles (SPIOs) are attractive. While Gd causes a positive contrast in T1-weighted MRI, SPIO or in general iron oxide-based contrast agents result in a negative contrast due to susceptibility artifact in T2*-weighted MRI sequences [9, 10]. SPIOs are available in different sizes, e.g., as ultrasmall SPIOs (USPIOs) or microparticles of iron oxide (MPIOs). Depending upon the size and formulation, particles can be loaded with different quantities of iron and therefore have variable effects on susceptibility. These artifacts appear as black signal extinctions in T2*-weighted MRI, and 1 μm-sized MPIOs can cause signal effect extending their effective diameter on the image by a factor of 50.

When choosing a strategy of PET or SPECT for molecular imaging, radionuclides are usually conjugated with the targeting antibody or peptide mimetic. For ultrasound molecular imaging, air-filled microbubbles can be used to cause imaging artifacts.

The selection of the optimal imaging technique is crucial in molecular imaging, depending on the localization and distribution of the imaging target in the pathophysiology of interest (also see next paragraph). Each imaging technology has its advantages and disadvantages. While PET and SPECT have a very high sensitivity for molecular probes marked with radionuclides (nanogram range), the spatial resolution is usually low (PET: 1–2 mm; SPECT: 0.3–1 mm). MRI has a better spatial resolution of 50–250 μm, thereby providing important anatomical information, while molecular probes can be detected in a micro- and milligram range [11].

1.4 Imaging Targets in Vascular Inflammation and Thrombosis

Atherosclerosis is a very complex disease, involving a large number of vascular receptors, cell types, and other processes [1, 5, 8]. Usually, atherosclerosis begins with endothelial activation. Inflammatory endothelial markers such as vascular cellular adhesion molecule (VCAM) and P-selectin can be found, but also platelets adhere at early stages of atherosclerotic plaque formation [2]. This vascular inflammation attracts macrophages, which adhere to the plaque surface and finally migrate into the plaque. In the course of inflammation progression, a lipid core may develop, and proteolysis and apoptosis further promote formation of a lipid-rich necrotic core, neovessels, and formation of a fibrous cap separating this process from the bloodstream. Plaque rupture and exposure of the inflammatory core is a critical event in the pathogenesis of acute vascular syndromes since aggregation of circulating platelets and fibrin can result in immediate vessel closure and therefore acute ischemia in the remote tissue, leading to myocardial infarction in the context of a coronary artery, and stroke when a cerebral vessel such as the carotid artery is occluded. As mentioned above, growth of an atherosclerotic plaque is not a continuous sequence, and also nonobstructive plaques can rupture, particularly where there is an accumulated lipid core and cap thinning associated with local inflammation.

1.5 What to Consider When Performing Imaging Approaches in Vascular Pathologies

Multiple studies have been published over the last years describing molecular imaging approaches in atherosclerosis, involving proof of feasibility or mechanistic studies. Important factors when evaluating such approaches are the following questions: does this study provide a diagnostic value? Does it even allow a prognostic value? Or can it help to guide therapies or evaluate an outcome benefit?

When thinking about imaging studies in atherosclerosis, it is also important to consider the stage at which the imaging is performed, and if the epitopes or processes of interest are exposed superficially, and therefore readily accessible to blood-borne agents (e.g., VCAM or P-selectin), or inside plaque (e.g., macrophages, apoptosis). In this context, it is then crucial to choose the sort of imaging technique and the preparation of the contrast agent itself. It is hypothesized that macrophages can phagocytose USPIOs in the bloodstream and enter the plaque carrying the particles and the accumulated USPIO can then provide a signal from "inside" the plaque [12]. By contrast, USPIO would potentially not provide sufficient contrast to image epitopes on the plaque surface in the flowing blood, since the iron load is too low. For this purpose, MPIOs would be more attractive: although they are exposed to high shear stress in the flowing blood, even single particles are detectable by MRI due to the large signal extinction caused [13].

Other thoughts involve the differentiation of noninvasive or invasive imaging. The latter one can be performed with IVUS or optical coherence tomography (OCT), which will be described further below. Finally, the question of whether the imaging is performed on an experimental level in animals or clinically in humans is important. Not all contrast agents described in well-recognized animal imaging studies are necessarily compatible for application in humans due to issues of toxicity or biocompatibility.

1.6 Imaging of Inflammation

In the following section, exemplary studies for imaging of inflammation will be described, using different contrast agent and imaging approaches.

1.6.1 Magnetic Resonance Imaging

As already mentioned above, microparticles of iron oxide (MPIOs) can deliver high payloads of iron toward endothelial epitopes.

In a study published by our group, we performed dual targeting of MPIOs with two different markers of inflammation, imitating leukocyte binding: VCAM and p-selectin [14]. The resulting VCAM/p-selectin-MPIO contrast agent was injected into Apolipoprotein E knockout mice ($ApoE^{-/-}$), which develop atherosclerotic plaques in the ascending aorta and aortic root. Dual-targeted MPIOs, injected intravenously in vivo, bound the aortic root endothelium and were quantifiable by MRI ex vivo. MPIOs were well tolerated in vivo by all mice, with sequestration in the spleen after 24 h. This approach allowed the design of a functional MRI probe for detecting endothelial-specific markers not only in atherosclerosis but in a range of vascular pathologies [15, 16].

Also ultrasmall superparamagnetic iron oxides were used for imaging of endothelial and intraplaque markers of atherosclerosis. In a study by Burtea et al., VCAM-1 and apoptotic cell-targeted peptides were conjugated to USPIO and assessed in $ApoE^{-/-}$ mice by MRI [17]. Plaques enhanced by VCAM-targeted USPIOs contained macrophages concentrated in the cap and a large necrotic core, whereas apoptosis-targeted USPIOs produced a negative enhancement of macrophage-rich plaques inside the plaque.

As discussed above, applications in humans would be desirable, also adding another dimension, such as monitoring of therapeutical effects. One example is the ATHEROMA study, which evaluated the effects of low-dose (10 mg) and high-dose (80 mg) atorvastatin on carotid artery plaque inflammation, as measured by USPIO-enhanced MRI [18]. Twenty patients completed the full 12 weeks of treatment in each group. A significant reduction from baseline in USPIO-defined inflammation was observed in the 80-mg group at both 6 weeks and 12 weeks, whereas there was no visible effect in the low-dose regimen. Interestingly, USPIO were cycled out of the plaque region in between the imaging time points. Unfortunately, to our knowledge, no further clinical studies were performed with this agent, although this molecular imaging strategy could have been a useful biomarker imaging strategy for screening and assessment of therapeutic response to anti-inflammatory interventions in patients with atherosclerotic lesions.

1.6.2 Hybrid Imaging Approaches

Underlying the strategy to combine two different imaging techniques is the idea of getting the best out of each technology. A study by Taqueti et al. investigated the potential of imaging the relationship between markers of inflammatory activation, plaque microvascularization, and vessel wall permeability [19]. Patients with carotid artery plaques were imaged using a multimodality approach combining (1) FDG positron emission tomography (FDG-PET), (2) dynamic contrast-enhanced magnetic resonance imaging (dce-MRI), and (3) histopathology after endarterectomy in 32 subjects with carotid artery stenosis. As a result, plaque regions with active inflammation, as determined by macrophage content and major histocompatibility complex class II expression, showed increased FDG-PET uptake. This correlated with increased microvascularization and permeability, as measured by dce-MRI. Interestingly, the correlation was independent of clinical symptoms and plaque luminal severity, which might therefore be an option for detecting nonobstructive but highly vulnerable/inflamed plaques. Larger studies are desirable to confirm and further strengthen such findings.

Another approach aiming for the detection of ruptured or high-risk coronary atherosclerotic plaques has been described by Joshi et al., combining PET and CT with the radioactive tracers ^{18}F-sodium fluoride ((18)F-NaF) and ^{18}F-fluorodeoxyglucose ((18)F-FDG) [20]. Invasive coronary angiography, ^{18}F-NaF, and ^{18}F-FDG PET-CT were performed in patients with myocardial infarction and stable angina, and tissue-to-background ratios of culprit and non-culprit coronary plaques of patients with acute myocardial infarction were evaluated. Figure 1.1a shows the PET-CT of a patient with acute ST-segment elevation myocardial infarction with proximal occlusion of the left anterior descending artery on invasive coronary angiography and intense focal 18F-fluoride uptake at the site of the culprit plaque but in remote myocardium. In contrast, the corresponding 18F-fluorodeoxyglucose PET-CT image shows no uptake at the site of the culprit plaque. Another example of a patient with anterior non-ST-segment elevation

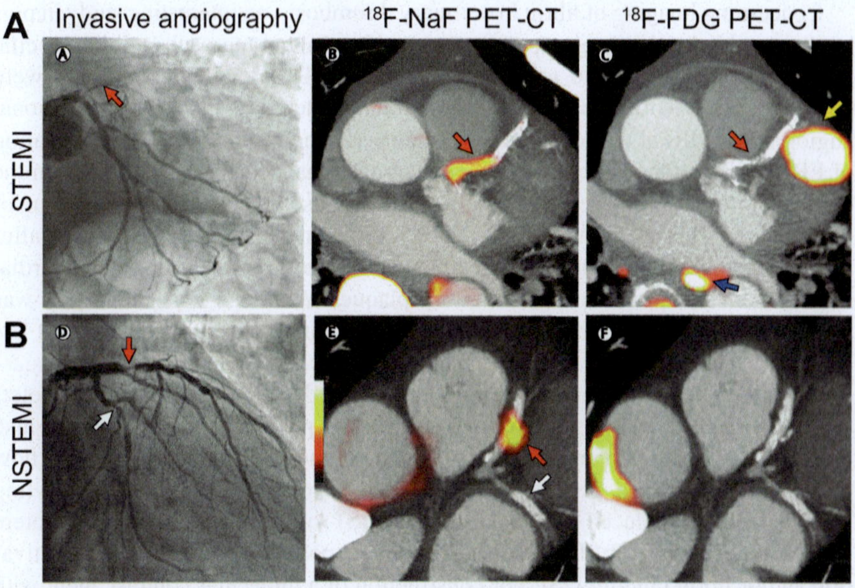

Fig. 1.1 (**a**) PET-CT of a patient with acute ST-segment elevation myocardial infarction with proximal occlusion of the left anterior descending artery on invasive coronary angiography and intense focal 18F-fluoride uptake at the site of the culprit plaque but in remote myocardium. The corresponding 18F-fluorodeoxyglucose PET-CT image shows no uptake at the site of the culprit plaque. (**b**) Example of a patient with anterior non-ST-segment elevation myocardial infarction with a culprit lesion (*red arrow*) and bystander non-culprit lesion demonstrates that only the culprit lesion had increased 18F-NaF uptake on PET-CT; the corresponding 18F-fluorodeoxyglucose PET-CT shows no uptake at either the culprit or the bystander stented lesion

myocardial infarction with a culprit lesion (red arrow) and bystander non-culprit lesion demonstrates that only the culprit lesion had increased 18F-NaF uptake on PET-CT; the corresponding 18F-fluorodeoxyglucose PET-CT shows no uptake at either the culprit or the bystander stented lesion (Fig. 1.1b). In this study, ^{18}F-NaF PET-CT was the first noninvasive imaging method to identify and localize ruptured and high-risk coronary plaque in a noninvasive way. This is also an interesting and exciting approach of how to combine two imaging modalities and get the best information from each: functional information by PET and anatomical information by CT.

1.7 Imaging of Thrombosis

Thrombosis after plaque rupture involves platelet activation and cross-linking of platelets with fibrin. Both therefore constitute a promising approach for molecular imaging of plaque rupture and atherothrombosis. Platelets are also involved into the inflammatory processes after ischemia caused by reperfusion.

Concerning imaging of fibrin in coronary thrombosis, an interesting study in pigs has been published by the group of Spuentrup [21]. In this approach, a fibrin-specific peptide was conjugated to gadolinium, called EP-2104R. Human thrombi were engineered ex vivo and delivered into coronary arteries of pigs by conventional angiography. MRI of the coronaries was then performed before and after application of EP2104R, and the gadolinium-specific signal enhancement was observed in the coronary arteries. Some years later, EP2104R was also applied in a mixed setting of clinical scenarios in humans [22]. MRI was again performed before and after application of EP2104R, thereby detecting ventricular thrombi, fibrinous pericardial effusions, or symptomatic carotid artery plaques. Although the latter study was published in 2009, so far no further or even routine application in humans has been described.

Our group has an interest in targeting activated platelets. Activated platelets can not only be found on ruptured plaques further promoting thrombosis but also on the surface of inflamed plaques, even those not resulting in coronary flow obstruction. We have previously developed a single-chain antibody targeting so-called ligand-induced binding sites (LIBS) of the activated glycoprotein IIb/IIIa receptor (GPIIb/IIIa), which becomes exposed only upon platelet activation and/or fibrin binding [23]. For construction of a molecular contrast agent with this very specific antibody, we have conjugated MPIO toward the LIBS, resulting in the LIBS-MPIO contrast agent. LIBS-MPIO has been studied by our group in various settings of murine vascular inflammation or atherothrombosis [24–28].

In one study, wall-adherent nonocclusive thrombosis was induced by ferric chloride in carotid arteries of mice, which simulated the situation of a ruptured plaque with thrombus formation [24]. MRI of the carotid arteries was performed before and after injection of LIBS-MPIO. After injection of the contrast agent, increasing signal extinction as the typical MPIO-induced effect was observed at the site of thrombosis, which was reversible after performing thrombolysis. Also in artificially induced plaque rupture in ApoE$^{-/-}$ mice, vascular thrombosis was detectable by this approach [28].

The technology was also transferred into human pathology. Patients with symptomatic carotid artery disease who underwent surgery were included in this study. The surgically removed endarterectomies were imaged before and after incubation with LIBS-MPIO, and the MPIO-induced signal void was well visible after incubation. MRI correlated well with results from histology, demonstrating specific binding of LIBS-MPIO at sites of platelet aggregation on the surface of these symptomatic plaques [24].

Of course, imaging of coronary thrombosis is the ultimate goal, but challenging in a mouse model due to the small size of the coronary vessels and the high heart rate in mice (ca. 600 bpm). We performed surgery in mice and exposed the left anterior descending artery (LAD) to ferric chloride, which again induced a wall-adherent, nonocclusive thrombosis in the vessel, similar to the approach described above with the carotid artery. Again, LIBS-MPIO was injected, but the heart removed thereafter and imaged by MRI ex vivo. MPIO-induced signal extinction was well visible inside the coronary artery and correlated well with the presence of

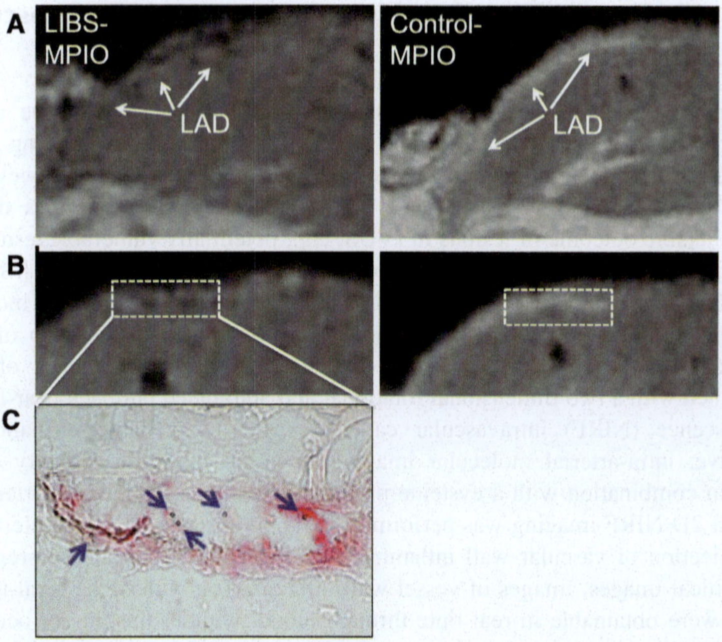

Fig. 1.2 Molecular imaging of platelets in coronary thrombosis. Nonocclusive thrombosis of the left anterior descending coronary artery was induced. LIBS-MPIO was injected, the heart removed and imaged by MRI ex vivo. MPIO-induced signal extinction was well visible inside the coronary artery and correlated well with the presence of local thrombi and platelet aggregation in histology (**a**/**c**); no signal was detectable in animals with control-MPIO injection (**b**)

local thrombi and platelet aggregation in histology (Fig. 1.2a,c); no signal was detectable in animals with control-MPIO injection (Fig. 1.2b).

Concerning inflammation in remote myocardium in the context of reperfusion after coronary ligation, LIBS-MPIO also helped to image the extent of the ischemia/reperfusion injury [29]. After ligation and reopening of the LAD, in vivo MRI of the heart and especially the area of ischemia was performed after injection of (1) LIBS-MPIO and (2) gadolinium. LIBS-MPIO-induced signal extinction correlated well with platelet aggregates on histology but also with platelet–neutrophil conjugated reflecting the severity of the myocardial inflammatory process. Gadolinium was used to detect necrotic myocardium in the same imaging approach, which worked well and confirmed the concept of a dual imaging study.

These studies seem as a promising strategy to target activated platelets in vascular thrombosis, since the antibody works in mouse and men and is less immunogenic due to its small molecular size. However, the MPIOs used for targeting are potentially toxic, and therefore research is ongoing toward the development of human compatible signal-giving elements. Magnetoliposomes appear as a promising strategy [30]; however, reliable in vivo applicability still has to be demonstrated.

1.8 Current Interventional Imaging Approaches of Vascular Wall Characterization

As already discussed above, intravascular ultrasound (IVUS) can be used to characterize coronary plaques during interventional procedures. Adding Virtual histology to IVUS can help to characterize plaque components. However, the predictive value of this technique is still not entirely clear. Studies describe unpredictable outcome of lesions in IVUS, e.g., potentially vulnerable lesions that transform into a potentially stable fibrofatty lesion, and vice versa [31]. Further, spatial resolution of IVUS is limited, which makes application for molecular imaging purposes difficult; adding another imaging technique on top of IVUS therefore seems reasonable. Jaffer et al. describe such an approach of IVUS combined with a two-dimensional rotational and automated pullback near-infrared fluorescence (NIRF) intravascular catheter apparatus, capable of nanomolar-sensitive, intra-arterial molecular imaging in larger diameter coronary arteries [32]. In combination with a cysteine protease-activatable imaging reporter, intra-arterial 2D NIRF imaging was performed in rabbit aortas with atherosclerosis for the detection of vascular wall inflammation, while IVUS provided co-registered anatomical images. Images of vessel wall inflammation with high signal-to-noise ratios were obtainable in real time through blood, without flushing or occlusion, revealing increased inflammation-regulated cysteine protease activity in atheromata and stent-induced arterial injury. Although this appears as a promising approach for molecular characterization of vascular pathologies, a transfer into human pathologies is still pending.

Another interventional imaging technique already used in clinical routine is the optical coherence tomography (OCT), using light to capture micrometer-resolution images from within optical scattering media, e.g., biological tissue. However, current concepts only allow the generation of anatomical information at a very good spatial resolution (ca. 15 μm). In one study, a dual-modality intra-arterial catheter for simultaneous microstructural and molecular imaging was applied, using a combination of optical frequency domain imaging (OFDI) and near-infrared fluorescence (NIRF) imaging [33]. By providing simultaneous molecular information in the context of the surrounding tissue microstructure, this new catheter was evaluated for investigating coronary atherosclerosis and identifying high-risk biological and structural coronary arterial plaques in rabbits. Two different targets were studied: fibrin and factors promoting plaque progression. After injection of a fluorochrome binding to fibrin, the combined imaging approach provided a molecular signal along the vascular wall with a very efficient signal-to-noise ratio, and correlation with findings in histology was excellent. The same was true for the inflammatory protease activity by using a cathepsin protease-activatable NIRF molecular beacon. Compared to the IVUS-NIRF approach, OFDI-NIRF provides higher resolution of potentially localized processes on the vascular wall, such as vascular thrombosis reflected by fibrin accumulation. Although very promising, transfer into human applicability is also pending.

1.9 Perspectives

Molecular imaging is a promising approach for characterizing biological processes and especially atherosclerosis, which presents numerous mechanistically important targets. Inflammation and thrombus formation as key events are reflected by a wide range of potential targets, e.g., inflammatory adhesion molecules, inflammatory cells and proteases, or fibrin and platelets. Molecular imaging of these processes is possible by applying single imaging techniques, such as MRI, or the combination of different imaging modalities, such as PET and CT. Currently, transfer into routine applications is ongoing but challenging, mainly due to the lack of human compatible or nontoxic signaling elements.

Compliance with Ethical Standards

Conflict of Interest: Constantin von zur Mühlen and Robin P. Choudhury declares that they have no conflict of interest.

Ethical Approval: This article does not contain any studies with human participants or animals performed by any of the authors.

References

1. Hansson GK. Inflammation, atherosclerosis, and coronary artery disease. N Engl J Med. 2005;352:1685–95.
2. Gawaz M, Stellos K, Langer HF. Platelets modulate atherogenesis and progression of atherosclerotic plaques via interaction with progenitor and dendritic cells. J Thromb Haemost. 2008;6(2):235–42.
3. Gawaz M, Langer H, May AE. Platelets in inflammation and atherogenesis. J Clin Invest. 2005;115:3378–84.
4. McGill HC, McMahan CA, Gidding SS. Are pediatricians responsible for prevention of adult cardiovascular disease? Nat Clin Pract Cardiovasc Med. 2009;6:10–1.
5. Choudhury RP, Fuster V, Fayad ZA. Molecular, cellular and functional imaging of atherothrombosis. Nat Rev Drug Discov. 2004;3:913–25.
6. Finn AV, Nakano M, Narula J, Kolodgie FD, Virmani R. Concept of vulnerable/unstable plaque. Arterioscler Thromb Vasc Biol. 2010;30:1282–92.
7. Stone GW, Maehara A, Lansky AJ, de Bruyne B, Cristea E, Mintz GS, Mehran R, McPherson J, Farhat N, Marso SP, Parise H, Templin B, White R, Zhang Z, Serruys PW. A prospective natural-history study of coronary atherosclerosis. N Engl J Med. 2011;364:226–35.
8. Choudhury RP, Fisher EA. Molecular imaging in atherosclerosis, thrombosis, and vascular inflammation. Arterioscler Thromb Vasc Biol. 2009;29:983–91.
9. Shapiro EM, Skrtic S, Koretsky AP. Sizing it up: cellular MRI using micron-sized iron oxide particles. Magn Reson Med. 2005;53:329–38.
10. Shapiro EM, Sharer K, Skrtic S, Koretsky AP. In vivo detection of single cells by MRI. Magn Reson Med. 2006;55:242–9.
11. Rudd JH, Fayad ZA. Imaging atherosclerotic plaque inflammation. Nat Clin Pract Cardiovasc Med. 2008;5(Suppl 2):S11–7.
12. Kooi ME, Cappendijk VC, Cleutjens KB, Kessels AG, Kitslaar PJ, Borgers M, Frederik PM, Daemen MJ, van Engelshoven JM. Accumulation of ultrasmall superparamagnetic particles of

iron oxide in human atherosclerotic plaques can be detected by in vivo magnetic resonance imaging. Circulation. 2003;107:2453–8.
13. Shapiro EM, Skrtic S, Sharer K, Hill JM, Dunbar CE, Koretsky AP. MRI detection of single particles for cellular imaging. Proc Natl Acad Sci USA. 2004;101:10901–6.
14. McAteer MASJ, Ali ZA, Warrick N, Bursill CA, von zur Mühlen C, Neubauer S, Channon KM, Choudhury RP. Magnetic resonance imaging of endothelial adhesion molecule expression in mouse atherosclerosis using dual-targeted microparticles of iron oxide. Arterioscler Thromb Vasc Biol. 2008;28(1):77–83.
15. McAteer MA, von Zur Muhlen C, Anthony DC, Sibson NR, Choudhury RP. Magnetic resonance imaging of brain inflammation using microparticles of iron oxide. Methods Mol Biol. 2011;680:103–15.
16. McAteer MA, Sibson NR, von Zur Muhlen C, Schneider JE, Lowe AS, Warrick N, Channon KM, Anthony DC, Choudhury RP. In vivo magnetic resonance imaging of acute brain inflammation using microparticles of iron oxide. Nat Med. 2007;13:1253–8.
17. Burtea C, Ballet S, Laurent S, Rousseaux O, Dencausse A, Gonzalez W, Port M, Corot C, Vander Elst L, Muller RN. Development of a magnetic resonance imaging protocol for the characterization of atherosclerotic plaque by using vascular cell adhesion molecule-1 and apoptosis-targeted ultrasmall superparamagnetic iron oxide derivatives. Arterioscler Thromb Vasc Biol. 2012;32:e36–48.
18. Tang TY, Howarth SP, Miller SR, Graves MJ, Patterson AJ, U-King-Im JM, Li ZY, Walsh SR, Brown AP, Kirkpatrick PJ, Warburton EA, Hayes PD, Varty K, Boyle JR, Gaunt ME, Zalewski A, Gillard JH. The atheroma (atorvastatin therapy: effects on reduction of macrophage activity) study. Evaluation using ultrasmall superparamagnetic iron oxide-enhanced magnetic resonance imaging in carotid disease. J Am Coll Cardiol. 2009;53(22):2039–50.
19. Taqueti VR, Di Carli MF, Jerosch-Herold M, Sukhova GK, Murthy VL, Folco EJ, Kwong RY, Ozaki CK, Belkin M, Nahrendorf M, Weissleder R, Libby P. Increased microvascularization and vessel permeability associate with active inflammation in human atheromata. Circ Cardiovasc Imaging. 2014;7:920–9.
20. Joshi NV, Vesey AT, Williams MC, Shah AS, Calvert PA, Craighead FH, Yeoh SE, Wallace W, Salter D, Fletcher AM, van Beek EJ, Flapan AD, Uren NG, Behan MW, Cruden NL, Mills NL, Fox KA, Rudd JH, Dweck MR, Newby DE. 18f-fluoride positron emission tomography for identification of ruptured and high-risk coronary atherosclerotic plaques: a prospective clinical trial. Lancet. 2014;383:705–13.
21. Spuentrup E, Buecker A, Katoh M, Wiethoff AJ, Parsons EC Jr, Botnar RM, Weisskoff RM, Graham PB, Manning WJ, Gunther RW. Molecular magnetic resonance imaging of coronary thrombosis and pulmonary emboli with a novel fibrin-targeted contrast agent. Circulation. 2005;111:1377–82.
22. Vymazal J, Spuentrup E, Cardenas-Molina G, Wiethoff AJ, Hartmann MG, Caravan P, Parsons ECJ. Thrombus imaging with fibrin-specific gadolinium-based mr contrast agent ep-2104r: results of a phase ii clinical study of feasibility. Investig Radiol. 2009;44 (11):697–704.
23. Schwarz M, Rottgen P, Takada Y, Le Gall F, Knackmuss S, Bassler N, Buttner C, Little M, Bode C, Peter K. Single-chain antibodies for the conformation-specific blockade of activated platelet integrin alphaiibbeta3 designed by subtractive selection from naive human phage libraries. FASEB J. 2004;18:1704–6.
24. von zur Muhlen C, von Elverfeldt D, Moeller JA, Choudhury RP, Paul D, Hagemeyer CE, Olschewski M, Becker A, Neudorfer I, Bassler N, Schwarz M, Bode C, Peter K. Magnetic resonance imaging contrast agent targeted toward activated platelets allows in vivo detection of thrombosis and monitoring of thrombolysis. Circulation. 2008;118:258–67.
25. von Zur Muhlen C, von Elverfeldt D, Choudhury RP, Ender J, Ahrens I, Schwarz M, Hennig J, Bode C, Peter K. Functionalized magnetic resonance contrast agent selectively binds to glycoprotein iib/iiia on activated human platelets under flow conditions and is detectable at clinically relevant field strengths. Mol Imaging. 2008;7:59–67.

26. von Zur MC, Sibson NR, Peter K, Campbell SJ, Wilainam P, Grau GE, Bode C, Choudhury RP, Anthony DC. A contrast agent recognizing activated platelets reveals murine cerebral malaria pathology undetectable by conventional MRI. J Clin Invest. 2008;118:1198–207.
27. Duerschmied D, Meißner M, Peter K, Neudorfer I, Roming F, Zirlik A, Bode C, von Elverfeldt D, von Zur Muhlen C. Molecular magnetic resonance imaging allows the detection of activated platelets in a new mouse model of coronary artery thrombosis. Invest Radiol. 2011;46(10):618–23.
28. von Elverfeldt D, von zur Muhlen C, Wiens K, Neudorfer I, Zirlik A, Meissner M, Tilly P, Charles AL, Bode C, Peter K, Fabre JE. In vivo detection of activated platelets allows characterizing rupture of atherosclerotic plaques with molecular magnetic resonance imaging in mice. PLoS One. 2012;7:e45008.
29. von Elverfeldt D, Maier A, Duerschmied D, Braig M, Witsch T, Wang X, Mauler M, Neudorfer I, Menza M, Idzko M, Zirlik A, Heidt T, Bronsert P, Bode C, Peter K, von Zur Muhlen C. Dual-contrast molecular imaging allows noninvasive characterization of myocardial ischemia/reperfusion injury after coronary vessel occlusion in mice by magnetic resonance imaging. Circulation. 2014;130:676–87.
30. Meier S, Putz G, Massing U, Hagemeyer CE, von Elverfeldt D, Meissner M, Ardipradja K, Barnert S, Peter K, Bode C, Schubert R, von zur Muhlen C. Immuno-magnetoliposomes targeting activated platelets as a potentially human-compatible MRI contrast agent for targeting atherothrombosis. Biomaterials. 2015;53:137–48.
31. Kubo T, Maehara A, Mintz GS, Doi H, Tsujita K, Choi SY, Katoh O, Nasu K, Koenig A, Pieper M, Rogers JH, Wijns W, Bose D, Margolis MP, Moses JW, Stone GW, Leon MB. The dynamic nature of coronary artery lesion morphology assessed by serial virtual histology intravascular ultrasound tissue characterization. J Am Coll Cardiol. 2010;55:1590–7.
32. Jaffer FA, Calfon MA, Rosenthal A, Mallas G, Razansky RN, Mauskapf A, Weissleder R, Libby P, Ntziachristos V. Two-dimensional intravascular near-infrared fluorescence molecular imaging of inflammation in atherosclerosis and stent-induced vascular injury. J Am Coll Cardiol. 2011;57:2516–26.
33. Yoo H, Kim JW, Shishkov M, Namati E, Morse T, Shubochkin R, McCarthy JR, Ntziachristos V, Bouma BE, Jaffer FA, Tearney GJ. Intra-arterial catheter for simultaneous microstructural and molecular imaging in vivo. Nat Med. 2011;17:1680–4.

In the Heat of the Artery: Inflammation as Trigger and Target of Atherosclerosis

2

Peter Stachon and Andreas Zirlik

Abstract

Abundant experimental and clinical work identifies inflammation of the vessel wall as a crucial factor in the development of atherosclerosis. Low-density lipoprotein (LDL) cholesterol initiates besides other factors endothelial activation. Monocytes and other immune cells invade the nascent lesion and create a pro-inflammatory milieu ultimately facilitating plaque rupture, the pathologic correlate of an acute coronary syndrome. We particularly shed light on the basic and clinical data implicating inflammation and immunity with this disease and its clinical sequelae. Furthermore, we comment on the large phase III CANTOS trial investigating the effect of an anti-inflammatory treatment with the IL-1b antibody canakinumab in over 10,000 patients with coronary heart disease and inflammatory status. This proof-of-concept trial showed for the first time that anti-inflammatory treatment may be a feasible and working option for patients with coronary heart disease. Finally, we end with an outlook on other promising targets for therapeutic intervention.

Contents

P. Stachon • A. Zirlik (✉)
Medical Faculty, Department of Cardiology and Angiology I, University Heart Center Freiburg, University of Freiburg, Hugstetter Street 55, 79106 Freiburg, Germany
e-mail: andreas.zirlik@universitaets-herzzentrum.de

A. Zirlik et al. (eds.), *Platelets, Haemostasis and Inflammation*,
Cardiac and Vascular Biology 5, https://doi.org/10.1007/978-3-319-66224-4_2

2.1 Introduction

Coronary heart disease is a multifactorial disorder triggered by modifiable and non-modifiable risk factors. Current treatment strategies include risk factor modification, platelet inhibition, and revascularization [1]. However, abundant experimental and clinical studies identify plaque inflammation and immune cell recruitment as crucial pathogenic factors in atherogenesis. This chapter discusses the current understanding of how inflammation initiates, promotes, and complicates the course of this devastating disease. Finally, we will shed some light on targeting inflammation clinically as a novel promising therapeutic approach [2–4] (Fig. 2.1).

2.2 Initiation of Atherosclerosis: Endothelial Cells and "Response to Injury"

Inflammation drives atherosclerotic plaque formation, growth, and vulnerability [5]. Upon initiation of atherogenesis, low-density lipoprotein cholesterol (LDL) accumulates in the arterial wall and is oxidized (ox) by reactive oxygen species, myeloperoxidase, or lipoperoxidase [6]. OxLDL enhances the expression of cell adhesion molecules such as E-selectin, vascular cell adhesion molecule 1 (VCAM-1), and intercellular cell adhesion molecule 1 (ICAM-1) on endothelial cells (EC) [7]. Thus, monocytes are attracted to the vessel wall, enter the intimal layer, and ingest oxLDL upon differentiation to macrophages. These steps lead to the formation of so-called "fatty streaks," the first manifestation of atherosclerosis. The concept of endothelial activation during atherogenesis is part of the "response-to-injury" hypothesis raised by Russel Ross and John Glomset in 1976 [8]. Beyond lipids, a variety of other factors elicit endothelial activation: arterial hypertension induces shear stress promoting expression of adhesion molecules. Inflammatory cytokines, e.g., arising from abdominal fat masses in obesity, may facilitate or directly activate the endothelium [9]. Other triggers include so-called advanced glycation end products (AGEs) forming in diabetic individuals and components of cigarette smoke or particulate matter from air pollution [10–12].

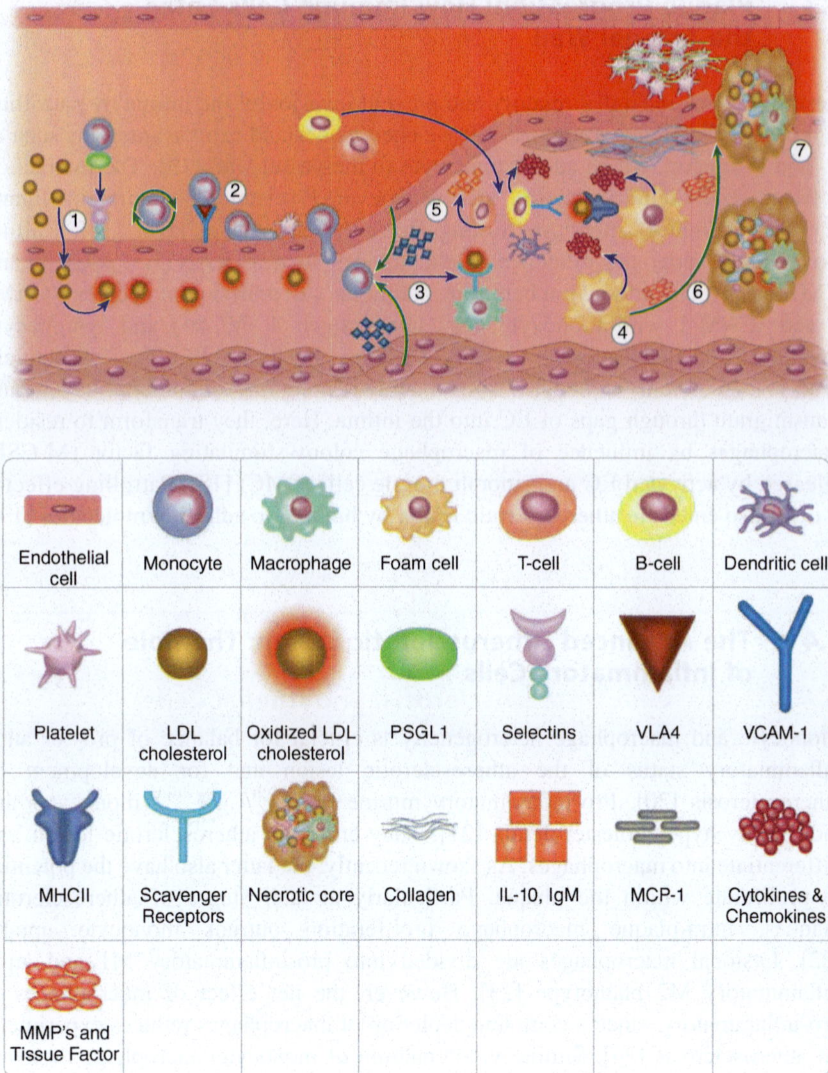

Fig. 2.1 LDL cholesterol enters the intimal wall and is oxidized to oxLDL. OxLDL activates endothelial cells, which express chemokines and cell adhesion molecules in order to attract monocytes (*1*). Due to the interaction of selectins, monocytes roll along the endothelium and adhere firmly by the interaction of VCAM-1 and VLA-1. Finally, they transmigrate into the intimal wall (*2*). Under the influence of endothelial and smooth muscle cells, derived M-CSF monocytes develop to macrophages, which uptake oxLDL and form lipid-loaded foam cells (*3*). These foam cells are inflammatory active and attract further leukocytes such as B- or T-cells by release of cytokines and chemokines (*4*). T- and B-cells are activated by antigen presentation. Th1 T-cells are pro-inflammatory; Th2 and T-regulatory cells are their anti-inflammatory counterparts (*5*). Release of IFNg and MMP does weaken the protective fibrotic cap (*6*). Finally, the plaque can rupture and prothrombotic components are released into the bloodstream. This results in a platelet activation and vessel occlusion (*7*)

2.3 Plaque Progression: How Immune Cells Enter the Arterial Wall

Resting EC are anti-inflammatory and prevent invasion of the intima by patrolling leukocytes. Upon activation, EC release chemokines and express integrins such as E- and P-selectin, which attract monocyte to the vessel wall [13]. The interaction between endothelial selectins and monocyte P-selectin glycoprotein ligand 1 (PSGL1) induces monocyte rolling along the endothelial layer [14]. Rolling slows down under the influence of chemokines such as RANTES, MCP-1, and CXCL5 [15]. The interaction between endothelial cell adhesion molecules VCAM-1 and ICAM-1 with monocyte very late antigen 4 (VLA4) and lymphocyte function-associated antigen 1 (LFA1) results in firm adhesion, respectively [16, 17]. Adherent monocytes spread, crawl along the endothelial layer, and transmigrate through gaps of EC into the intima. Here, they transform to resident macrophages by influence of macrophage colony-stimulating factor (M-CSF) released by activated EC and smooth muscle cells (SMC) [18]. Patrolling effector T cells also enter the atherosclerotic lesion by binding to adhesion molecules [19].

2.4 The Advanced Atherosclerotic Lesion: The Role of Inflammatory Cells

Monocyte and macrophage heterogeneity is crucial for balance of pro- or anti-inflammatory status of the atherosclerotic lesion and for development of atherosclerosis [20]. Pro-inflammatory murine Gr-1^{high}/Ly6C^{high} monocytes are induced by hypercholesterolemia [21]. They enter the atherosclerotic lesion and differentiate into macrophages. As shown recently, the latter also have the potential to proliferate within the plaque. Particularly at later stages of atherosclerotic plaques, intra-plaque macrophage proliferation outruns monocyte uptake [22]. Lesional macrophages are divided into pro-inflammatory M1 and anti-inflammatory M2 phenotype [23]. However, the net effect of macrophages is pro-inflammatory, since a complete depletion of macrophages reduces experimental atherosclerosis [24]. Similarly, attenuation of monocyte/macrophage turnover and activation, e.g., by inhibition of spleen tyrosine kinase (SYK), slows de novo atherogenesis [25, 26]. The macrophages are activated by oxLDL, which they take up via scavenger receptors SR-A, SR-B1, CD36, or lectin-type oxLDL receptor 1 (LOX-1) and form foam cells [27]. If the intracellular cholesterol concentrations increase, microcrystals are formed and activate inflammation via the inflammasome [28]. Foam cells propagate plaque inflammation and promote further leukocyte uptake via direct pro-inflammatory effector functions, e.g., pro-inflammatory cytokine expression as well as induction of a T-cell-dependent adaptive immune response via antigen presentation [29]. However, foam cell formation is reversible since cholesterol efflux via ATP-binding cassette transporter A1 (ABCA-1) and G1 (ABCA-G1) can reduce it [30]. If LDL cholesterol uptake dominates, macrophages are eventually overloaded with cholesterol and become apoptotic [31]. The cell debris, pro-inflammatory components such as cholesterol crystals, matrix metalloproteinases (MMP), and tissue factor form the necrotic core [32].

About 10% of cells within the plaque are CD3+ and CD4+ T-helper cells. Most of these T cells are pro-inflammatory T_h1 cells, which are generated under the influence of IL-12 [33]. After activation by antigen-presenting cells (APC) such as dendritic cells or macrophages T_h1 cells release interferon γ (IFNγ), interleukin 2 (IL-2), IL-3, and tumor necrosis factor α and β (TNFα and β) and aggravate plaque inflammation by activating macrophages, SMC, and EC [34, 35]. However, T_h2 and T-regulatory cells are the anti-inflammatory counterpart of T_h1 cells by releasing IL-1, IL-5, IL-10, and transforming growth factor β (TGFβ) [36]. The precise antigens leading to T-cell activation by APCs are under investigation. LDL cholesterol and heat shock proteins (i.e., HSP60) have been suggested as likely auto-antigens [37, 38]. Infiltrating mast cells support the pro-inflammatory milieu by releasing pro-inflammatory mediators and enzymes [39]. Divergent B-cell subsets have pro- and anti-atherogenic properties. B1 cells produce atheroprotective Immunoglobulin M (IgM) [40]. However, depletion of B2 cells attenuates atherosclerosis indicating a pro-inflammatory effect [41]. The imbalance between pro- and anti-inflammatory cells leads to a growing and maturing of the atherosclerotic lesion. Furthermore, repair mechanisms are suppressed, and the plaque becomes vulnerable.

2.5 The Plaque Extracellular Matrix: The Pro-inflammatory Micro-milieu for Cell–Cell Communication

The extracellular matrix contains a pro-inflammatory micro-milieu with a diversity of cytokines, chemokines, enzymes, and damage-associated molecular patterns (DAMP) released by activated or dying cells but also protective factors such as stabilizing collagen fibers. Consequently, there is a growing interest in the biology of the extracellular matrix [42]. Various cytokines orchestrate the inflammation by supporting the cell–cell communication within the plaque but also within the whole vascular system since atherosclerosis is a systemic disease. Macrophage-derived TNFα, IL-1β, and IL-6 activate endothelial cells to express integrins, cell adhesion molecules, and chemokines [43]. Furthermore, macrophages produce IL-12-promoting T_h1 differentiation of T cells [37, 44]. T_h1 cells are further activated by co-stimulatory factors. A crucial co-stimulatory factor is CD40 ligand (CD40L)-promoting inflammation classically via CD40 and alternatively via the leukocyte integrin Mac-1 [45, 46]. The latter interaction appears to be particularly relevant for CD40L-induced recruitment of inflammatory cells and the development of atherosclerosis [47]. Similarly, other costimulatory molecules as well as adaptor proteins downstream of CD40L and the TNFα/IL-1 receptor superfamily potently affect vascular inflammation and atherosclerosis [48–51]. Activated T cells support the inflammation by the release of IL-2 and IFNγ. The inflammatory activation leads to a release of chemokines such as IL-8 and MCP-1 recruiting more leukocytes to the vessel wall perpetuating vascular inflammation.

However, macrophages and T_h2 cells also release anti-inflammatory cytokines such as IL-10, IL-33, IL1-receptor antagonist, or TGFβ with the ability to dampen the inflammatory response. Therefore, it is mainly a disbalance of pro-inflammatory and anti-inflammatory cytokines favoring a pro-inflammatory extracellular milieu that exacerbates atherosclerosis. Beyond cytokines, chemokines, and stabilizing collagen, the lesional extracellular matrix contains a variety of other molecules affecting the inflammatory response. Cholesterol microcrystals, which develop after accumulation of LDL cholesterol, activate pattern recognition receptors (PRR) such as Toll-like or scavenger receptors and activate the inflammasome [28, 52]. The inflammasome activates Caspase-1, and pro-IL1β is cleaved to active IL-1β, which is secreted. However, the inflammasome can be co-activated by other factors. DAMPs such as extracellular adenosine-3′-phosphate (ATP) released by dying cells bind to the purinergic receptor X7 (P2X7), thereby propagating inflammasome activation [53]. A recent study showed that deficiency of the P2X7 receptor inhibits the lesional NLRP3 inflammasome activation and reduces atherosclerosis [54]. Similarly overall inhibition of the inflammasome attenuates experimental atherosclerosis. IL-1b can activate itself and promotes further IL-1b release turning on a vicious inflammatory circle [55]. Moreover, other purinergic receptors such as P2Y2 and P2Y6, which bind to extracellular nucleotides as danger signals, promote atherosclerosis [56, 57]. During this process, lesional cells, mostly macrophages, are dying by necrosis or apoptosis. The resulting cell debris mingled with components of the extracellular matrix forms the necrotic core [31].

2.6 Inflammation Drives Atherosclerotic Complications: Plaque Rupture and Endothelial Erosion

A fibrous cap formed by EC, collagen fibers, and SMC separates the procoagulant necrotic core from circulation. Collagen fibers produced by SMC are crucial to the integrity of this cap. The breakdown of the fibrous cap leads to a vulnerable plaque and can promote plaque rupture [58]. Consequently, prothrombotic molecules including tissue factor, phospholipids, and further pro-thrombotic material are released to the bloodstream, activate platelets, which form a thrombus, and ultimately occlude the vessel [59]. If the inflammation within a nascent atherosclerotic lesion maintains, foam cells are activated by IL1β and secret matrix metalloproteinases (MMP), enzymes degrading collagen fibers. Furthermore, IL-12 derived by foam cells induces T-cell transformation into T_h1-effector cells releasing IFNγ [60]. The latter has two effects on SMC: First, it dampens plaque-stabilizing collagen synthesis. Second, it inhibits SMC proliferation further limiting the source of collagen [61]. Thus, the fibrous cap thins and the plaque can rupture with potential life-threatening consequences. These data directly causally link increased inflammation within atherosclerotic lesions with plaque rupture and its unfavorable clinical sequela [62, 63]. Accordingly, in the PROSPECT trial following up 697 patients suffering from an acute coronary syndrome with IVUS identified thin-cap atheroma along with plaque burden and a luminal area smaller than 4.00 mm^2 as major predictor of clinical events (REF). Beyond plaque rupture, erosions within the endothelial layer may promote thrombocyte activation and vessel occlusion by the exposure of prothrombotic plaque material to the blood stream. The erosions occur even in younger patients with relatively small atherosclerotic lesions [64]. It is speculated, that activation of endothelial Toll-like receptor 2 (TLR2) results in endothelial apoptosis after ligation of extracellular hyaluronan or components of Gram-positive bacteria [59, 65, 66].

2.7 From Bench to Bedside: Clinical Perspective of Anti-inflammatory Treatment of Atherosclerosis

Two decades of experimental work clearly identify unresolved inflammation as driving force behind atherosclerosis. In accord, large clinical trials associate an increased inflammatory status in humans, e.g., as indicated by the elevation of high sensitive C-reactive protein (hsCRP), with cardiovascular events [67]. Yet, we still lack a genuine anti-inflammatory or immune-modulatory treatment option for atherosclerosis beyond lipid-lowering therapies. The time is ripe for stringent translation of experimental work to clinical development of such anti-inflammatory and/or immune-modulatory treatments combating atherosclerosis. In that respect we need to recognize that most the experimental data available have been acquired in mice. While extremely valuable, by no means are these data guaranteed to hold up their promise in humans and therefore no single mouse experiment can replace a rigorous clinical investigation [68].

The large statins trials provide convincing evidence that reduction of systemic inflammation reduces cardiovascular events: indeed, treatment with rosuvastatin decreases levels of hsCRP, an effect most likely independent of the lipid-lowering effect of statins [69]. However, several clinical trials challenging potentially anti-inflammatory treatment options failed: The selective inhibitor for lipoprotein-associated phospholipase A2, darapladib, which has large experimental evidence on its side to drive inflammation in atherosclerosis, did not reduce cardiovascular events in the SOLID-TIMI 52 and STABILITY trial [70, 71]. Similarly, strategies of increasing high-density lipoproteins by CETP [72] inhibition failed to translate into reduction of cardiovascular events. Epidemiological data also suggest that cyclooxygenase (COX) [73] inhibition may even be associated with cardiovascular harm.

Some promising data arise from smaller trials: In the BLAST study, 225 patients undergoing effective coronary angioplasty received an alendronate-loaded liposome treatment (LABR-312) attenuating macrophage biology. In subjects with elevated monocyte counts, this treatment resulted in lower late lumen loss [74]. Colchicine inhibits expression of cell adhesion molecule, the inflammasome, and inflammatory cytokines through the inhibition of tubulin polymerization. It is a widely used anti-inflammatory drug approved for acute use in patients with gout and pericarditis or chronically in patients with Familial Mediterranean Fever. An open-label pilot study found a relative reduction of hsCRP at 60% in patients with CHD [75]. Accordingly, a retrospective analysis indicated that the continuous treatment with colchicine in patients with gout or Familial Mediterranean Fever reduces myocardial infarctions [76, 77]. The prospective Low Dose Colchicine (LoDoCo) trial demonstrated a reduction of cardiovascular events in patients with stable CHD with a hazard ratio of 0.33 [78]. Potential benefits are also observed in patients with ST-segment elevation myocardial infarction (STEMI), but clinical end point trials are missing [79].

Large trials testing the inflammatory hypothesis are currently under way: The IL-1β antibody canakinumab reduces level of hsCRP, IL-6, and fibrinogen without influencing the lipid metabolism [80]. The canakinumab anti-inflammatory thrombosis outcome study (CANTOS) enrolled over 10,000 participants with coronary heart disease and elevated hsCRP and tests for cardiovascular events in patients treated with canakinumab or placebo. The results were presented in Summer 2017: Treatment with Canakinumab could reduce cardiovascular events and mortality. The positive results of the CANTOS trial induce a paradigm shift in the treatment of patients with coronary heart disease. The inflammatory nature of atherosclerosis can be addressed directly in patients with increased inflammatory status [81]. Therefore, it is a step toward an individualized therapy for patients with coronary heart disease. It is well known that treatment with methotrexate reduces cardiovascular events in patients with rheumatoid arthritis [81]. The aim of the Cardiovascular Inflammation Reduction Trial (CIRT) is to evaluate whether low-dose methotrexate can reduce cardiovascular events among patients with a cardiovascular inflammatory status such as type 2 diabetes or metabolic syndrome. The CIRT trial is currently recruiting; results will be available in 2019 [82].

2.8 Conclusion

Atherosclerosis is a chronic inflammatory disease. Inflammation drives all steps of atherogenesis from initiation, plaque progression, and plaque rupture. Despite its crucial role in atherosclerosis, a genuine anti-inflammatory therapy is not established in atherosclerosis. However, several promising large clinical endpoint trials investigating IL1β-antibody, methotrexate, or colchicine in patients with CHD are ongoing. These will likely provide us with novel, valuable, tailored treatment options.

Compliance with Ethical Standards

Conflict of Interest: Peter Stachon and Andreas Zirlik declares that they have no conflict of interest.

Ethical Approval: This article does not contain any studies with human participants or animals performed by any of the authors.

References

1. Smith SC Jr, Benjamin EJ, Bonow RO, Braun LT, Creager MA, Franklin BA, Gibbons RJ, Grundy SM, Hiratzka LF, Jones DW, Lloyd-Jones DM, Minissian M, Mosca L, Peterson ED, Sacco RL, Spertus J, Stein JH, Taubert KA, World Heart Federation, The Preventive Cardiovascular Nurses Association. Aha/accf secondary prevention and risk reduction therapy for patients with coronary and other atherosclerotic vascular disease: 2011 update: a guideline from the american heart association and american college of cardiology foundation. Circulation. 2011;124:2458–73.
2. Libby P. Inflammation in atherosclerosis. Nature. 2002;420:868–74.
3. Lichtman AH, Binder CJ, Tsimikas S, Witztum JL. Adaptive immunity in atherogenesis: new insights and therapeutic approaches. J Clin Invest. 2013;123:27–36.
4. Libby P, Lichtman AH, Hansson GK. Immune effector mechanisms implicated in atherosclerosis: from mice to humans. Immunity. 2013;38:1092–104.
5. Wolf D, Stachon P, Bode C, Zirlik A. Inflammatory mechanisms in atherosclerosis. Hamostaseologie. 2014;34:63–71.
6. Hansson GK, Hermansson A. The immune system in atherosclerosis. Nat Immunol. 2011;12:204–12.
7. Cybulsky MI, Gimbrone MA Jr. Endothelial expression of a mononuclear leukocyte adhesion molecule during atherogenesis. Science. 1991;251:788–91.
8. Ross R, Glomset JA. The pathogenesis of atherosclerosis (first of two parts). N Engl J Med. 1976;295:369–77.
9. Zirlik A, Lutgens E. An inflammatory link in atherosclerosis and obesity.Co-stimulatory molecules. Hamostaseologie. 2015;35:272–8.
10. Aronson D, Rayfield EJ. How hyperglycemia promotes atherosclerosis: molecular mechanisms. Cardiovasc Diabetol. 2002;1:1.
11. Hagiwara E, Takahashi KI, Okubo T, Ohno S, Ueda A, Aoki A, Odagiri S, Ishigatsubo Y. Cigarette smoking depletes cells spontaneously secreting th(1) cytokines in the human airway. Cytokine. 2001;14:121–6.
12. Marchini T, Wolf D, Michel NA, Mauler M, Dufner B, Hoppe N, Beckert J, Jackel M, Magnani N, Duerschmied D, Tasat D, Alvarez S, Reinohl J, von Zur MC, Idzko M, Bode C, Hilgendorf I, Evelson P, Zirlik A. Acute exposure to air pollution particulate matter aggravates experimental myocardial infarction in mice by potentiating cytokine secretion from lung macrophages. Basic Res Cardiol. 2016;111:44.
13. Ley K, Laudanna C, Cybulsky MI, Nourshargh S. Getting to the site of inflammation: the leukocyte adhesion cascade updated. Nat Rev Immunol. 2007;7:678–89.
14. McEver RP, Cummings RD. Role of psgl-1 binding to selectins in leukocyte recruitment. J Clin Invest. 1997;100:S97–103.
15. von Hundelshausen P, Koenen RR, Sack M, Mause SF, Adriaens W, Proudfoot AE, Hackeng TM, Weber C. Heterophilic interactions of platelet factor 4 and rantes promote monocyte arrest on endothelium. Blood. 2005;105:924–30.
16. Chan JR, Hyduk SJ, Cybulsky MI. Chemoattractants induce a rapid and transient upregulation of monocyte alpha4 integrin affinity for vascular cell adhesion molecule 1 which mediates arrest: an early step in the process of emigration. J Exp Med. 2001;193:1149–58.
17. Sigal A, Bleijs DA, Grabovsky V, van Vliet SJ, Dwir O, Figdor CG, van Kooyk Y, Alon R. The lfa-1 integrin supports rolling adhesions on icam-1 under physiological shear flow in a permissive cellular environment. J Immunol. 2000;165:442–52.

18. Smith JD, Trogan E, Ginsberg M, Grigaux C, Tian J, Miyata M. Decreased atherosclerosis in mice deficient in both macrophage colony-stimulating factor (op) and apolipoprotein e. Proc Natl Acad Sci USA. 1995;92:8264–8.
19. Hansson GK, Libby P. The immune response in atherosclerosis: a double-edged sword. Nat Rev Immunol. 2006;6:508–19.
20. Swirski FK, Pittet MJ, Kircher MF, Aikawa E, Jaffer FA, Libby P, Weissleder R. Monocyte accumulation in mouse atherogenesis is progressive and proportional to extent of disease. Proc Natl Acad Sci USA. 2006;103:10340–5.
21. Swirski FK, Libby P, Aikawa E, Alcaide P, Luscinskas FW, Weissleder R, Pittet MJ. Ly-6chi monocytes dominate hypercholesterolemia-associated monocytosis and give rise to macrophages in atheromata. J Clin Invest. 2007;117:195–205.
22. Robbins CS, Hilgendorf I, Weber GF, Theurl I, Iwamoto Y, Figueiredo JL, Gorbatov R, Sukhova GK, Gerhardt LM, Smyth D, Zavitz CC, Shikatani EA, Parsons M, van Rooijen N, Lin HY, Husain M, Libby P, Nahrendorf M, Weissleder R, Swirski FK. Local proliferation dominates lesional macrophage accumulation in atherosclerosis. Nat Med. 2013;19:1166–72.
23. Ley K, Miller YI, Hedrick CC. Monocyte and macrophage dynamics during atherogenesis. Arterioscler Thromb Vasc Biol. 2011;31:1506–16.
24. Stoneman V, Braganza D, Figg N, Mercer J, Lang R, Goddard M, Bennett M. Monocyte/macrophage suppression in cd11b diphtheria toxin receptor transgenic mice differentially affects atherogenesis and established plaques. Circ Res. 2007;100:884–93.
25. Lindau A, Hardtner C, Hergeth SP, Blanz KD, Dufner B, Hoppe N, Anto-Michel N, Kornemann J, Zou J, Gerhardt LM, Heidt T, Willecke F, Geis S, Stachon P, Wolf D, Libby P, Swirski FK, Robbins CS, McPheat W, Hawley S, Braddock M, Gilsbach R, Hein L, von zur Muhlen C, Bode C, Zirlik A. Hilgendorf I Atheroprotection through syk inhibition fails in established disease when local macrophage proliferation dominates lesion progression. Basic Res Cardiol. 2016;111:20.
26. Stachon P, Ahrens I, Bode C, Zirlik A. Dual pathway therapy in acute coronary syndrome. J Thromb Thrombolysis. 2016;42:254–60.
27. Greaves DR, Gordon S. The macrophage scavenger receptor at 30 years of age: current knowledge and future challenges. J Lipid Res. 2009;50(Suppl):S282–6.
28. Duewell P, Kono H, Rayner KJ, Sirois CM, Vladimer G, Bauernfeind FG, Abela GS, Franchi L, Nunez G, Schnurr M, Espevik T, Lien E, Fitzgerald KA, Rock KL, Moore KJ, Wright SD, Hornung V, Latz E. Nlrp3 inflammasomes are required for atherogenesis and activated by cholesterol crystals. Nature. 2010;464:1357–61.
29. Wolf D, Zirlik A, Ley K. Beyond vascular inflammation—recent advances in understanding atherosclerosis. Cell Mol Life Sci. 2015;72:3853–69.
30. Yvan-Charvet L, Ranalletta M, Wang N, Han S, Terasaka N, Li R, Welch C, Tall AR. Combined deficiency of abca1 and abcg1 promotes foam cell accumulation and accelerates atherosclerosis in mice. J Clin Invest. 2007;117:3900–8.
31. Libby P, Tabas I, Fredman G, Fisher EA. Inflammation and its resolution as determinants of acute coronary syndromes. Circ Res. 2014;114:1867–79.
32. Shah PK, Falk E, Badimon JJ, Fernandez-Ortiz A, Mailhac A, Villareal-Levy G, Fallon JT, Regnstrom J, Fuster V. Human monocyte-derived macrophages induce collagen breakdown in fibrous caps of atherosclerotic plaques. Potential role of matrix-degrading metalloproteinases and implications for plaque rupture. Circulation. 1995;92:1565–9.
33. Robertson AK, Hansson GK. T cells in atherogenesis: for better or for worse? Arterioscler Thromb Vasc Biol. 2006;26:2421–32.
34. Buono C, Binder CJ, Stavrakis G, Witztum JL, Glimcher LH, Lichtman AH. T-bet deficiency reduces atherosclerosis and alters plaque antigen-specific immune responses. Proc Natl Acad Sci USA. 2005;102:1596–601.
35. Buono C, Come CE, Stavrakis G, Maguire GF, Connelly PW, Lichtman AH. Influence of interferon-gamma on the extent and phenotype of diet-induced atherosclerosis in the ldlr-deficient mouse. Arterioscler Thromb Vasc Biol. 2003;23:454–60.

36. Hori S, Nomura T, Sakaguchi S. Control of regulatory t cell development by the transcription factor foxp3. Science. 2003;299:1057–61.
37. Hermansson A, Ketelhuth DF, Strodthoff D, Wurm M, Hansson EM, Nicoletti A, Paulsson-Berne G, Hansson GK. Inhibition of t cell response to native low-density lipoprotein reduces atherosclerosis. J Exp Med. 2010;207:1081–93.
38. Ketelhuth DF, Hansson GK. Cellular immunity, low-density lipoprotein and atherosclerosis: break of tolerance in the artery wall. Thromb Haemost. 2011;106:779–86.
39. Bot I, de Jager SC, Zernecke A, Lindstedt KA, van Berkel TJ, Weber C, Biessen EA. Perivascular mast cells promote atherogenesis and induce plaque destabilization in apolipoprotein e-deficient mice. Circulation. 2007;115:2516–25.
40. Binder CJ, Horkko S, Dewan A, Chang MK, Kieu EP, Goodyear CS, Shaw PX, Palinski W, Witztum JL, Silverman GJ. Pneumococcal vaccination decreases atherosclerotic lesion formation: molecular mimicry between streptococcus pneumoniae and oxidized ldl. Nat Med. 2003;9:736–43.
41. Kyaw T, Tay C, Hosseini H, Kanellakis P, Gadowski T, MacKay F, Tipping P, Bobik A, Toh BH. Depletion of b2 but not b1a b cells in baff receptor-deficient apoe mice attenuates atherosclerosis by potently ameliorating arterial inflammation. PLoS One. 2012;7:e29371.
42. Nilsson J, Hansson GK. The changing face of atherosclerotic plaque inflammation. J Intern Med. 2015;278:430–2.
43. Ait-Oufella H, Taleb S, Mallat Z, Tedgui A. Recent advances on the role of cytokines in atherosclerosis. Arterioscler Thromb Vasc Biol. 2011;31:969–79.
44. Stemme S, Faber B, Holm J, Wiklund O, Witztum JL, Hansson GK. T lymphocytes from human atherosclerotic plaques recognize oxidized low density lipoprotein. Proc Natl Acad Sci USA. 1995;92:3893–7.
45. Wolf D, Hohmann JD, Wiedemann A, Bledzka K, Blankenbach H, Marchini T, Gutte K, Zeschky K, Bassler N, Hoppe N, Rodriguez AO, Herr N, Hilgendorf I, Stachon P, Willecke F, Duerschmied D, von zur Muhlen C, Soloviev DA, Zhang L, Bode C, Plow EF, Libby P, Peter K, Zirlik A. Binding of cd40l to mac-1's i-domain involves the eqlkksktl motif and mediates leukocyte recruitment and atherosclerosis—but does not affect immunity and thrombosis in mice. Circ Res. 2011;109:1269–79.
46. Zirlik A, Maier C, Gerdes N, MacFarlane L, Soosairajah J, Bavendiek U, Ahrens I, Ernst S, Bassler N, Missiou A, Patko Z, Aikawa M, Schonbeck U, Bode C, Libby P, Peter K. Cd40 ligand mediates inflammation independently of cd40 by interaction with mac-1. Circulation. 2007;115:1571–80.
47. Schonbeck U, Sukhova GK, Shimizu K, Mach F, Libby P. Inhibition of cd40 signaling limits evolution of established atherosclerosis in mice. Proc Natl Acad Sci USA. 2000;97:7458–63.
48. Missiou A, Rudolf P, Stachon P, Wolf D, Varo N, Aichele P, Colberg C, Hoppe N, Ernst S, Munkel C, Walter C, Sommer B, Hilgendorf I, Nakano H, Bode C, Zirlik A. Traf5 deficiency accelerates atherogenesis in mice by increasing inflammatory cell recruitment and foam cell formation. Circ Res. 2010;107:757–66.
49. Stachon P, Missiou A, Walter C, Varo N, Colberg C, Wolf D, Buchner M, von Zur MC, Zirlik K, Bode C, Zirlik A. Tumor necrosis factor receptor associated factor 6 is not required for atherogenesis in mice and does not associate with atherosclerosis in humans. PLoS One. 2010;5:e11589.
50. Missiou A, Kostlin N, Varo N, Rudolf P, Aichele P, Ernst S, Munkel C, Walter C, Stachon P, Sommer B, Pfeifer D, Zirlik K, MacFarlane L, Wolf D, Tsitsikov E, Bode C, Libby P, Zirlik A. Tumor necrosis factor receptor-associated factor 1 (traf1) deficiency attenuates atherosclerosis in mice by impairing monocyte recruitment to the vessel wall. Circulation. 2010;121:2033–44.
51. Zirlik A, Bavendiek U, Libby P, MacFarlane L, Gerdes N, Jagielska J, Ernst S, Aikawa M, Nakano H, Tsitsikov E, Schonbeck U. Traf-1, -2, -3, -5, and -6 are induced in atherosclerotic plaques and differentially mediate proinflammatory functions of cd40l in endothelial cells. Arterioscler Thromb Vasc Biol. 2007;27:1101–7.

52. Rajamaki K, Lappalainen J, Oorni K, Valimaki E, Matikainen S, Kovanen PT, Eklund KK. Cholesterol crystals activate the nlrp3 inflammasome in human macrophages: a novel link between cholesterol metabolism and inflammation. PLoS One. 2010;5:e11765.
53. Tschopp J, Schroder K. Nlrp3 inflammasome activation: the convergence of multiple signalling pathways on ros production? Nat Rev Immunol. 2010;10:210–5.
54. Stachon P, et al. P2X7 deficiency blocks lesional inflammasome activity and ameliorates atherosclerosis in mice. Circulation. 2017;135(25):2524–33. https://doi.org/10.1161/CIRCULATIONAHA.117.027400.
55. Libby P, Ordovas JM, Birinyi LK, Auger KR, Dinarello CA. Inducible interleukin-1 gene expression in human vascular smooth muscle cells. J Clin Invest. 1986;78:1432–8.
56. Stachon P, Geis S, Peikert A, Heidenreich A, Michel NA, Unal F, Hoppe N, Dufner B, Schulte L, Marchini T, Cicko S, Ayata K, Zech A, Wolf D, Hilgendorf I, Willecke F, Reinohl J, von Zur MC, Bode C, Idzko M, Zirlik A. Extracellular atp induces vascular inflammation and atherosclerosis via purinergic receptor y2 in mice. Arterioscler Thromb Vasc Biol. 2016;36:1577–86.
57. Stachon P, Peikert A, Michel NA, Hergeth S, Marchini T, Wolf D, Dufner B, Hoppe N, Ayata CK, Grimm M, Cicko S, Schulte L, Reinohl J, von zur Muhlen C, Bode C, Idzko M, Zirlik A. P2y6 deficiency limits vascular inflammation and atherosclerosis in mice. Arterioscler Thromb Vasc Biol. 2014;34:2237–45.
58. Falk E, Shah PK, Fuster V. Coronary plaque disruption. Circulation. 1995;92:657–71.
59. Hansson GK, Libby P, Tabas I. Inflammation and plaque vulnerability. J Intern Med. 2015;278:483–93.
60. Frostegard J, Ulfgren AK, Nyberg P, Hedin U, Swedenborg J, Andersson U, Hansson GK. Cytokine expression in advanced human atherosclerotic plaques: dominance of pro-inflammatory (th1) and macrophage-stimulating cytokines. Atherosclerosis. 1999;145:33–43.
61. Amento EP, Ehsani N, Palmer H, Libby P. Cytokines and growth factors positively and negatively regulate interstitial collagen gene expression in human vascular smooth muscle cells. Arteriosclerosis Thromb. 1991;11:1223–30.
62. Galis ZS, Sukhova GK, Lark MW, Libby P. Increased expression of matrix metalloproteinases and matrix degrading activity in vulnerable regions of human atherosclerotic plaques. J Clin Invest. 1994;94:2493–503.
63. Stone GW, Maehara A, Lansky AJ, de Bruyne B, Cristea E, Mintz GS, Mehran R, McPherson J, Farhat N, Marso SP, Parise H, Templin B, White R, Zhang Z, Serruys PW. A prospective natural-history study of coronary atherosclerosis. N Engl J Med. 2011;364:226–35.
64. Farb A, Burke AP, Tang AL, Liang TY, Mannan P, Smialek J, Virmani R. Coronary plaque erosion without rupture into a lipid core. A frequent cause of coronary thrombosis in sudden coronary death. Circulation. 1996;93:1354–63.
65. Quillard T, Araujo HA, Franck G, Shvartz E, Sukhova G, Libby P. Tlr2 and neutrophils potentiate endothelial stress, apoptosis and detachment: implications for superficial erosion. Eur Heart J. 2015;36:1394–404.
66. Williams H, Johnson JL, Carson KG, Jackson CL. Characteristics of intact and ruptured atherosclerotic plaques in brachiocephalic arteries of apolipoprotein e knockout mice. Arterioscler Thromb Vasc Biol. 2002;22:788–92.
67. Ridker PM. Establishing a clinical basis for hscrp in the prevention and treatment of cardiovascular disease. Clin Chem. 2010;56:1186–7.
68. Libby P. Murine "model" monotheism: an iconoclast at the altar of mouse. Circ Res. 2015;117:921–5.
69. Ridker PM, Danielson E, Fonseca FA, Genest J, Gotto AM Jr, Kastelein JJ, Koenig W, Libby P, Lorenzatti AJ, Macfadyen JG, Nordestgaard BG, Shepherd J, Willerson JT, Glynn RJ, Group JTS. Reduction in c-reactive protein and ldl cholesterol and cardiovascular event

rates after initiation of rosuvastatin: a prospective study of the jupiter trial. Lancet. 2009;373:1175–82.
70. Stability Investigators, White HD, Held C, Stewart R, Tarka E, Brown R, Davies RY, Budaj A, Harrington RA, Steg PG, Ardissino D, Armstrong PW, Avezum A, Aylward PE, Bryce A, Chen H, Chen MF, Corbalan R, Dalby AJ, Danchin N, De Winter RJ, Denchev S, Diaz R, Elisaf M, Flather MD, Goudev AR, Granger CB, Grinfeld L, Hochman JS, Husted S, Kim HS, Koenig W, Linhart A, Lonn E, Lopez-Sendon J, Manolis AJ, Mohler ER 3rd, Nicolau JC, Pais P, Parkhomenko A, Pedersen TR, Pella D, Ramos-Corrales MA, Ruda M, Sereg M, Siddique S, Sinnaeve P, Smith P, Sritara P, Swart HP, Sy RG, Teramoto T, Tse HF, Watson D, Weaver WD, Weiss R, Viigimaa M, Vinereanu D, Zhu J, Cannon CP, Wallentin L. Darapladib for preventing ischemic events in stable coronary heart disease. N Engl J Med. 2014;370:1702–11.
71. O'Donoghue ML, Braunwald E, White HD, Lukas MA, Tarka E, Steg PG, Hochman JS, Bode C, Maggioni AP, Im K, Shannon JB, Davies RY, Murphy SA, Crugnale SE, Wiviott SD, Bonaca MP, Watson DF, Weaver WD, Serruys PW, Cannon CP, Investigators S-T, Steen DL. Effect of darapladib on major coronary events after an acute coronary syndrome: the solid-timi 52 randomized clinical trial. JAMA. 2014;312:1006–15.
72. Kosmas CE, DeJesus E, Rosario D, Vittorio TJ. Cetp inhibition: past failures and future hopes. Clin Med Insights Cardiol. 2016;10:37–42.
73. Howes LG. Selective cox-2 inhibitors, nsaids and cardiovascular events – is celecoxib the safest choice? Ther Clin Risk Manag. 2007;3:831–45.
74. Banai S, Finkelstein A, Almagor Y, Assali A, Hasin Y, Rosenschein U, Apruzzese P, Lansky AJ, Kume T, Edelman ER. Targeted anti-inflammatory systemic therapy for restenosis: the biorest liposomal alendronate with stenting study (blast)-a double blind, randomized clinical trial. Am Heart J. 2013;165:234–40. e231
75. Nidorf M, Thompson PL. Effect of colchicine (0.5 mg twice daily) on high-sensitivity c-reactive protein independent of aspirin and atorvastatin in patients with stable coronary artery disease. Am J Cardiol. 2007;99:805–7.
76. Crittenden DB, Lehmann RA, Schneck L, Keenan RT, Shah B, Greenberg JD, Cronstein BN, Sedlis SP, Pillinger MH. Colchicine use is associated with decreased prevalence of myocardial infarction in patients with gout. J Rheumatol. 2012;39:1458–64.
77. Sari I, Yuksel A, Kozaci D, Selcuk S, Gokce G, Yildiz Y, Demirel H, Sop G, Alacacioglu A, Gunay N, Akkoc N. The effect of regular colchicine treatment on biomarkers related with vascular injury in newly diagnosed patients with familial mediterranean fever. Inflammation. 2012;35:1191–7.
78. Nidorf SM, Eikelboom JW, Budgeon CA, Thompson PL. Low-dose colchicine for secondary prevention of cardiovascular disease. J Am Coll Cardiol. 2013;61:404–10.
79. Deftereos S, Giannopoulos G, Angelidis C, Alexopoulos N, Filippatos G, Papoutsidakis N, Sianos G, Goudevenos J, Alexopoulos D, Pyrgakis V, Cleman MW, Manolis AS, Tousoulis D, Lekakis J. Anti-inflammatory treatment with colchicine in acute myocardial infarction: a pilot study. Circulation. 2015;132:1395–403.
80. Ridker PM, Howard CP, Walter V, Everett B, Libby P, Hensen J, Thuren T, Group CPI. Effects of interleukin-1beta inhibition with canakinumab on hemoglobin a1c, lipids, c-reactive protein, interleukin-6, and fibrinogen: a phase iib randomized, placebo-controlled trial. Circulation. 2012;126:2739–48.
81. Kisiel B, Kruszewski R, Juszkiewicz A, Raczkiewicz A, Bachta A, Tlustochowicz M, Staniszewska-Varga J, Klos K, Duda K, Boguslawska-Walecka R, Ploski R, Tlustochowicz W. Methotrexate, cyclosporine a, and biologics protect against atherosclerosis in rheumatoid arthritis. J Immunol Res. 2015;2015:759610.
82. Ridker PM. Testing the inflammatory hypothesis of atherothrombosis: scientific rationale for the cardiovascular inflammation reduction trial (cirt). J Thromb Haemost. 2009;7(Suppl 1):332–9.

Vaccination to Prevent Cardiovascular Disease

3

Dennis Wolf, Teresa Gerhardt, and Klaus Ley

Abstract

Atherosclerosis and attendant cardiovascular disease account for most deaths worldwide. Despite considerable progress in understanding disease mechanisms, the only causal treatment strategy is lowering LDL cholesterol. Atherosclerosis is a complex disorder that is initiated and maintained by dyslipidemia, inflammation, and (auto-) immunity. This complex interplay is orchestrated by cells of the innate and adaptive immune system. It has been proposed that modulating the aberrant immune response by vaccination against specific antigens could become a new causal treatment for atherosclerosis. Immune-modulatory therapies have been explored using strategies designed to dampen autoimmunity by immuno-modulators, tolerize for auto-antigens, remove potentially harmful antigens, or neutralize proteins that participate in atherogenesis by active and passive immunization. Beneficial effects of such strategies have been shown in animal models of atherosclerosis and other cardiovascular diseases. This chapter categorizes and summarizes the efforts that have been made in the last decades to design a vaccine to prevent cardiovascular disease.

Contents

D. Wolf • T. Gerhardt
Division of Inflammation Biology, La Jolla Institute for Allergy and Immunology, 9420 Athena Circle Drive, La Jolla, CA 92037, USA

K. Ley (✉)
Division of Inflammation Biology, La Jolla Institute for Allergy and Immunology, 9420 Athena Circle Drive, La Jolla, CA 92037, USA

Department of Bioengineering, University of California San Diego, La Jolla, CA, USA
e-mail: klaus@lji.org

A. Zirlik et al. (eds.), *Platelets, Haemostasis and Inflammation*,
Cardiac and Vascular Biology 5, https://doi.org/10.1007/978-3-319-66224-4_3

3.1 Atherosclerosis: The Driving Pathology in Cardiovascular Disease

Cardiovascular events account for over 30% of all deaths, more than all forms of cancer combined [1]. The most common underlying cause of all these events is atherosclerosis, a systemic immune-inflammatory and fibro-proliferative disease, in which the intima of medium-sized and large arteries is focally thickened by a fibro-fatty plaque containing immune cells [2]. Neovascularization and hemorrhage destabilize plaque and lead to calcification, which promotes the rupture of the cap overlying atherosclerotic lesions. Plaque rupture, ulceration, and erosion can trigger sudden thrombosis and occlusion, which precipitate acute life-threatening clinical manifestations, such as acute coronary syndrome, myocardial infarction, occlusion of peripheral arties, and stroke [3].

Elevated levels of serum cholesterol, especially low-density lipoprotein (LDL) cholesterol, represent a major risk factor for initiation and progression of atherosclerosis [4–6]. The primary protein constituent of the LDL particle is apolipoprotein B (ApoB)-100, which organizes a core of cholesterol esters and triglycerides and stabilizes a shell of phospholipids and free cholesterol. ApoB-100 contains the binding site for the LDL receptor (LDLR) in the liver [7].

During early atherosclerosis, monocyte-derived macrophages accumulate in the developing atherosclerotic lesion. In the plaque, LDL and ApoB-100 become susceptible to modification and oxidation [8, 9]. Modified lipids are recognized by Toll-like and other innate receptors and initiate a cascade of inflammatory and pro-atherogenic events that include the development of foam cells, fatty streaks, and ultimately complex atherosclerotic lesions [7, 10]. Both innate and adaptive immune cells produce inflammatory cytokines and proteases that destabilize the plaque and drive thinning of the plaque cap, lesion progression, and the risk of rupture [11].

Several players, of which two are of great therapeutic interest, maintain the homeostasis of LDL cholesterol: HMG-CoA reductase is the rate-limiting enzyme of endogenous cholesterol biosynthesis [12]. By competitive inhibition of this enzyme, statins effectively block cholesterol synthesis in the liver, which reduces overall cardiovascular mortality. Statins are now recommended in primary and secondary prevention [12–15]. Proprotein convertase subtilisin/kexin type 9 (PCSK9) is a serine protease expressed in hepatocytes, which acts by enhancing LDLR degradation and removal from the cell surface [16]. Blocking PCSK9 increases LDL clearance and effectively reduces plasma cholesterol levels [17]. Despite the success of lipid-lowering treatment [13], cardiovascular disease has frequently been observed to occur in individuals without hyperlipidemia or the traditional risk factors such as age, smoking, hypertension, and diabetes [4, 18]. Based on these epidemiologic data, it is clear that attention must be focused on alternative pathophysiological processes beyond lipid homeostasis.

3.2 Rationale for an Atherosclerosis Vaccine

Atherosclerosis has been studied for over 100 years [19]. The adaptive immune component of atherosclerosis was discovered in 1986, when Hansson et al. described $CD4^+$ T-helper cells in the plaque [20]. However, immune modulation has not yet translated to promising clinical prevention strategies. The potency of anti-inflammatory therapies has best been shown for statins, which have many pleiotropic anti-inflammatory effects [17, 21] and are now recommended for individuals at high risk for cardiovascular disease, even in the absence of clinical atherosclerosis [22]. Several anti-inflammatory therapies have been shown to be effective in animal models of atherosclerosis [23]. In humans, an antibody to IL-1β has successfully been tested in clinical trials [24]. However, despite their concomitant immune-modulating effects, anti-inflammatory therapies are not capable of specifically targeting autoimmunity and still hold the risk to dampen host defense or to promote cancer.

Lymphocytes are found even in the healthy aortic wall in mice [25]. Therefore, it is reasonable to suspect that lymphocytes initiate the immune response in atherosclerosis [26]. In particular, T-lymphocytes, which appear in early stages of atherosclerosis and account for over 10% of all cells found in human plaque [20, 27], seem to be early key modulators of atherosclerosis [28, 29] (Table 3.1). Certain self- and non-self antigens, such as ApoB-100 [30] or heat-shock proteins (HSPs) [31], have been proposed as candidate antigens in the plaque that trigger $CD4^+$ T cell interaction with antigen-presenting cells in the aortic wall [32]. Modulating these early events by antigen-specific immunization or immune modulation could represent a causal treatment for atherosclerosis [33, 34]—a concept first tested in rabbits by Gero et al. in 1959 [35]. Such immune-modulatory approaches (Table 3.2) could represent antigen-specific therapies that would not affect host defense [11].

Table 3.1 Lymphocyte subsets in atherosclerosis

Lymphocyte	Lineage	Markers	Effector cytokines	Role in atherosclerosis
$CD4^+$ T cells	T_H1	T-bet	IFN-γ, TNF, IL-2, IL-3	Pro-atherogenic [87, 158, 159]
	T_H2	GATA3	IL-4, IL-5, IL-13	Controversial [29, 88]
	T_H17	ROR-γT	IL-17A+F, IL-21, IL-22	Controversial [89–92]
	T_{reg}	FoxP3, CD25, CTLA-4	IL-10, TGF-β	Protective [160–164]
	T_R1	Unknown ($FoxP3^-$ $CD25^-$)	IL-10, TGF-β IFN-γ, IL-5+	Protective [162, 165]
	T_{FH}	Bcl6, CXCR5, PD-1	IL-21	Not tested
	Subset	*Cell markers*	*Antibodies*	
B cells	B1	$B220^{low}$, $CD11b^+$, $CD23^-$	IgM, IL-10	Protective [106, 107]
	B2	$B220^+$, $CD23^+$, $CD11b^-$	IgG	Pro-atherogenic [108–111]

T_H T-helper, T_{reg} T-regulatory cell, T_{FH} T-follicular-helper cell, *IgM* Immunoglobulin M, *IgG* Immunoglobulin G

Table 3.2 Modulation of the immune system

Tolerance	Immunosuppression	Active immunization	Passive immunization
– Non-reactivity of the immune system toward a foreign antigen or self-antigen – Can prevent autoimmune disease – Shaped in thymus (central tolerance) or in peripheral tissue (peripheral tolerance)	– Therapeutic reduction of immune activity by drugs, plasmapheresis, or radiation – Clinically used in organ transplantation and autoimmune disease	– Induction of immunity through an antigen – Elicits a humoral response by antibodies targeting the antigen – Induces a cellular ($CD4^+$/$CD8^+$ T cell) response – Antigen-specific memory	– Induction of immunity through delivery of antigen-specific antibodies or blood products containing antibodies (sera, ivIG) – Generates no cellular response – Effects are immediate

ivIG intravenous immunoglobulin

3.3 Atherosclerosis-Associated Antigens

Among all cells of the adaptive immune system that can be found in the atherosclerotic plaque, $CD4^+$ T-helper (T_H) cells predominate, while $CD8^+$ cytotoxic T cells and B cells are a minority of plaque lymphocytes [25]. Lymphocytes are activated by specific antigens that can be recognized by unique surface molecules, the T cell receptor (TCR), and the B cell receptor (BCR), respectively [36]. The BCR is essentially a cell surface version of the antibody made by that B cell and can recognize any structure, protein, lipid, carbohydrate, and even chemicals [36]. By contrast, the TCR must recognize antigen-derived peptides presented by specific cell surface molecules, MHC-II for $CD4^+$ T cells and MHC-I for $CD8^+$ T cells. MHC-I is expressed on almost all cells and MHC-II is expressed on antigen-presenting cells, such as dendritic cells, macrophages, and B cells in the atherosclerotic plaque or the draining lymph nodes [37]. Several reports have strengthened the importance of specific antigen recognition in atherosclerosis. T-helper cells interact frequently with antigen-presenting cells in the plaque [32]. T cells found in human coronary arteries and mouse lesions showed a restricted repertoire of TCRs [38, 39], indicating that only T cells accumulate and proliferate in the plaque that are specific for atherosclerosis-associated antigens. The challenge in recent years was to identify specific antigens that initiate the immune response in the plaque [34]. Exogenous antigens stem from bacteria, viruses, or other invading organisms and have spawned the infectious hypothesis of atherosclerosis [40], which was not supported by clinical trials using various antibiotics. An immune reaction to self-antigens, endogenously expressed by the host organism or generated by cell death, gave rise to the autoimmune hypothesis [33, 41] (Fig. 3.1). It is currently unknown whether the autoimmune response in atherosclerosis is pro- or anti-atherogenic.

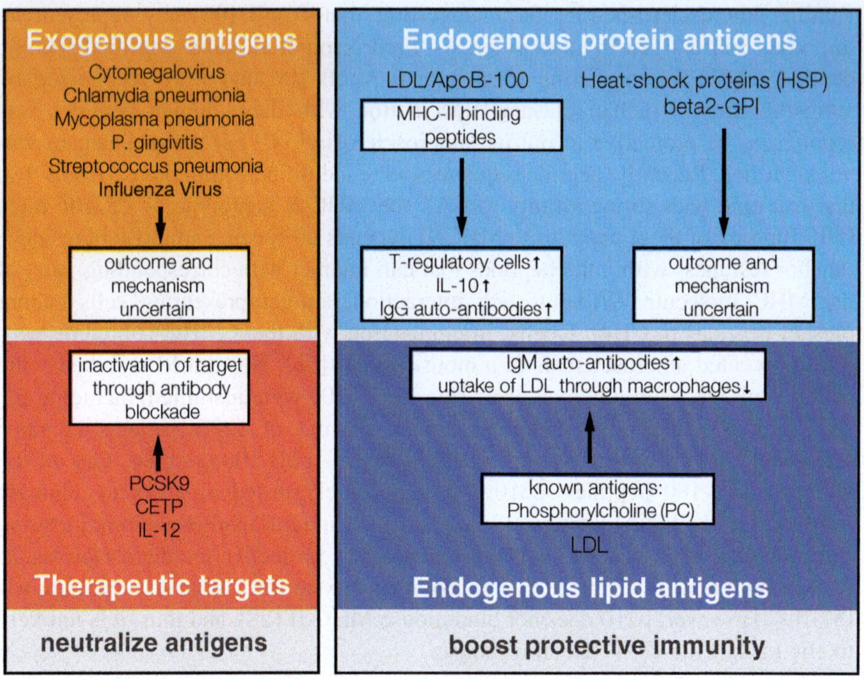

Fig. 3.1 Proposed cardiovascular vaccination strategies

3.4 Modulating Autoimmunity to Endogenous Protein Antigens

LDL/ApoB-100 Of all particles that accumulate in the plaque, LDL is most highly positively correlated with adverse clinical outcomes [2] including coronary atherosclerosis [42]. LDL is a large heterogeneous particle with a diameter of 25 nm [6]. Upon deposition in the atherosclerotic lesions, LDL is modified by oxidative processes and eventually taken up by lesional macrophages, which promotes their transition into foam cells [43, 44]. LDL was suggested to contain auto-epitopes recognized by T cells. Indeed, T cells resident in the plaque specifically respond and proliferate when LDL is presented by APCs [32]. Unexpectedly, vaccination with either native or modified (oxidized) LDL was shown to prevent atherosclerosis in different animal species. Successful vaccination with LDL has been described for various adjuvant combinations, dosages, and routes [45–52]. The protective effect of immunizing against LDL appears to encompass two fundamental mechanisms: (1) the recognition of peptide antigens derived from ApoB-100 by $CD4^+$ T cells and (2) the recognition of (modified) lipid epitopes that induce B cells, which differentiate into plasma cells and produce antibodies to modified LDL.

$CD4^+$ T cells recognize peptides from autoantigens that are presented on Major Histocompatibility Complex (MHC)-II but not the antigen itself or lipid epitopes. ApoB-100 contains the most plausible candidates for the primary immunogenic peptide epitopes. Indeed, vaccination with ApoB-100 alone was effective in preventing atherosclerosis [53, 54], suggesting that ApoB-100 may represent an important self-antigen in atherosclerosis. Fredrikson et al. have identified specific peptide sequences originating from human ApoB-100 that were recognized by autoantibodies in human plasma. Vaccination with these human peptides was reported to be protective in murine atherosclerosis [55], but the mechanism was not identified. Recently, peptide sequences were identified within mouse ApoB-100 that can bind with strong affinity to I-A^b, the MHC-II haplotype of C57Bl6 mice [56]. Binding of these peptides to MHC-II depends on specific amino acid residues (anchor residues) within the peptides that can interact with corresponding parts of the MHC molecule. Without such interaction, antigen-presenting cells cannot present peptides to $CD4^+$ T cells. Immunization with these MHC-II-binding peptides prevented atherosclerosis in a mouse model [56]. Now, at least five specific peptide epitopes within human or murine ApoB-100 were identified, including p2, p3, p6, p143, and p210. Notably, human plasma of CVD patients contains antibodies that cross-reacted with some of these ApoB-100 peptides. One of the human ApoB-100 peptides, p210, was extensively tested in different adjuvant formulations, including intranasal, subcutaneous, and intraperitoneal immunization [53]. Immunization against p210 was shown to be protective in atherosclerosis, as well as angiotensin II-induced hypertension, aneurysm formation, and renal fibrosis [57–59]. However, p210 does not bind mouse MHC-II [28], and thus, it is unlikely to elicit a specific $CD4^+$ T cell response.

Heat-Shock Proteins (HSP) Another group of self-antigens that may trigger the autoimmune response in atherosclerotic lesions are heat-shock proteins [60]. HSPs are intracellular chaperones that are expressed upon cellular stress, cold, UV light, and changes in pH [61]. They are highly conserved throughout different organisms [61]. In humans, HSP60 antibody titers correlate with cardiovascular disease [62, 63]. The efficacy of immunization against HSPs remains controversial: Immunization against HSP60/65 increased atherosclerosis in several reports [64–69], while others reported atheroprotection [70–73]. Interestingly, human HSP60 and bacterial HSP65 share some B cell epitopes [74], giving rise to the hypothesis that molecular mimicry could support cross-reactivity between infectious epitopes and self-epitopes. Such cross-reactivity was previously shown for epitopes within *Streptococcus pneumoniae* and oxidation-specific epitopes in mice [75]. A systematic approach to screen for immunogenic peptides derived from HSPs has not been described so far. Only recently, a 24-amino acid peptide within HSP60, termed p266, has been reported to accelerate atherosclerosis [69].

β2-Glycoprotein I (β2GPI) β2GPI—the target of anti-cardiolipin antibodies [76]—is a plasma protein that is the principal auto-antigen in patients with the anti-phospholipid syndrome [77], a hypercoagulation disorder that accompanies systemic lupus erythematosus. β2GPI is found in human atherosclerotic lesions [78]. The outcome of vaccination against β2GPI is controversial: Immunization against β2GPI accelerated early atherosclerosis in LDL receptor-deficient mice [79–81], while it was protective in another study [82]. Notably, lymphocytes from mice immunized with β2GPI promoted lesion formation in $LDLR^{-/-}$ mice [83].

3.4.1 Atheroprotective Mechanisms

In principle, two different mechanisms have been proposed to mediate atheroprotection following immunization: a humoral antibody-driven response initiated by B cells or a cellular protective immune response by $CD4^+$ T cells.

Autoantibodies induced by vaccination can neutralize the target antigen. If the antigen is a cytokine, antibodies can increase or decrease its activity at the receptor. Antibodies can also promote or diminish antigen uptake by macrophages in the context of atherosclerosis [108, 109]. An increase of autoantibodies has been observed in some but not in all atheroprotective immunization protocols. For example, Freigang et al. observed that immunization with malondialdehyde (MDA)-modified LDL as well as native LDL protected from atherosclerosis in $LDLR^{-/-}$ mice. However, only MDA-modified LDL induced an IgG1 and IgG2 antibody response, while antibodies targeting native LDL were not found in LDL-immunized mice [45]. Vaccination with ApoB-100-derived peptides protected from atherosclerosis and induced IgG antibodies against the peptides. However, these antibodies did not cross-react with intact LDL [56]. Another study reported atheroprotection by a mix of different ApoB-100-derived peptides without induction of autoantibodies against these peptides [84], suggesting that such antibodies are not required for atheroprotection. If this is true, antibodies may just be biomarkers of vaccination and not causal actors in atheroprotection: Upon intracellular processing of the antigen, anti-peptide antibodies are generated, but the relevant peptide sequence is not accessible in the native protein. As such antibodies would exhibit no biological function. Indeed, vaccination against HSPs induced antibodies, but IgG antibodies did not correlate with the disease outcome after vaccination. Interestingly, direct transfer of IgGs induced by immunization enhanced atherosclerosis, while vaccination strategies that resulted in athero-protection did also show enhanced IgG antibody titers [53, 64]. Of note, it has been shown that IgG antibodies against LDL can promote LDL aggregation and enhance uptake of LDL into macrophages, which may provide a link between autoantibody generation and disease progression [85]. Antibodies directed against a peptide epitope of HSP60 bound to endothelial cells and caused increased cytotoxicity [69]. The direct effect of specific IgG antibodies against peptide epitopes from ApoB-100 has not been tested systematically.

The alternative hypothesis is that atheroprotection following immunization may be due to atheroprotective cytokines secreted by antigen-specific $CD4^+$ T-helper cells. This is supported by the observation that immunization with oxLDL generated T cells that did only respond to native LDL but not to oxLDL [54]. $CD4^+$ T-helper cells are the main lymphocytes residing in the atherosclerotic plaque [25]. The majority of lesional T cells are $CD4^+$ and show distinct lineage commitment into Type 1 T-helper (T_H1), Type 2 T-helper (T_H2), Type 17 T-helper (T_H17), or T-regulatory helper (T_{reg}) cells (an overview of T-helper cell lineages is provided in [86], Table 3.1). T_H1 cells and the main T_H1-derived cytokine IFN-γ clearly exacerbate atherosclerosis [28, 87]. T_H2 responses, mainly based on the T_H2 signature cytokine IL-4, antagonize T_H1 pro-atherogenic effects and protect against early lesion formation [29], although in some studies, depletion of IL-4 has been suggested to be atheroprotective [88]. Studies on the role of IL-17-producing T_H17 cells have yielded inconsistent results. Blockade or genetic deletion of IL-17 reduced atherosclerosis and induced a stable plaque phenotype [89–91], while in other reports IL-17 deficiency accelerated plaque formation [92]. T_{regs} secrete IL-10, TGF-β, and other anti-inflammatory cytokines. T_{regs} are the main mechanism that curbs autoimmunity [93], dramatically demonstrated by the rampant, lethal autoimmunity in mice [94] or humans [95] with defective T_{regs}. T_{regs} are beneficial in atherosclerosis and the prototypic T_{reg} cytokines TGF-β and IL-10 show anti-inflammatory and atheroprotective effects [96]. The finding that much fewer T_{reg} cells are found in all stages of atherosclerotic lesions than in other chronic inflammatory diseases [97] has led to the hypothesis that local breach of self-tolerance against potential antigens in the plaque and a loss of balance between protective T_{regs} and pro-atherogenic T-effector cells may be promoting factors of inflammation and lesion progression [28, 97].

Atheroprotection following immunization has been proposed to skew the polarization of T cells into distinct, favorable T_H lineages, such as into T_{regs} [28]. Wigren et al. reported increased percentage of $CD25^+$ $FoxP3^+$ T_{reg} cells in the spleen. FoxP3 is the signature transcription factor defining T_{regs}. Splenic T cells following immunization with the ApoB-100 peptide p210 showed T_{reg} polarization [98]. Depletion of Treg cells abolished the atheroprotective effect of a set of ApoB-100 peptides in another study [84]. The exact cellular and molecular mechanisms by which T_{reg} cells can protect from atherosclerosis in the context of immunization, as well as their antigen specificity, are unknown. It has been reported that IL-10 secreted by T_{regs} protects from atherosclerosis [99]. Consistent with this finding, immunization with the ApoB-100 peptides p3 and p6 induced IL-10 expression in the atherosclerotic aorta [56]. Other studies reported a decrease in T_H1 and T_H2 cytokines upon vaccination [84], which could also be a result of increased T_{reg}-mediated effector T cell suppression. A recent report indicated that immunization in autoimmune disease induces T_R1 cells and $CD4^+$ T-helper cells that express IL-10 but not the T_{reg} marker FoxP3 [100]. Vaccination with HSP60 induces both T_{regs} and T_R1 cells in the spleen and cervical lymph nodes of mice [73]. This effect was accompanied by increased levels of circulating IL-10 and TGF-β and decreased secretion of IL-17 and IFN-γ. Notably, inhibition of IL-10 abrogated the atheroprotective effect [73]. It is unknown which specific pathways induce the activation of T_{reg} cells in the context of vaccination. It was proposed that Fms-like tyrosine kinase 3 (Flt3)-expressing tolerogenic dendritic cells drive the generation of Treg cells in atherosclerosis [101–103], an effect thought to be dependent on expression of TGFβ and retinoic acid by dendritic cells [104]. Injection of ApoB-100-loaded dendritic cells treated with IL-10 protected from atherosclerosis and induced a T_{reg} response in lymphoid organs, corroborating this concept [54]. Taken together, these findings identify T_{regs} and T_R1 cells, as well as the atheroprotective cytokine IL-10, as the best candidates to cause atheroprotection after vaccination.

Antigen-specific T cell responses to protein antigens during the natural course of atherosclerosis or after vaccination have not been studied. All published data stem either from analyzing bulk T cells at different locations or from indirect evidence, mainly by restimulation protocols, in which $CD4^+$ T cells were isolated from atherosclerosis-prone animals, atherosclerotic lesions, or after vaccination with the proposed antigens and exposed to antigen-loaded APCs to activate and expand the responding T cells. These classical strategies, albeit capable of identifying a part of the auto-reactive T cell repertoire, are neither effective in detecting T cells with a low-affinity TCR that are less likely to proliferate (likely T_{regs}), nor can they identify the natural phenotype of these cells without prior stimulation in vitro. It is likely that only a tiny proportion of all T cells are specific for a given antigen [105]. T cell-mediated atheroprotection has been shown to be important early in atherosclerosis in mice [26]. Thus, it is unclear whether manipulating response cell immunity in patients with already established atherosclerosis would be beneficial.

3.5 Modulating the Autoimmune Response to Endogenous (Modified) Lipid Antigens

B cells figure in all stages of atherosclerosis, but their effect on lesion formation is complex. B cells have been reported to protect from atherosclerosis [106, 107] or to promote atherosclerosis [108–111], depending on the particular B cell subset studied. Plasma cells derived from B cells secrete antigen-specific IgM or IgG antibodies that can bind to the antigen and neutralize it. It has been reported that treatment with infused polyclonal immunoglobulin preparations (ivIg) protects from murine atherosclerosis, although the mechanism is unclear [112]. It has been demonstrated that B1 cells, which can be defined by expression of certain surface markers (Table 3.1), are capable of inducing a protective humoral immune response that can protect from atherosclerosis and do not require participation of T cells [113]. The protective response of B1 cells depends on the generation of IgM antibodies that can cross-react with lipid B cell epitopes within LDL. Such antibodies help clearing LDL from the plaque and inhibit uptake of LDL into macrophages [113, 114]. Oxidation-specific (neo-) epitopes that are generated by oxidation of LDL, such as phosphorylcholine (PC) and oxidized cholesterol, have been identified as primary targets of these antibodies [115]. The majority of reports suggested that IgM but not IgG antibodies that require T cell help carry out this protective response. Notably, the majority of all IgM antibodies found in mice and humans target oxidation-specific epitopes [115]. In humans, IgM and in some studies also IgG antibodies to modified LDL correlate negatively with disease [116–118], although other studies have reported a positive correlation between IgM and IgG antibodies with the extent of atherosclerotic lesions [119]. Antibodies against oxidation-specific epitopes protected mice from atherosclerosis [120, 121]. Vaccination against oxLDL, MDA-modified LDL, or *Streptococcus pneumoniae*, which shows molecular mimicry with oxidized LDL [75], generated antibodies specific for PC. Anti-PC antibodies generated after vaccination with PC protected from atherosclerosis [122]. Some of these neo-epitopes within oxLDL are also found in apoptotic cells. In fact, some antibodies against oxLDL can also bind to apoptotic cells and vice versa [123]. Injection of apoptotic cells that bear similar neo-epitopes as found in oxLDL induced oxidation-specific antibodies in the spleen and protected from atherosclerosis in ApoE$^{-/-}$ mice [123]. B1a, B1b, and marginal zone B cells (MZB) have been proposed as origin of this protective IgM response [124, 125].

3.6 Clearance of Potential Exogenous Antigens

A considerable number of different pathogens were identified in human and mouse atherosclerotic lesions, including bacteria, such as *Chlamydia pneumoniae*, *Mycoplasma pneumoniae*, and *Porphyromonas gingivalis*, as well as viruses like *Cytomegalovirus* (CMV), *Hepatitis C Virus* (HCV), *Human Immunodeficiency Virus* (HIV), or *Human Papillomavirus* (HPV) [126, 127]. Infectious diseases caused by some of these pathogens have been linked to CVD in some epidemiological studies.

For instance, infection with the varicella-zoster virus (VZV) that causes chickenpox and herpes zoster upon reactivation is clinically linked to stroke [128, 129]. Whether vaccination against VZV alters cardiovascular outcome is not known. Notably, clinical trials that neutralized *Chlamydia* infection by an antibiotic treatment have failed to translate into a measurable cardiovascular disease protection [130]. Influenza is clinically associated with increased risk for CVD, in particular for acute myocardial infarction and overall cardiovascular mortality [131]. It has been proposed that enhanced plaque inflammation following influenza accounts for this increase in CVD [132, 133]. Interestingly, a recent meta-analysis shows a reduction of cardiovascular events following influenza vaccination within 1 year [134]. These findings have been confirmed in further case-controlled trials [135, 136]. Based on these findings, influenza vaccination is now recommended as secondary prevention for patients with heart disease [137].

Whether these results suggest a causal link between influenza and T cell immunity is unclear. However, it is possible that TCRs or BCRs cross-react with epitopes on pathogens that have similarities to self-antigens [105], which could elicit an immune response against a self-antigen, that was initiated by the foreign antigen. Vaccination against *S. pneumoniae* suppressed atherosclerosis in a mouse model [75]. Interestingly, anti-oxLDL antibodies are also found more frequently in cardiovascular healthy individuals after pneumococcal vaccination [138]. Whether this antibody response is beneficial in protecting from cardiovascular disease has been controversial [139]. Particularly, one study found no protection from acute myocardial infarction and stroke within 30 days after pneumococcal vaccination [140]. It has also been proposed that immune activation caused by infection or other autoimmune disease could trigger an immune response to atherosclerosis antigens by indirect pathways.

3.7 Vaccination to Neutralize Protein-Effector Functions

The classic concept of vaccination is to induce a specific immune response that is capable of clearing the infectious organisms or neutralizing the relevant antigen, often a toxin [36]. This concept has been translated into cardiovascular vaccination strategies to neutralize potential pro-atherogenic mediators that interfere with lipid metabolism or TCR functioning.

Blocking TCR-MHC II Interaction Hermansson et al. recently identified that $CD4^+$ T cells generated by immunization against native LDL showed oligoclonal usage of TCR variable (V) β segments as determined by TCR sequencing [141]. Hybridomas generated from these T cells were enriched for the TCRBV31 segment. The V segment is involved in TCR interaction with MHC-II and is therefore needed for priming and expansion of T cells by antigen-presenting cells. Immunization with a TCRBV31 peptide resulted in antibodies specific for this TCR and protected from atherosclerosis, likely by eliminating atherosclerosis-specific T cell clones.

Neutralizing Interleukin-12 (IL-12) IL-12 is a cytokine secreted by macrophages and other antigen-presenting cells that activates T cells and induces an atherogenic T_H1 polarization. Immunization of LDLR$^{-/-}$ mice with an IL-12-adjuvant complex resulted in IL-12 specific, neutralizing antibodies that blocked IL-12 downstream signaling, diminished IFN-γ production in T cells, and decreased atherosclerotic lesions with a clinically more favorable stable plaque phenotype in mice [142].

PCSK9 Proprotein convertase subtilisin/kexin type 9 (PCSK9) is a protein that helps to degrade LDL receptor. Its expression is negatively correlated with cardiovascular events and individuals with a loss-of-function mutation are greatly protected [16]. In a recent study, neutralizing antibodies to PCSK9 were induced in mice and primates by vaccination with virus-like particles that displayed PCSK9-derived peptides [143]. However, vaccination did not reduce LDL cholesterol.

CETP Cholesteryl ester transfer protein (CETP) is an enzyme critically involved in cholesterol loading and transfer from high-density lipoproteins (HDL) to LDL and very-low-density lipoproteins (VLDL), which have a strong pro-atherogenic potential, while HDL negatively correlates with disease. Antibodies directed against CETP, induced by vaccination with a CETP fusion protein with a fragment of tetanus toxin can inhibit cholesterol loading of VLDL and LDL. CETP antibody titers are negatively correlated with atherosclerosis. Preclinical studies have shown that such antibodies are atheroprotective and increase HDL cholesterol [144, 145].

3.8 Considerations for Translational Strategies

Vaccination is a powerful tool in primary prevention of infectious diseases. Recent advances in checkpoint inhibitors suggest that vaccines could also effectively prevent or treat cancer. However, current vaccination approaches against atherosclerosis remain at the preclinical level. Many questions that are fundamental to ultimately translate these findings into a human vaccine remain unresolved. Development of a vaccine to prevent cardiovascular disease requires a stepwise approach of (1) identifying of the exact epitopes within antigens that modify atherosclerosis, (2) defining the atheroprotective mechanisms with respect to their cellular origins, and (3) specifying immune-tolerizing adjuvants, doses, and routes of antigen delivery that could be translated to humans.

Immune-Tolerizing Adjuvants and Routes It has been demonstrated that the effect of vaccination is greatly dependent on the choice of adjuvants. Many clinically used vaccines are adjuvanted by aluminum salts (alum). Traditional vaccination protocols in rodents use a combination of complete Freud's adjuvant (CFA), which contains heat-inactivated *Mycobacterium tuberculosis* in an emulsion with mineral oil for prime injection, and incomplete Freud's adjuvant (IFA), which does not contain *M. tuberculosis*, for subsequent booster injections [34]. It has been shown that some adjuvants, including IFA and the clinically widely used alum, may foster an atheroprotective response alone, even without a specific antigen [98, 146]. This effect was thought to be partially caused by enhanced autoantibody generation and a yet undefined immune mechanism that depends on the site of injection [147]. Also, booster injections have been placed in the peritoneal cavity of mice and other animals, a strategy that is not translatable to humans. The route of antigen delivery seems to be important: nasal and oral application of an HSP65 vaccine was atheroprotective [71, 72], while intra- or subcutaneous injection promoted lesion formation [64, 65]. Alternative routes have also been exploited in vaccination against oxLDL and an ApoB-100 fusion protein, including oral and nasal antigen delivery [53, 148].

Multivalent Vaccines Several antigens have been successfully tested in animal models of atherosclerosis. It has been proposed that multivalent vaccines that incorporate not one but instead several antigens could help to design a broader and more efficient vaccine [149], because they may induce synergistic, beneficial effects. In particular, simultaneous immunization with two ApoB-100 and HSP60 peptides was more effective than one of the peptides alone [150], albeit the mechanisms were not identified. Efforts have also been made to incorporate antigenic peptides from HSPs, ApoB-100, and β2GPI into one multivalent vaccine [151]. Recently, it has been proposed that delivery of peptides embedded in MHC-II molecules, rather than the peptide alone, may be more potent to trigger a T cell specific responses capable of inducing a protective T_R1 response in autoimmune disease [100], but this approach has not been tested in atherosclerosis.

DNA Vaccines The delivery of antigens not through the antigen itself but by delivery of DNA plasmids that encode for the antigen that is de novo expressed in target cells has been tested in animal models of autoimmunity, including EAE and rheumatoid arthritis [152–154]. In the context of atherosclerosis, DNA vaccines have been shown to induce a specific response toward vascular endothelial growth factor 2 (VEGF2), which is expressed in stressed endothelial cells. DNA immunization induced $CD8^+$ cytotoxic T cells that neutralized VEGF-2 expressing endothelial cells and protected from atherosclerosis [155].

Passive Immunization in Clinical Trials It has been proposed that some species of autoantibodies against modified LDL may be atheroprotective. A human IgG1 antibody specific for an oxidation-specific epitope from ApoB-100, termed p45, was protective in mouse atherosclerosis [120, 121]. This antibody, MLDL1278a, showed anti-inflammatory properties in obese primates [156] and was later tested in the multicenter, randomized GLACIER trial (Goal of Oxidized LDL and Activated Macrophage Inhibition by Exposure to a Recombinant Antibody). The primary endpoint in this study was arterial wall inflammation quantified by positron emission tomography (PET) imaging with 18F-fluorodeoxyglucose (FDG). Despite promising experimental evidence in animals, the trial failed to show an effect in its primary endpoint [157].

3.9 Conclusion

Atherosclerosis is the main underlying pathologic process that drives CAD, MI, stroke, and PAD. Its initiation and progression in now understood as a complex interplay of innate and adaptive immunity that engages humoral and cellular immunity. Adaptive immune responses to foreign antigens and to self-antigens have been reported. Several antigens that could trigger such autoimmune response in disease-prone animals and individuals have been identified. Vaccination against such antigens was atheroprotective in different species, albeit precise mechanisms and antigen specificity have not been identified. T-regulatory cells, the atheroprotective cytokine IL-10, and antigen-specific IgM antibodies are candidates that may confer atheroprotection. The first clinical trial to test passive immunization in humans has failed. These results should not discourage but instead drive future work to define epitopes in humans, clarify the mechanisms that underlie atheroprotective vaccination, and design vaccination protocols that can be translated to clinical practice.

Compliance with Ethical Standards

Conflict of Interest: Dennis Wolf, Teresa Gerhardt, and Klaus Ley declare that they have no conflict of interest.

Ethical Approval: This article does not contain any studies with human participants or animals performed by any of the authors.

References

1. Lozano R, Naghavi M, Foreman K, Lim S, Shibuya K, Aboyans V, Abraham J, Adair T, Aggarwal R, Ahn SY, Alvarado M, Anderson HR, Anderson LM, Andrews KG, Atkinson C, Baddour LM, Barker-Collo S, Bartels DH, Bell ML, Benjamin EJ, Bennett D, Bhalla K, Bikbov B, Bin Abdulhak A, Birbeck G, Blyth F, Bolliger I, Boufous S, Bucello C, Burch M, Burney P, Carapetis J, Chen H, Chou D, Chugh SS, Coffeng LE, Colan SD, Colquhoun S, Colson KE, Condon J, Connor MD, Cooper LT, Corriere M, Cortinovis M, de Vaccaro KC, Couser W, Cowie BC, Criqui MH, Cross M, Dabhadkar KC, Dahodwala N, De Leo D, Degenhardt L, Delossantos A, Denenberg J, Des Jarlais DC, Dharmaratne SD, Dorsey ER, Driscoll T, Duber H, Ebel B, Erwin PJ, Espindola P, Ezzati M, Feigin V, Flaxman AD, Forouzanfar MH, Fowkes FG, Franklin R, Fransen M, Freeman MK, Gabriel SE, Gakidou E, Gaspari F, Gillum RF, Gonzalez-Medina D, Halasa YA, Haring D, Harrison JE, Havmoeller R, Hay RJ, Hoen B, Hotez PJ, Hoy D, Jacobsen KH, James SL, Jasrasaria R, Jayaraman S, Johns N, Karthikeyan G, Kassebaum N, Keren A, Khoo JP, Knowlton LM, Kobusingye O, Koranteng A, Krishnamurthi R, Lipnick M, Lipshultz SE, Ohno SL, Mabweijano J, MacIntyre MF, Mallinger L, March L, Marks GB, Marks R, Matsumori A, Matzopoulos R, Mayosi BM, McAnulty JH, McDermott MM, McGrath J, Mensah GA, Merriman TR, Michaud C, Miller M, Miller TR, Mock C, Mocumbi AO, Mokdad AA, Moran A, Mulholland K, Nair MN, Naldi L, Narayan KM, Nasseri K, Norman P, O'Donnell M, Omer SB, Ortblad K, Osborne R, Ozgediz D, Pahari B, Pandian JD, Rivero AP, Padilla RP, Perez-Ruiz F, Perico N, Phillips D, Pierce K, Pope CA 3rd, Porrini E,

Pourmalek F, Raju M, Ranganathan D, Rehm JT, Rein DB, Remuzzi G, Rivara FP, Roberts T, De Leon FR, Rosenfeld LC, Rushton L, Sacco RL, Salomon JA, Sampson U, Sanman E, Schwebel DC, Segui-Gomez M, Shepard DS, Singh D, Singleton J, Sliwa K, Smith E, Steer A, Taylor JA, Thomas B, Tleyjeh IM, Towbin JA, Truelsen T, Undurraga EA, Venketasubramanian N, Vijayakumar L, Vos T, Wagner GR, Wang M, Wang W, Watt K, Weinstock MA, Weintraub R, Wilkinson JD, Woolf AD, Wulf S, Yeh PH, Yip P, Zabetian A, Zheng ZJ, Lopez AD, Murray CJ, AlMazroa MA, Memish ZA. Global and regional mortality from 235 causes of death for 20 age groups in 1990 and 2010: a systematic analysis for the Global Burden of Disease Study 2010. Lancet. 2012;380:2095–128.

2. Libby P. Inflammation in atherosclerosis. Nature. 2002;420:868–74.
3. Falk E. Pathogenesis of atherosclerosis. J Am Coll Cardiol. 2006;47:C7–C12.
4. Mozaffarian D, Benjamin EJ, Go AS, Arnett DK, Blaha MJ, Cushman M, de Ferranti S, Despres JP, Fullerton HJ, Howard VJ, Huffman MD, Judd SE, Kissela BM, Lackland DT, Lichtman JH, Lisabeth LD, Liu S, Mackey RH, Matchar DB, McGuire DK, Mohler ER 3rd, Moy CS, Muntner P, Mussolino ME, Nasir K, Neumar RW, Nichol G, Palaniappan L, Pandey DK, Reeves MJ, Rodriguez CJ, Sorlie PD, Stein J, Towfighi A, Turan TN, Virani SS, Willey JZ, Woo D, Yeh RW, Turner MB, American Heart Association Statistics C, Stroke Statistics S. Heart disease and stroke statistics—2015 update: a report from the American Heart Association. Circulation. 2015;131:e29–322.
5. Tabas I, Williams KJ, Boren J. Subendothelial lipoprotein retention as the initiating process in atherosclerosis: update and therapeutic implications. Circulation. 2007;116:1832–44.
6. Libby P, Aikawa M, Schonbeck U. Cholesterol and atherosclerosis. Biochim Biophys Acta. 2000;1529:299–309.
7. Glass CK, Witztum JL. Atherosclerosis. Cell. 2001;104:503–16.
8. de Winther MP, van Dijk KW, Havekes LM, Hofker MH. Macrophage scavenger receptor class A: a multifunctional receptor in atherosclerosis. Arterioscler Thromb Vasc Biol. 2000;20:290–7.
9. Miller YI, Choi S-H, Fang L, Tsimikas S. Lipoprotein modification and macrophage uptake: role of pathologic cholesterol transport in atherogenesis. In: Harris RJ, editor. Cholesterol binding and cholesterol transport proteins: structure and function in health and disease. Dordrecht: Springer; 2010. p. 229–51.
10. Mitra S, Goyal T, Mehta JL, Oxidized LDL. LOX-1 and atherosclerosis. Cardiovasc Drugs Ther. 2011;25:419–29.
11. Ketelhuth DF, Hansson GK. Adaptive response of T and B cells in atherosclerosis. Circ Res. 2016;118:668–78.
12. Istvan ES, Deisenhofer J. Structural mechanism for statin inhibition of HMG-CoA reductase. Science. 2001;292:1160–4.
13. Pedersen TR. The success story of LDL cholesterol lowering. Circ Res. 2016;118:721–31.
14. Downs JR, Clearfield M, Weis S, Whitney E, Shapiro DR, Beere PA, Langendorfer A, Stein EA, Kruyer W, Gotto AM Jr. Primary prevention of acute coronary events with lovastatin in men and women with average cholesterol levels: results of AFCAPS/TexCAPS. Air Force/ Texas Coronary Atherosclerosis Prevention Study. JAMA. 1998;279:1615–22.
15. Sacks FM, Pfeffer MA, Moye LA, Rouleau JL, Rutherford JD, Cole TG, Brown L, Warnica JW, Arnold JM, Wun CC, Davis BR, Braunwald E. The effect of pravastatin on coronary events after myocardial infarction in patients with average cholesterol levels. Cholesterol and Recurrent Events Trial investigators. N Engl J Med. 1996;335:1001–9.
16. Kwon HJ, Lagace TA, McNutt MC, Horton JD, Deisenhofer J. Molecular basis for LDL receptor recognition by PCSK9. Proc Natl Acad Sci USA. 2008;105:1820–5.
17. Shapiro MD, Fazio S. From lipids to inflammation: new approaches to reducing atherosclerotic risk. Circ Res. 2016;118:732–49.
18. Greenland P, Knoll MD, Stamler J, Neaton JD, Dyer AR, Garside DB, Wilson PW. Major risk factors as antecedents of fatal and nonfatal coronary heart disease events. JAMA. 2003;290:891–7.

19. Virchow R. Die Cellularpathologie in ihrer Begründung auf physiologische und pathologische Gewebelehre. Berlin: Verlag von August Hirschwald; 1859.
20. Jonasson L, Holm J, Skalli O, Bondjers G, Hansson GK. Regional accumulations of T cells, macrophages, and smooth muscle cells in the human atherosclerotic plaque. Arteriosclerosis. 1986;6:131–8.
21. Back M, Hansson GK. Anti-inflammatory therapies for atherosclerosis. Nat Rev Cardiol. 2015;12:199–211.
22. Stone NJ, Robinson JG, Lichtenstein AH, Bairey Merz CN, Blum CB, Eckel RH, Goldberg AC, Gordon D, Levy D, Lloyd-Jones DM, McBride P, Schwartz JS, Shero ST, Smith SC Jr, Watson K, Wilson PW, American College of Cardiology/American Heart Association Task Force on Practice G. 2013 ACC/AHA guideline on the treatment of blood cholesterol to reduce atherosclerotic cardiovascular risk in adults: a report of the American College of Cardiology/American Heart Association Task Force on Practice Guidelines. J Am Coll Cardiol. 2014;63:2889–934.
23. Tabas I, Glass CK. Anti-inflammatory therapy in chronic disease: challenges and opportunities. Science. 2013;339:166–72.
24. Ridker PM, Everett BM, Thuren T, MacFadyen JG, Chang WH, Ballantyne C, Fonseca F, Nicolau J, Koenig W, Anker SD, Kastelein JJP, Cornel JH, Pais P, Pella D, Genest J, Cifkova R, Lorenzatti A, Forster T, Kobalava Z, Vida-Simiti L, Flather M, Shimokawa H, Ogawa H, Dellborg M, Rossi PRF, Troquay RPT, Libby P, Glynn RJ, CANTOS Trial Group. Antiinflammatory therapy with canakinumab for atherosclerotic disease. N Engl J Med. 2017;377(12):1119–31.
25. Galkina E, Kadl A, Sanders J, Varughese D, Sarembock IJ, Ley K. Lymphocyte recruitment into the aortic wall before and during development of atherosclerosis is partially L-selectin dependent. J Exp Med. 2006;203:1273–82.
26. Song L, Leung C, Schindler C. Lymphocytes are important in early atherosclerosis. J Clin Investig. 2001;108:251–9.
27. Ketelhuth DF, Hansson GK. Cellular immunity, low-density lipoprotein and atherosclerosis: break of tolerance in the artery wall. Thromb Haemost. 2011;106:779–86.
28. Tse K, Tse H, Sidney J, Sette A, Ley K. T cells in atherosclerosis. Int Immunol. 2013;25:615–22.
29. Mallat Z, Taleb S, Ait-Oufella H, Tedgui A. The role of adaptive T cell immunity in atherosclerosis. J Lipid Res. 2009;50(Suppl):S364–9.
30. Nilsson J, Bjorkbacka H, Fredrikson GN. Apolipoprotein B100 autoimmunity and atherosclerosis – disease mechanisms and therapeutic potential. Curr Opin Lipidol. 2012;23:422–8.
31. Kilic A, Mandal K. Heat shock proteins: pathogenic role in atherosclerosis and potential therapeutic implications. Autoimmune Dis. 2012;2012:9.
32. Koltsova EK, Garcia Z, Chodaczek G, Landau M, McArdle S, Scott SR, von Vietinghoff S, Galkina E, Miller YI, Acton ST, Ley K. Dynamic T cell-APC interactions sustain chronic inflammation in atherosclerosis. J Clin Invest. 2012;122:3114–26.
33. Ley K. 2015 Russell Ross Memorial Lecture in vascular biology: protective autoimmunity in atherosclerosis. Arterioscler Thromb Vasc Biol. 2016;36:429–38.
34. Kimura T, Tse K, Sette A, Ley K. Vaccination to modulate atherosclerosis. Autoimmunity. 2015;48:152–60.
35. Gero S, Gergely J, Jakab L, Szekely J, Virag S, Farkas K, Czuppon A. Inhibition of cholesterol atherosclerosis by immunisation with beta-lipoprotein. Lancet. 1959;2:6–7.
36. Adler R. Janeway's immunobiology. Choice: Curr Rev Acad Libr. 2008;45:1793–4.
37. Ley K. The second touch hypothesis: T cell activation, homing and polarization. F1000Res. 2014;3:37.
38. Paulsson G, Zhou X, Tornquist E, Hansson GK. Oligoclonal T cell expansions in atherosclerotic lesions of apolipoprotein E-deficient mice. Arterioscler Thromb Vasc Biol. 2000;20:10–7.

39. De Palma R, Del Galdo F, Abbate G, Chiariello M, Calabro R, Forte L, Cimmino G, Papa MF, Russo MG, Ambrosio G, Giombolini C, Tritto I, Notaristefano S, Berrino L, Rossi F, Golino P. Patients with acute coronary syndrome show oligoclonal T-cell recruitment within unstable plaque: evidence for a local, intracoronary immunologic mechanism. Circulation. 2006;113:640–6.
40. Alviar CL, Echeverri JG, Jaramillo NI, Figueroa CJ, Cordova JP, Korniyenko A, Suh J, Paniz-Mondolfi A. Infectious atherosclerosis: is the hypothesis still alive? A clinically based approach to the dilemma. Med Hypotheses. 2011;76:517–21.
41. Wick G, Perschinka H, Millonig G. Atherosclerosis as an autoimmune disease: an update. Trends Immunol. 2001;22:665–9.
42. Colantonio LD, Bittner V, Reynolds K, Levitan EB, Rosenson RS, Banach M, Kent ST, Derose SF, Zhou H, Safford MM, Muntner P. Association of serum lipids and coronary heart disease in contemporary observational studies. Circulation. 2016;133:256–64.
43. Palinski W, Rosenfeld ME, Yla-Herttuala S, Gurtner GC, Socher SS, Butler SW, Parthasarathy S, Carew TE, Steinberg D, Witztum JL. Low density lipoprotein undergoes oxidative modification in vivo. Proc Natl Acad Sci USA. 1989;86:1372–6.
44. Kruth HS, Jones NL, Huang W, Zhao B, Ishii I, Chang J, Combs CA, Malide D, Zhang WY. Macropinocytosis is the endocytic pathway that mediates macrophage foam cell formation with native low density lipoprotein. J Biol Chem. 2005;280:2352–60.
45. Freigang S, Horkko S, Miller E, Witztum JL, Palinski W. Immunization of LDL receptor-deficient mice with homologous malondialdehyde-modified and native LDL reduces progression of atherosclerosis by mechanisms other than induction of high titers of antibodies to oxidative neoepitopes. Arterioscler Thromb Vasc Biol. 1998;18:1972–82.
46. Palinski W, Miller E, Witztum JL. Immunization of low density lipoprotein (LDL) receptor-deficient rabbits with homologous malondialdehyde-modified LDL reduces atherogenesis. Proc Natl Acad Sci USA. 1995;92:821–5.
47. Ameli S, Hultgardh-Nilsson A, Regnstrom J, Calara F, Yano J, Cercek B, Shah PK, Nilsson J. Effect of immunization with homologous LDL and oxidized LDL on early atherosclerosis in hypercholesterolemic rabbits. Arterioscler Thromb Vasc Biol. 1996;16:1074–9.
48. George J, Afek A, Gilburd B, Levkovitz H, Shaish A, Goldberg I, Kopolovic Y, Wick G, Shoenfeld Y, Harats D. Hyperimmunization of apo-E-deficient mice with homologous malondialdehyde low-density lipoprotein suppresses early atherogenesis. Atherosclerosis. 1998;138:147–52.
49. Zhou X, Caligiuri G, Hamsten A, Lefvert AK, Hansson GKLDL. Immunization induces T-cell-dependent antibody formation and protection against atherosclerosis. Arterioscler Thromb Vasc Biol. 2001;21:108–14.
50. Chyu KY, Reyes OS, Zhao X, Yano J, Dimayuga P, Nilsson J, Cercek B, Shah PK. Timing affects the efficacy of LDL immunization on atherosclerotic lesions in apo E (-/-) mice. Atherosclerosis. 2004;176:27–35.
51. Zhou X, Robertson AK, Rudling M, Parini P, Hansson GK. Lesion development and response to immunization reveal a complex role for CD4 in atherosclerosis. Circ Res. 2005;96:427–34.
52. Zhong Y, Wang X, Ji Q, Mao X, Tang H, Yi G, Meng K, Yang X, Zeng Q. CD4+LAP + and CD4 +CD25 +Foxp3 + regulatory T cells induced by nasal oxidized low-density lipoprotein suppress effector T cells response and attenuate atherosclerosis in ApoE-/- mice. J Clin Immunol. 2012;32:1104–17.
53. Klingenberg R, Lebens M, Hermansson A, Fredrikson GN, Strodthoff D, Rudling M, Ketelhuth DF, Gerdes N, Holmgren J, Nilsson J, Hansson GK. Intranasal immunization with an apolipoprotein B-100 fusion protein induces antigen-specific regulatory T cells and reduces atherosclerosis. Arterioscler Thromb Vasc Biol. 2010;30:946–52.
54. Hermansson A, Johansson DK, Ketelhuth DF, Andersson J, Zhou X, Hansson GK. Immunotherapy with tolerogenic apolipoprotein B-100-loaded dendritic cells attenuates atherosclerosis in hypercholesterolemic mice. Circulation. 2011;123:1083–91.

55. Fredrikson GN, Hedblad B, Berglund G, Alm R, Ares M, Cercek B, Chyu KY, Shah PK, Nilsson J. Identification of immune responses against aldehyde-modified peptide sequences in apoB associated with cardiovascular disease. Arterioscler Thromb Vasc Biol. 2003;23:872–8.
56. Tse K, Gonen A, Sidney J, Ouyang H, Witztum JL, Sette A, Tse H, Ley K. Atheroprotective vaccination with MHC-II restricted peptides from ApoB-100. Front Immunol. 2013;4:493.
57. Honjo T, Chyu KY, Dimayuga PC, Lio WM, Yano J, Trinidad P, Zhao X, Zhou J, Cercek B, Shah PK. Immunization with an ApoB-100 related peptide vaccine attenuates angiotensin-II induced hypertension and renal fibrosis in mice. PLoS One. 2015;10:e0131731.
58. Honjo T, Chyu KY, Dimayuga PC, Yano J, Lio WM, Trinidad P, Zhao X, Zhou J, Chen S, Cercek B, Arditi M, Shah PK. ApoB-100-related peptide vaccine protects against angiotensin II-induced aortic aneurysm formation and rupture. J Am Coll Cardiol. 2015;65:546–56.
59. Chyu KY, Zhao X, Dimayuga PC, Zhou J, Li X, Yano J, Lio WM, Chan LF, Kirzner J, Trinidad P, Cercek B, Shah PK. CD8+ T cells mediate the athero-protective effect of immunization with an ApoB-100 peptide. PLoS One. 2012;7:e30780.
60. Wick C. Tolerization against atherosclerosis using heat shock protein 60. Cell Stress Chaperones. 2015; 201–211.
61. Wick G, Jakic B, Buszko M, Wick MC, Grundtman C. The role of heat shock proteins in atherosclerosis. Nat Rev Cardiol. 2014;11:516–29.
62. Zhu J, Quyyumi AA, Rott D, Csako G, Wu H, Halcox J, Epstein SE. Antibodies to human heat-shock protein 60 are associated with the presence and severity of coronary artery disease: evidence for an autoimmune component of atherogenesis. Circulation. 2001;103:1071–5.
63. Kervinen H, Huittinen T, Vaarala O, Leinonen M, Saikku P, Manninen V, Manttari M. Antibodies to human heat shock protein 60, hypertension and dyslipidemia. A study of joint effects on coronary risk. Atherosclerosis. 2003;169:339–44.
64. Xu Q, Dietrich H, Steiner HJ, Gown AM, Schoel B, Mikuz G, Kaufmann SH, Wick G. Induction of arteriosclerosis in normocholesterolemic rabbits by immunization with heat shock protein 65. Arterioscler Thromb. 1992;12:789–99.
65. George J, Shoenfeld Y, Afek A, Gilburd B, Keren P, Shaish A, Kopolovic J, Wick G, Harats D. Enhanced fatty streak formation in C57BL/6J mice by immunization with heat shock protein-65. Arterioscler Thromb Vasc Biol. 1999;19:505–10.
66. Zhang Y, Xiong Q, Hu X, Sun Y, Tan X, Zhang H, Lu Y, Liu J. A novel atherogenic epitope from Mycobacterium tuberculosis heat shock protein 65 enhances atherosclerosis in rabbit and LDL receptor-deficient mice. Heart Vessels. 2012;27:411–8.
67. Afek A, George J, Gilburd B, Rauova L, Goldberg I, Kopolovic J, Harats D, Shoenfeld Y. Immunization of low-density lipoprotein receptor deficient (LDL-RD) mice with heat shock protein 65 (HSP-65) promotes early atherosclerosis. J Autoimmun. 2000;14:115–21.
68. George J, Afek A, Gilburd B, Shoenfeld Y, Harats D. Cellular and humoral immune responses to heat shock protein 65 are both involved in promoting fatty-streak formation in LDL-receptor deficient mice. J Am Coll Cardiol. 2001;38:900–5.
69. Xiong Q, Feng J, Zhang Y, Sun Y, Lu Y, Li T, Zhang X, Cao R, Jin L, Wu J. Promotion of atherosclerosis in high cholesterol diet-fed rabbits by immunization with the P277 peptide. Immunol Lett. 2016;170:80–7.
70. Klingenberg R, Ketelhuth DF, Strodthoff D, Gregori S, Hansson GK. Subcutaneous immunization with heat shock protein-65 reduces atherosclerosis in Apoe(-)/(-) mice. Immunobiology. 2012;217:540–7.
71. Long J, Lin J, Yang X, Yuan D, Wu J, Li T, Cao R, Liu J. Nasal immunization with different forms of heat shock protein-65 reduced high-cholesterol-diet-driven rabbit atherosclerosis. Int Immunopharmacol. 2012;13:82–7.
72. van Puijvelde GH, van Es T, van Wanrooij EJ, Habets KL, de Vos P, van der Zee R, van Eden W, van Berkel TJ, Kuiper J. Induction of oral tolerance to HSP60 or an HSP60-peptide

activates T cell regulation and reduces atherosclerosis. Arterioscler Thromb Vasc Biol. 2007;27:2677–83.
73. Zhong Y, Tang H, Wang X, Zeng Q, Liu Y, Zhao XI, Yu K, Shi H, Zhu R, Mao X. Intranasal immunization with heat shock protein 60 induces CD4(+) CD25(+) GARP(+) and type 1 regulatory T cells and inhibits early atherosclerosis. Clin Exp Immunol. 2016;183:452–68.
74. Perschinka H, Mayr M, Millonig G, Mayerl C, van der Zee R, Morrison SG, Morrison RP, Xu Q, Wick G. Cross-reactive B-cell epitopes of microbial and human heat shock protein 60/65 in atherosclerosis. Arterioscler Thromb Vasc Biol. 2003;23:1060–5.
75. Binder CJ, Horkko S, Dewan A, Chang MK, Kieu EP, Goodyear CS, Shaw PX, Palinski W, Witztum JL, Silverman GJ. Pneumococcal vaccination decreases atherosclerotic lesion formation: molecular mimicry between Streptococcus pneumoniae and oxidized LDL. Nat Med. 2003;9:736–43.
76. Tsutsumi A, Matsuura E, Ichikawa K, Fujisaku A, Mukai M, Kobayashi S, Koike T. Antibodies to beta 2-glycoprotein I and clinical manifestations in patients with systemic lupus erythematosus. Arthritis Rheum. 1996;39:1466–74.
77. Kandiah DA, Krilis SA. Beta 2-glycoprotein I. Lupus. 1994;3:207–12.
78. George J, Harats D, Gilburd B, Afek A, Levy Y, Schneiderman J, Barshack I, Kopolovic J, Shoenfeld Y. Immunolocalization of beta2-glycoprotein I (apolipoprotein H) to human atherosclerotic plaques: potential implications for lesion progression. Circulation. 1999;99:2227–30.
79. George J, Afek A, Gilburd B, Blank M, Levy Y, Aron-Maor A, Levkovitz H, Shaish A, Goldberg I, Kopolovic J, Harats D, Shoenfeld Y. Induction of early atherosclerosis in LDL-receptor-deficient mice immunized with beta2-glycoprotein I. Circulation. 1998;98:1108–15.
80. Dunoyer-Geindre S, Kwak BR, Pelli G, Roth I, Satta N, Fish RJ, Reber G, Mach F, Kruithof EK, de Moerloose P. Immunization of LDL receptor-deficient mice with beta2-glycoprotein 1 or human serum albumin induces a more inflammatory phenotype in atherosclerotic plaques. Thromb Haemost. 2007;97:129–38.
81. Afek A, George J, Shoenfeld Y, Gilburd B, Levy Y, Shaish A, Keren P, Janackovic Z, Goldberg I, Kopolovic J, Harats D. Enhancement of atherosclerosis in beta-2-glycoprotein I-immunized apolipoprotein E-deficient mice. Pathobiology. 1999;67:19–25.
82. De Haro J, Esparza L, Bleda S, Varela C, Sanchez C, Acin F. Attenuation of early atherosclerotic lesions by immunotolerance with beta2 glycoprotein I and the immunomodulatory effectors interleukin 2 and 10 in a murine model. J Vasc Surg. 2015;62:1625–31.
83. George J, Harats D, Gilburd B, Afek A, Shaish A, Kopolovic J, Shoenfeld Y. Adoptive transfer of beta(2)-glycoprotein I-reactive lymphocytes enhances early atherosclerosis in LDL receptor-deficient mice. Circulation. 2000;102:1822–7.
84. Herbin O, Ait-Oufella H, Yu W, Fredrikson GN, Aubier B, Perez N, Barateau V, Nilsson J, Tedgui A, Mallat Z. Regulatory T-cell response to apolipoprotein B100-derived peptides reduces the development and progression of atherosclerosis in mice. Arterioscler Thromb Vasc Biol. 2012;32:605–12.
85. Khoo JC, Miller E, Pio F, Steinberg D, Witztum JL. Monoclonal antibodies against LDL further enhance macrophage uptake of LDL aggregates. Arterioscler Thromb. 1992;12:1258–66.
86. Wolf D, Zirlik A, Ley K. Beyond vascular inflammation—recent advances in understanding atherosclerosis. Cell Mol Life Sci. 2015;72:3853–69.
87. Buono C, Binder CJ, Stavrakis G, Witztum JL, Glimcher LH, Lichtman AH. T-bet deficiency reduces atherosclerosis and alters plaque antigen-specific immune responses. Proc Natl Acad Sci USA. 2005;102:1596–601.
88. King VL, Szilvassy SJ, Daugherty A. Interleukin-4 deficiency decreases atherosclerotic lesion formation in a site-specific manner in female LDL receptor-/- mice. Arterioscler Thromb Vasc Biol. 2002;22:456–61.

89. Smith E, Prasad KM, Butcher M, Dobrian A, Kolls JK, Ley K, Galkina E. Blockade of interleukin-17A results in reduced atherosclerosis in apolipoprotein E-deficient mice. Circulation. 2010;121:1746–55.
90. Gao Q, Jiang Y, Ma T, Zhu F, Gao F, Zhang P, Guo C, Wang Q, Wang X, Ma C, Zhang Y, Chen W, Zhang L. A critical function of Th17 proinflammatory cells in the development of atherosclerotic plaque in mice. J Immunol. 2010;185:5820–7.
91. Gistera A, Robertson AK, Andersson J, Ketelhuth DF, Ovchinnikova O, Nilsson SK, Lundberg AM, Li MO, Flavell RA, Hansson GK. Transforming growth factor-beta signaling in T cells promotes stabilization of atherosclerotic plaques through an interleukin-17-dependent pathway. Sci Transl Med. 2013;5:196ra100.
92. Danzaki K, Matsui Y, Ikesue M, Ohta D, Ito K, Kanayama M, Kurotaki D, Morimoto J, Iwakura Y, Yagita H, Tsutsui H, Uede T. Interleukin-17A deficiency accelerates unstable atherosclerotic plaque formation in apolipoprotein E-deficient mice. Arterioscler Thromb Vasc Biol. 2012;32:273–80.
93. Sakaguchi S, Yamaguchi T, Nomura T, Ono M. Regulatory T cells and immune tolerance. Cell. 2008;133:775–87.
94. Sakaguchi S, Sakaguchi N, Asano M, Itoh M, Toda M. Immunologic self-tolerance maintained by activated T cells expressing IL-2 receptor alpha-chains (CD25). Breakdown of a single mechanism of self-tolerance causes various autoimmune diseases. J Immunol. 1995;155:1151–64.
95. Bacchetta R, Passerini L, Gambineri E, Dai M, Allan SE, Perroni L, Dagna-Bricarelli F, Sartirana C, Matthes-Martin S, Lawitschka A, Azzari C, Ziegler SF, Levings MK, Roncarolo MG. Defective regulatory and effector T cell functions in patients with FOXP3 mutations. J Clin Invest. 2006;116:1713–22.
96. Pastrana JL, Sha X, Virtue A, Mai J, Cueto R, Lee IA, Wang H, Yang X-F. Regulatory T cells and atherosclerosis. J Clin Exp Cardiol. 2012;2012:002.
97. de Boer OJ, van der Meer JJ, Teeling P, van der Loos CM, van der Wal AC. Low numbers of FOXP3 positive regulatory T cells are present in all developmental stages of human atherosclerotic lesions. PLoS One. 2007;2:e779.
98. Wigren M, Bengtsson D, Duner P, Olofsson K, Bjorkbacka H, Bengtsson E, Fredrikson GN, Nilsson J. Atheroprotective effects of Alum are associated with capture of oxidized LDL antigens and activation of regulatory T cells. Circ Res. 2009;104:e62–70.
99. Mallat Z, Ait-Oufella H, Tedgui A. Regulatory T-cell immunity in atherosclerosis. Trends Cardiovasc Med. 2007;17:113–8.
100. Clemente-Casares X, Blanco J, Ambalavanan P, Yamanouchi J, Singha S, Fandos C, Tsai S, Wang J, Garabatos N, Izquierdo C, Agrawal S, Keough MB, Yong VW, James E, Moore A, Yang Y, Stratmann T, Serra P, Santamaria P. Expanding antigen-specific regulatory networks to treat autoimmunity. Nature. 2016;530:434–440.
101. Choi JH, Cheong C, Dandamudi DB, Park CG, Rodriguez A, Mehandru S, Velinzon K, Jung IH, Yoo JY, GT O, Steinman RM. Flt3 signaling-dependent dendritic cells protect against atherosclerosis. Immunity. 2011;35:819–31.
102. Choi JH, Do Y, Cheong C, Koh H, Boscardin SB, YS O, Bozzacco L, Trumpfheller C, Park CG, Steinman RM. Identification of antigen-presenting dendritic cells in mouse aorta and cardiac valves. J Exp Med. 2009;206:497–505.
103. Subramanian M, Thorp E, Hansson GK, Tabas I. Treg-mediated suppression of atherosclerosis requires MYD88 signaling in DCs. J Clin Invest. 2013;123:179–88.
104. Coombes JL, Siddiqui KR, Arancibia-Carcamo CV, Hall J, Sun CM, Belkaid Y, Powrie F. A functionally specialized population of mucosal CD103+ DCs induces Foxp3+ regulatory T cells via a TGF-beta and retinoic acid-dependent mechanism. J Exp Med. 2007;204:1757–64.
105. Nelson RW, Beisang D, Tubo NJ, Dileepan T, Wiesner DL, Nielsen K, Wuthrich M, Klein BS, Kotov DI, Spanier JA, Fife BT, Moon JJ, Jenkins MK. T cell receptor cross-reactivity between similar foreign and self peptides influences naive cell population size and autoimmunity. Immunity. 2015;42:95–107.

106. Kyaw T, Tay C, Krishnamurthi S, Kanellakis P, Agrotis A, Tipping P, Bobik A, Toh BH. B1a B lymphocytes are atheroprotective by secreting natural IgM that increases IgM deposits and reduces necrotic cores in atherosclerotic lesions. Circ Res. 2011;109:830–40.
107. Lewis MJ, Malik TH, Ehrenstein MR, Boyle JJ, Botto M, Haskard DO. Immunoglobulin M is required for protection against atherosclerosis in low-density lipoprotein receptor-deficient mice. Circulation. 2009;120:417–26.
108. Kyaw T, Tay C, Khan A, Dumouchel V, Cao A, To K, Kehry M, Dunn R, Agrotis A, Tipping P, Bobik A, Toh BH. Conventional B2 B cell depletion ameliorates whereas its adoptive transfer aggravates atherosclerosis. J Immunol. 2010;185:4410–9.
109. Kyaw T, Tay C, Hosseini H, Kanellakis P, Gadowski T, MacKay F, Tipping P, Bobik A, Toh BH. Depletion of B2 but not B1a B cells in BAFF receptor-deficient ApoE mice attenuates atherosclerosis by potently ameliorating arterial inflammation. PLoS One. 2012;7:e29371.
110. Ait-Oufella H, Herbin O, Bouaziz JD, Binder CJ, Uyttenhove C, Laurans L, Taleb S, Van Vre E, Esposito B, Vilar J, Sirvent J, Van Snick J, Tedgui A, Tedder TF, Mallat Z. B cell depletion reduces the development of atherosclerosis in mice. J Exp Med. 2010;207:1579–87.
111. Sage AP, Tsiantoulas D, Baker L, Harrison J, Masters L, Murphy D, Loinard C, Binder CJ, Mallat Z. BAFF receptor deficiency reduces the development of atherosclerosis in mice—brief report. Arterioscler Thromb Vasc Biol. 2012;32:1573–6.
112. Nicoletti A, Kaveri S, Caligiuri G, Bariety J, Hansson GK. Immunoglobulin treatment reduces atherosclerosis in apo E knockout mice. J Clin Invest. 1998;102:910–8.
113. Tsiantoulas D, Diehl CJ, Witztum JL, Binder CJ. B cells and humoral immunity in atherosclerosis. Circ Res. 2014;114:1743–56.
114. Shaw PX, Horkko S, Chang MK, Curtiss LK, Palinski W, Silverman GJ, Witztum JL. Natural antibodies with the T15 idiotype may act in atherosclerosis, apoptotic clearance, and protective immunity. J Clin Invest. 2000;105:1731–40.
115. Chou MY, Fogelstrand L, Hartvigsen K, Hansen LF, Woelkers D, Shaw PX, Choi J, Perkmann T, Backhed F, Miller YI, Horkko S, Corr M, Witztum JL, Binder CJ. Oxidation-specific epitopes are dominant targets of innate natural antibodies in mice and humans. J Clin Invest. 2009;119:1335–49.
116. Tsimikas S, Brilakis ES, Lennon RJ, Miller ER, Witztum JL, McConnell JP, Kornman KS, Berger PB. Relationship of IgG and IgM autoantibodies to oxidized low density lipoprotein with coronary artery disease and cardiovascular events. J Lipid Res. 2007;48:425–33.
117. Ravandi A, Boekholdt SM, Mallat Z, Talmud PJ, Kastelein JJ, Wareham NJ, Miller ER, Benessiano J, Tedgui A, Witztum JL, Khaw KT, Tsimikas S. Relationship of IgG and IgM autoantibodies and immune complexes to oxidized LDL with markers of oxidation and inflammation and cardiovascular events: results from the EPIC-Norfolk Study. J Lipid Res. 2011;52:1829–36.
118. Bjorkbacka H, Alm R, Persson M, Hedblad B, Nilsson J, Fredrikson GN. Low levels of apolipoprotein B-100 autoantibodies are associated with increased risk of coronary events. Arterioscler Thromb Vasc Biol. 2016;36:765–771.
119. Tsimikas S, Palinski W, Witztum JL. Circulating autoantibodies to oxidized LDL correlate with arterial accumulation and depletion of oxidized LDL in LDL receptor-deficient mice. Arterioscler Thromb Vasc Biol. 2001;21:95–100.
120. Schiopu A, Frendeus B, Jansson B, Soderberg I, Ljungcrantz I, Araya Z, Shah PK, Carlsson R, Nilsson J, Fredrikson GN. Recombinant antibodies to an oxidized low-density lipoprotein epitope induce rapid regression of atherosclerosis in apobec-1(-/-)/low-density lipoprotein receptor(-/-) mice. J Am Coll Cardiol. 2007;50:2313–8.
121. Schiopu A, Bengtsson J, Soderberg I, Janciauskiene S, Lindgren S, Ares MP, Shah PK, Carlsson R, Nilsson J, Fredrikson GN. Recombinant human antibodies against aldehyde-modified apolipoprotein B-100 peptide sequences inhibit atherosclerosis. Circulation. 2004;110:2047–52.

122. Caligiuri G, Khallou-Laschet J, Vandaele M, Gaston AT, Delignat S, Mandet C, Kohler HV, Kaveri SV, Nicoletti A. Phosphorylcholine-targeting immunization reduces atherosclerosis. J Am Coll Cardiol. 2007;50:540–6.
123. Grasset EK, Duhlin A, Agardh HE, Ovchinnikova O, Hagglof T, Forsell MN, Paulsson-Berne G, Hansson GK, Ketelhuth DF, Karlsson MC. Sterile inflammation in the spleen during atherosclerosis provides oxidation-specific epitopes that induce a protective B-cell response. Proc Natl Acad Sci USA. 2015;112:E2030–8.
124. Rosenfeld SM, Perry HM, Gonen A, Prohaska TA, Srikakulapu P, Grewal S, Das D, McSkimming C, Taylor AM, Tsimikas S, Bender TP, Witztum JL, McNamara CA. B-1b cells secrete atheroprotective IgM and attenuate atherosclerosis. Circ Res. 2015;117:e28–39.
125. Hosseini H, Li Y, Kanellakis P, Tay C, Cao A, Tipping P, Bobik A, Toh BH, Kyaw T. Phosphatidylserine liposomes mimic apoptotic cells to attenuate atherosclerosis by expanding polyreactive IgM producing B1a lymphocytes. Cardiovasc Res. 2015;106:443–452.
126. Lawson JS, Glenn WK, Tran DD, Ngan CC, Duflou JA, Whitaker NJ. Identification of human papilloma viruses in atheromatous coronary artery disease. Front Cardiovasc Med. 2015;2:17.
127. Rosenfeld ME, Campbell LA. Pathogens and atherosclerosis: update on the potential contribution of multiple infectious organisms to the pathogenesis of atherosclerosis. Thromb Haemost. 2011;106:858–67.
128. Lin HC, Chien CW, Ho JD. Herpes zoster ophthalmicus and the risk of stroke: a population-based follow-up study. Neurology. 2010;74:792–7.
129. Kang JH, Ho JD, Chen YH, Lin HC. Increased risk of stroke after a herpes zoster attack: a population-based follow-up study. Stroke. 2009;40:3443–8.
130. Anderson JL, Muhlestein JB, Carlquist J, Allen A, Trehan S, Nielson C, Hall S, Brady J, Egger M, Horne B, Lim T. Randomized secondary prevention trial of azithromycin in patients with coronary artery disease and serological evidence for Chlamydia pneumoniae infection: the azithromycin in coronary artery disease: elimination of myocardial infection with chlamydia (ACADEMIC) study. Circulation. 1999;99:1540–7.
131. Hebsur S, Vakil E, Oetgen WJ, Kumar PN, Lazarous DF. Influenza and coronary artery disease: exploring a clinical association with myocardial infarction and analyzing the utility of vaccination in prevention of myocardial infarction. Rev Cardiovasc Med. 2014;15:168–75.
132. Haidari M, Wyde PR, Litovsky S, Vela D, Ali M, Casscells SW, Madjid M. Influenza virus directly infects, inflames, and resides in the arteries of atherosclerotic and normal mice. Atherosclerosis. 2010;208:90–6.
133. Naghavi M, Wyde P, Litovsky S, Madjid M, Akhtar A, Naguib S, Siadaty MS, Sanati S, Casscells W. Influenza infection exerts prominent inflammatory and thrombotic effects on the atherosclerotic plaques of apolipoprotein E-deficient mice. Circulation. 2003;107:762–8.
134. Udell JA, Zawi R, Bhatt DL, Keshtkar-Jahromi M, Gaughran F, Phrommintikul A, Ciszewski A, Vakili H, Hoffman EB, Farkouh ME, Cannon CP. Association between influenza vaccination and cardiovascular outcomes in high-risk patients: a meta-analysis. JAMA. 2013;310:1711–20.
135. Siriwardena AN, Asghar Z, Coupland CC. Influenza and pneumococcal vaccination and risk of stroke or transient ischaemic attack-matched case control study. Vaccine. 2014;32:1354–61.
136. Macintyre CR, Heywood AE, Kovoor P, Ridda I, Seale H, Tan T, Gao Z, Katelaris AL, Siu HW, Lo V, Lindley R, Dwyer DE. Ischaemic heart disease, influenza and influenza vaccination: a prospective case control study. Heart. 2013;99:1843–8.
137. Davis MM, Taubert K, Benin AL, Brown DW, Mensah GA, Baddour LM, Dunbar S, Krumholz HM, American Heart A, American College of C, American Association of C, Pulmonary R, American Association of Critical Care N, American Association of Heart Failure N, American Diabetes A, Association of Black Cardiologists I, Heart Failure Society of A, Preventive Cardiovascular Nurses A, American Academy of Nurse P, Centers for

Disease C, Prevention, the Advisory Committee on I. Influenza vaccination as secondary prevention for cardiovascular disease: a science advisory from the American Heart Association/American College of Cardiology. J Am Coll Cardiol. 2006;48:1498–502.
138. Suthers B, Hansbro P, Thambar S, McEvoy M, Peel R, Attia J. Pneumococcal vaccination may induce anti-oxidized low-density lipoprotein antibodies that have potentially protective effects against cardiovascular disease. Vaccine. 2012;30:3983–5.
139. Worzella SL, Hayney MS. Inflammatory chronic diseases: preventable by vaccines? J Am Pharm Assoc. 2014;54:446–8.
140. Vila-Corcoles A, Ochoa-Gondar O, Rodriguez-Blanco T, de Diego C, Satue E, Group ES. Ineffectiveness of pneumococcal vaccination in cardiovascular prevention: the CAPAMIS study. JAMA Intern Med. 2013;173:1918–20.
141. Hermansson A, Ketelhuth DF, Strodthoff D, Wurm M, Hansson EM, Nicoletti A, Paulsson-Berne G, Hansson GK. Inhibition of T cell response to native low-density lipoprotein reduces atherosclerosis. J Exp Med. 2010;207:1081–93.
142. Hauer AD, Uyttenhove C, de Vos P, Stroobant V, Renauld JC, van Berkel TJ, van Snick J, Kuiper J. Blockade of interleukin-12 function by protein vaccination attenuates atherosclerosis. Circulation. 2005;112:1054–62.
143. Crossey E, Amar MJ, Sampson M, Peabody J, Schiller JT, Chackerian B, Remaley AT. A cholesterol-lowering VLP vaccine that targets PCSK9. Vaccine. 2015;33:5747–55.
144. Gaofu Q, Jun L, Xin Y, Wentao L, Jie W, Xiuyun Z, Jingjing L. Vaccinating rabbits with a cholesteryl ester transfer protein (CETP) B-Cell epitope carried by heat shock protein-65 (HSP65) for inducing anti-CETP antibodies and reducing aortic lesions in vivo. J Cardiovasc Pharmacol. 2005;45:591–8.
145. Rittershaus CW, Miller DP, Thomas LJ, Picard MD, Honan CM, Emmett CD, Pettey CL, Adari H, Hammond RA, Beattie DT, Callow AD, Marsh HC, Ryan US. Vaccine-induced antibodies inhibit CETP activity in vivo and reduce aortic lesions in a rabbit model of atherosclerosis. Arterioscler Thromb Vasc Biol. 2000;20:2106–12.
146. Khallou-Laschet J, Tupin E, Caligiuri G, Poirier B, Thieblemont N, Gaston AT, Vandaele M, Bleton J, Tchapla A, Kaveri SV, Rudling M, Nicoletti A. Atheroprotective effect of adjuvants in apolipoprotein E knockout mice. Atherosclerosis. 2006;184:330–41.
147. Engman C, Wen Y, Meng WS, Bottino R, Trucco M, Giannoukakis N. Generation of antigen-specific Foxp3+ regulatory T-cells in vivo following administration of diabetes-reversing tolerogenic microspheres does not require provision of antigen in the formulation. Clin Immunol. 2015;160:103–23.
148. van Puijvelde GH, Hauer AD, de Vos P, van den Heuvel R, van Herwijnen MJ, van der Zee R, van Eden W, van Berkel TJ, Kuiper J. Induction of oral tolerance to oxidized low-density lipoprotein ameliorates atherosclerosis. Circulation. 2006;114:1968–76.
149. Xia M, Chen D, Endresz V, Lantos I, Szabo A, Kakkar V, Lu X. Modulation of recombinant antigenic constructs containing multi-epitopes towards effective reduction of atherosclerotic lesion in B6;129S-Ldlr(tm1Her)Apob(tm2Sgy)/J mice. PLoS One. 2015;10:e0123393.
150. Mundkur L, Mukhopadhyay R, Samson S, Varma M, Kale D, Chen D, Shivaprasad S, Sivanandan H, Soman V, Lu X, Kakkar VV. Mucosal tolerance to a combination of ApoB and HSP60 peptides controls plaque progression and stabilizes vulnerable plaque in Apob (tm2Sgy)Ldlr(tm1Her)/J mice. PLoS One. 2013;8:e58364.
151. Karkhah A, Amani J. A potent multivalent vaccine for modulation of immune system in atherosclerosis: an in silico approach. Clin Exp Vaccine Res. 2016;5:50–9.
152. Walczak A, Szymanska B, Selmaj K. Differential prevention of experimental autoimmune encephalomyelitis with antigen-specific DNA vaccination. Clin Neurol Neurosurg. 2004;106:241–5.
153. Kutzler MA, Weiner DB. DNA vaccines: ready for prime time? Nat Rev Genet. 2008;9:776–88.
154. Shen Y, Chen J, Zhang X, Wu X, Human XQ. TNF-alpha gene vaccination prevents collagen-induced arthritis in mice. Int Immunopharmacol. 2007;7:1140–9.

155. Hauer AD, van Puijvelde GH, Peterse N, de Vos P, van Weel V, van Wanrooij EJ, Biessen EA, Quax PH, Niethammer AG, Reisfeld RA, van Berkel TJ, Kuiper J. Vaccination against VEGFR2 attenuates initiation and progression of atherosclerosis. Arterioscler Thromb Vasc Biol. 2007;27:2050–7.
156. Li S, Kievit P, Robertson AK, Kolumam G, Li X, von Wachenfeldt K, Valfridsson C, Bullens S, Messaoudi I, Bader L, Cowan KJ, Kamath A, van Bruggen N, Bunting S, Frendeus B, Grove KL. Targeting oxidized LDL improves insulin sensitivity and immune cell function in obese Rhesus macaques. Mol Metab. 2013;2:256–69.
157. Lehrer-Graiwer J, Singh P, Abdelbaky A, Vucic E, Korsgren M, Baruch A, Fredrickson J, van Bruggen N, Tang MT, Frendeus B, Rudd JH, Hsieh F, Ballantyne CM, Ghoshhajra B, Rosenson RS, Koren M, Roth EM, Duprez DA, Fayad ZA, Tawakol AA. FDG-PET imaging for oxidized LDL in stable atherosclerotic disease: a phase II study of safety, tolerability, and anti-inflammatory activity. JACC Cardiovasc Imaging. 2015;8:493–4.
158. Buono C, Come CE, Stavrakis G, Maguire GF, Connelly PW, Lichtman AH. Influence of interferon-gamma on the extent and phenotype of diet-induced atherosclerosis in the LDLR-deficient mouse. Arterioscler Thromb Vasc Biol. 2003;23:454–60.
159. Gupta S, Pablo AM, Jiang X, Wang N, Tall AR, Schindler C. IFN-gamma potentiates atherosclerosis in ApoE knock-out mice. J Clin Invest. 1997;99:2752–61.
160. Mallat Z, Besnard S, Duriez M, Deleuze V, Emmanuel F, Bureau MF, Soubrier F, Esposito B, Duez H, Fievet C, Staels B, Duverger N, Scherman D, Tedgui A. Protective role of interleukin-10 in atherosclerosis. Circ Res. 1999;85:e17–24.
161. Robertson AK, Rudling M, Zhou X, Gorelik L, Flavell RA, Hansson GK. Disruption of TGF-beta signaling in T cells accelerates atherosclerosis. J Clin Invest. 2003;112:1342–50.
162. Mallat Z, Gojova A, Brun V, Esposito B, Fournier N, Cottrez F, Tedgui A, Groux H. Induction of a regulatory T cell type 1 response reduces the development of atherosclerosis in apolipoprotein E-knockout mice. Circulation. 2003;108:1232–7.
163. Ait-Oufella H, Salomon BL, Potteaux S, Robertson AK, Gourdy P, Zoll J, Merval R, Esposito B, Cohen JL, Fisson S, Flavell RA, Hansson GK, Klatzmann D, Tedgui A, Mallat Z. Natural regulatory T cells control the development of atherosclerosis in mice. Nat Med. 2006;12:178–80.
164. Klingenberg R, Gerdes N, Badeau RM, Gistera A, Strodthoff D, Ketelhuth DF, Lundberg AM, Rudling M, Nilsson SK, Olivecrona G, Zoller S, Lohmann C, Luscher TF, Jauhiainen M, Sparwasser T, Hansson GK. Depletion of FOXP3+ regulatory T cells promotes hypercholesterolemia and atherosclerosis. J Clin Invest. 2013;123:1323–34.
165. Ait-Oufella H, Horvat B, Kerdiles Y, Herbin O, Gourdy P, Khallou-Laschet J, Merval R, Esposito B, Tedgui A, Mallat Z. Measles virus nucleoprotein induces a regulatory immune response and reduces atherosclerosis in mice. Circulation. 2007;116:1707–13.

Platelets as Regulators of Thrombosis and Inflammation

4

Daniel Duerschmied and Steffen Massberg

Abstract

This chapter describes features of blood platelets that serve in the regulation of thrombosis and inflammation. Classically, platelets have been known for decades to promote hemostasis of wounds and arterial thrombosis, in particular atherothrombosis following atherosclerotic plaque rupture. More recently, the importance of platelets for the development of deep vein thrombosis has been recognized. But platelets also link thrombosis and inflammation in "immunothrombosis" within microvessels. In a collaborative "effort" of several cell types, plasma proteins, and neutrophil extracellular traps, platelets orchestrate the recognition, trapping, and killing of pathogens. Immunothrombosis also occurs in microvessels during ischemia–reperfusion injury, e.g., following myocardial infarction. In acute and chronic inflammation, platelets further cooperate with neutrophils, monocytes, and lymphocytes without clot formation to promote physiological—and pathophysiological—responses to pathogens or auto-antigens.

D. Duerschmied, MD (✉)
Department of Cardiology and Angiology I, Heart Center, University of Freiburg, Hugstetter Str. 55, 79106 Freiburg, Germany

Internal Medicine III – Intensive Care Medicine, University Hospital of Freiburg, Freiburg, Germany
e-mail: daniel.duerschmied@universitaets-herzzentrum.de

S. Massberg
Medizinische Klinik und Poliklinik I, Klinikum der Universität, Ludwig-Maximilians-Universität, Munich, Germany

DZHK (German Centre for Cardiovascular Research), Munich Heart Alliance, Munich, Germany

A. Zirlik et al. (eds.), *Platelets, Haemostasis and Inflammation*,
Cardiac and Vascular Biology 5, https://doi.org/10.1007/978-3-319-66224-4_4

Contents

4.1 Introduction

A quarter million platelets circulate in 1 μl of human blood (more than 600,000 μl^{-1} in murine blood). Considering that the mean volume of a platelet is 10 fl [1], up to one-fifth of the entire blood volume is actually platelet volume—divided into a large number of single small cells. In fact, platelets are the smallest and most numerous cells in the blood. They originate from bone marrow megakaryocytes [2, 3] and do not need a cell nucleus to regulate thrombosis and inflammation (among other functions). This chapter explains how platelets adhere to the vessel wall, form aggregates, and release soluble factors in venous and arterial thrombosis—and in acute and chronic inflammation. A table lists features that platelets need for these tasks. This collection of known platelet features may serve as a reference for researchers to further investigate immunothrombotic platelet functions or possible therapeutic targets.

4.2 Platelets Drive Arterial Thrombosis

Thrombus formation after vessel wall injury in arteries (or veins) is driven by platelets. This is important not only for hemostasis in bleeding wounds but also in atherothrombosis (Fig. 4.1). Circulating platelets adhere to the site of injury, become activated, and secrete soluble factors [4, 5]. Only if these autocrine agonists amplify activation, platelets then recruit other platelets to form a growing thrombus that needs to be stabilized. In myocardial infarction, platelet activation occurs at the site of atherosclerotic plaque rupture (and where a stent is implanted during percutaneous coronary intervention). The exposed plaque component collagen first induces platelet adhesion and activation, followed by tissue factor-driven coagulation in flow niches downstream of platelet aggregates [6]. Platelet

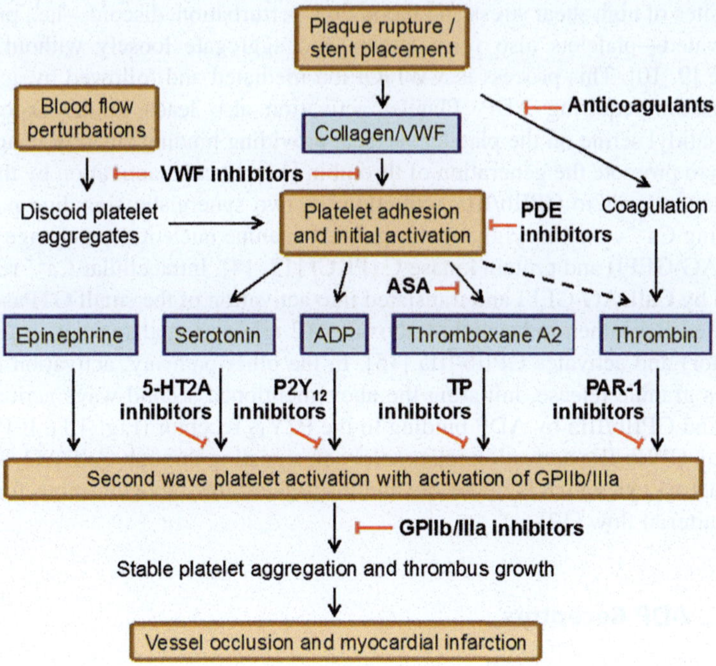

Fig. 4.1 Atherothrombosis. Platelets drive thrombus formation in acute coronary syndrome and are the primary therapeutic target. A near-occlusive intracoronary plaque creates blood flow perturbations, inducing formation of discoid platelet aggregates in a von Willebrand factor (VWF)-dependent manner. Once a plaque ruptures or a stent is implanted, collagen from the extracellular matrix is exposed and binds unfolded plasma VWF. Circulating platelets adhere to collagen-bound VWF and get activated by binding to collagen. Activated platelets release the depicted autocrine agonists and promote the formation of thrombin, initiating a second wave of platelet activation. Stable platelet aggregation via activated glycoprotein (GP) IIb/IIIa requires adenosine diphosphate (ADP) binding to P2Y12 receptor, which can be inhibited by clopidogrel. *PDE* phosphodiesterase, *ASA* acetylsalicylic acid, 5HT2A—serotonin receptor, TP—thromboxane A2 receptor, PAR1—thrombin receptor

glycoprotein (GP)Ibα binding to collagen-bound von Willebrand factor (VWF) constitutes the primary adhesion mechanism for platelets under arterial shear conditions [5, 7]. Subsequent binding of GPVI to collagen initiates platelet activation [8].

4.2.1 Platelet Signaling

Initially, activated platelets release thromboxane A2 (after cyclooxygenase activation) and ADP as mediators for a second wave of platelet activation. Enhanced platelet activation then induces a conformational change of the cell-adhesion molecule integrin αIIbβ3 (GPIIb/IIIa). In its active conformation, GPIIb/IIIa is the key molecule for platelet–platelet binding (via fibrinogen, VWF, and fibronectin) and stabilization of platelet aggregates [4].

At sites of high shear stress and blood flow perturbation, discoid—i.e., primarily unactivated—platelets also form tethers and aggregate loosely without plaque rupture [9, 10]. This process is VWF/GPIbα mediated and followed by aggregate stabilization requiring ADP. Platelet activation also leads to the exposure of phosphatidyl serine on the platelet surface, providing binding sites for coagulation factors to promote the generation of thrombin [11]. Platelet activation by thrombin or collagen leads to GPIIb/IIIa activation via two synergistic signaling pathways involving Ca^{2+}- and diacylglycerol-regulated guanine nucleotide exchange factor I (CalDAG-GEFI) and protein kinase C (PKC) [12–14]. Intracellular Ca^{2+} release is sensed by CalDAG-GEFI and translated into activation of the small GTPase Rap1. Activated Rap1 then induces thromboxane A2 release (another positive feedback mediator) and activates GPIIb/IIIa [15]. In the other pathway, activation of PKC induces granule release, initiating the abovementioned second-wave activation of Rap1 and GPIIb/IIIa by ADP binding to the $P2Y_{12}$ receptor (Fig. 4.1). If $P2Y_{12}$ is blocked, GPIIb/IIIa activation relies solely on signaling through CalDAG-GEFI for Rap1 activation, which is reversible and insufficient for stable thrombus formation under arterial flow [12].

4.2.2 ADP Receptors

Resting platelets store ADP in their dense granules at very high concentrations (650 mM) and, once activated, produce an ADP-rich environment [4, 16]. In the second-wave reaction, ADP then potentiates platelet aggregation by binding to both the $P2Y_1$ and $P2Y_{12}$ receptor [17, 18]. This mechanism ensures sustained platelet activation, which is crucial for the development of stable thrombi under arterial flow conditions [12, 17]. The Gαq-coupled $P2Y_1$ receptor amplifies initial platelet activation via Ca^{2+} mobilization and is important for the first phase of thrombus formation [19, 20]. Positive feedback through the Gαi-coupled $P2Y_{12}$ receptor is critical for the formation of stable platelet aggregates, and $P2Y_{12}$ antagonists effectively inhibit arterial thrombus formation [18, 21]. Activation of $P2Y_{12}$ by

ADP inhibits the formation of cyclic adenosine monophosphate (cAMP) by adenylyl cyclase [22]. Consequently, cAMP-dependent protein kinase (PKA)-mediated phosphorylation of vasodilator-stimulated phosphoprotein (VASP), a negative modulator of GPIIb/IIIa activation, is inhibited. VASP phosphorylation is not lowered by ADP stimulation when $P2Y_{12}$ is blocked. Further signaling events downstream of $P2Y_{12}$ are primarily mediated by phosphatidyl inositol-3 kinase (PI-3K). The $P2Y_{12}$ receptor hence amplifies platelet secretion, platelet aggregation, and platelet procoagulant activity.

4.2.3 Thrombin Receptors

The serine protease thrombin activates its receptors indirectly by enzymatic cleavage of a silencing domain, which allows subsequent binding of its unmasked ligand site to the receptor body ("autoactivation") [23, 24]. Human platelets express the Gq protein-coupled protease-activated receptors (PAR)-1 and PAR-4 (murine platelets express PAR-3 and PAR-4). It was understood around the turn of the millennium that PAR-mediated platelet activation by thrombin—in particular via PAR-1 in human—is a central feature of hemostasis [25, 26]. Interestingly, PAR-1 can be activated by 100-fold lower concentrations of thrombin than PAR-4 and is hence considered a more sensitive—albeit weaker—mediator of platelet activation [27, 28]. PAR-1 signaling via Ca^{2+} and CalDAG-GEFI is transient and unable to induce sustained GPIIb/IIIa activation unless amplified by the ADP feedback via the Gi-coupled $P2Y_{12}$ receptor and PKC activation in the parallel pathway.

4.2.4 Antithrombotic Strategies

Figure 4.1 also lists antiplatelet strategies to prevent arterial thrombosis. The primary indication for platelet inhibitors is (prevention of) atherothrombosis in coronary, cerebral, or peripheral arterial disease. Only acetylsalicylic acid (ASA) is also recommended for patients with deep vein thrombosis or pulmonary embolism, who have discontinued anticoagulation and would otherwise not receive any further antithrombotic treatment for secondary prevention (class IIb recommendation) [29]. Approved antiplatelet drugs are the cyclooxygenase inhibitor ASA, the phosphodiesterase inhibitor cilostazol, the serotonin receptor antagonist sarpogrelate (approved in Asia), ADP receptor antagonists (ticlopidine, clopidogrel, ticagrelor, prasugrel, and cangrelor), the thrombin receptor antagonist vorapaxar, and GPIIb/IIIa inhibitors (eptifibatide, tirofiban, and abciximab).

4.3 Platelets Promote Venous Thrombosis

The phase III clinical trials WARFASA and ASPIRE revealed a net clinical benefit of ASA for patients with venous thromboembolism after discontinuation of plasmatic anticoagulation [30, 31]. Platelet inhibition effectively prevented recurrent venous thromboembolism. This illustrates how important platelets are for thrombus formation in veins. Disturbed venous blood flow activates the endothelium, and in a cooperative effort, endothelial cells, platelets, monocytes, and neutrophils interact with tissue factors, VWF, Factor XII, and neutrophil extracellular traps (NETs) to initiate and promote thrombus formation [32]. Platelets interact with VWF and leukocytes via the key adhesion receptor GPIbα, a component of the GPIb/V/IX complex [33, 34]. This supports leukocyte recruitment and stimulates NET formation by neutrophils. In addition, after endothelial injury, ADP-dependent platelet activation similar to arterial thrombosis is also required for thrombus formation in veins [35].

4.4 Platelets Mediate Immunothrombosis

A hypothetical physiological form of thrombosis has been described recently and termed "immunothrombosis" [36]. Immunothrombosis supports the innate immune response against pathogens in microvessels (Fig. 4.2). Pathogens are recognized, compartmentalized, and trapped inside intact vessels to prevent pathogen spreading and invasion. Ultimately, pathogens are killed within immunothrombi. An adaptive immune response and immune memory are also facilitated by immunothrombosis and platelets transport pathogens to remote lymphoid organs.

4.5 Platelets Release Soluble Factors in Thrombosis and Inflammation

Platelets have a number of prothrombotic and inflammatory features ranging from secretable factors to stably or variably expressed surface receptors (summarized in Table 4.1). Rapid secretion of a wide array of mediators following exocytosis of α granules, dense granules, and lysosomes upon activation is a unique feature of platelets. This enables an almost instantaneous response to stimuli in the affected vasculature. Of note, platelets not only contain preformed secretable factors but are also capable of newly synthesizing mediators such as interleukin (IL)-1β following signal-dependent splicing of pre-mRNA [37].

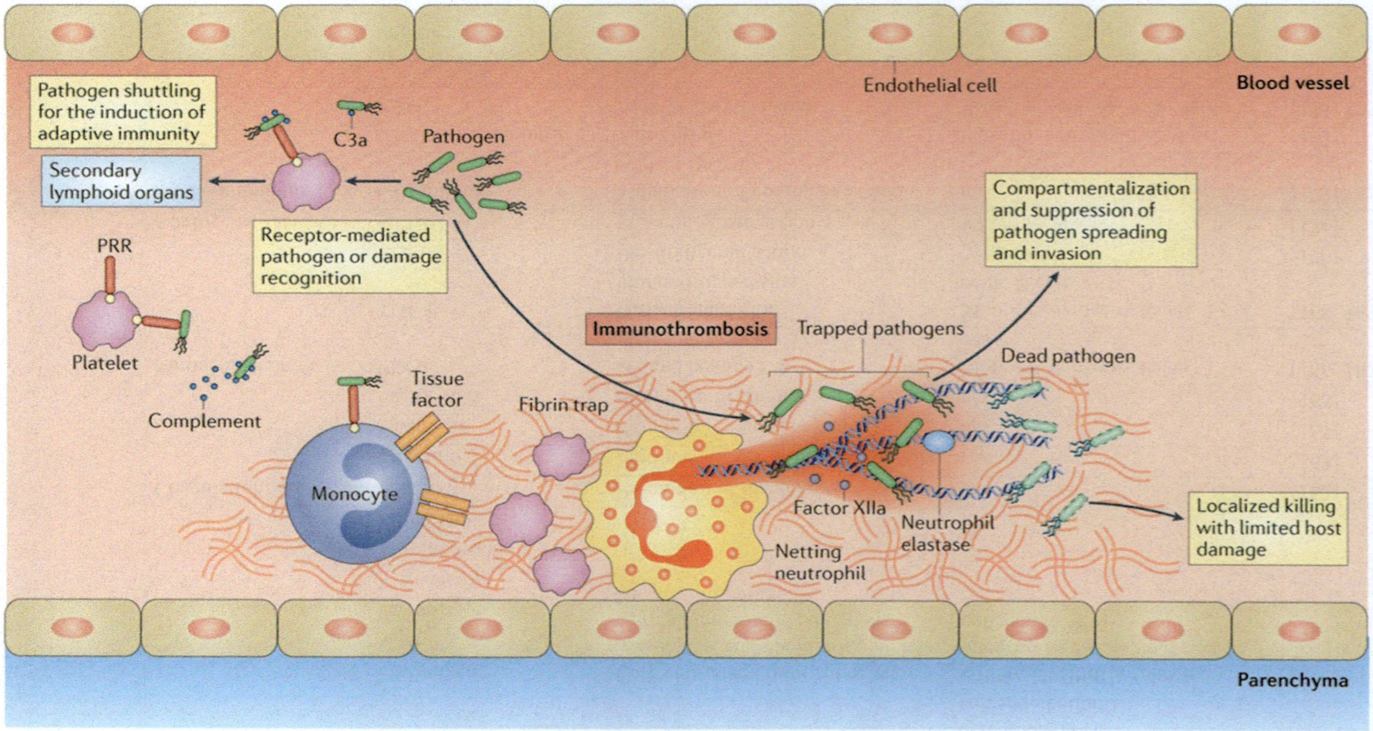

Fig. 4.2 Immunothrombosis. Platelets interact with endothelial cells, monocytes, and neutrophils not only to promote deep vein thrombosis but also to facilitate pathogen capture, presentation, and neutralization to fight infection (reprinted with permission from [36]; PRR, pattern-recognition receptor)

Table 4.1 Prothrombotic and immune-modulatory platelet components (reviewed in [38, 41, 42, 74, 95, 151])

	Superfamily	Molecule	Associated with	Effector cell/ligand	Ref.
α-granules	Adhesion molecules	P-selectin	Cell–cell interactions (leukocytes, endothelium)	PSGL-1 (on neutrophils, monocytes, microparticles, Th1 cells), unknown (on endothelium)	[38, 152, 153]
		Fibrinogen VWF Fibronectin Vitronectin Thrombospondin-1 PECAM-1	Cell–cell interactions (platelets, leukocytes, endothelium), platelet aggregation around pathogens, bacterial trapping	GPIbα, GPIIb/IIIa (platelets), integrins (platelets, leukocytes), collagen (VWF)	[41, 154–161]
	Coagulation factors	Factor V	Promotion of coagulation	n/a	[162]
		Protein S	Fibrinolysis	n/a	[163]
		Factor XI	Promotion of coagulation	n/a	[164]
		Factor XIII	Fibrin stabilization	n/a	[165]
	Mitogenic factors	PDGF	Wound healing	Monocytes, macrophages, T cells	[166, 167]
		TGF-β	Growth inhibition, immunosuppression	Monocytes, macrophages, T cells, B cells	[168, 169]
		EGF	Pro-mitogenic action	[a]	[170]
	Angiogenic factors	VEGF	Pro-angiogenic action	VEGF receptors	[171]
	Protease inhibitors	C1 inhibitor	Modulation of complement and contact system	n/a	[172]
		α2-plasmin inhibitor	Fibrinolysis	Plasmin, neutrophil elastase	[173]
		PAI-1	Antifibrinolytic action	Tissue plasminogen activator	[174, 175]

	Antimicrobial peptides	Thrombocidin 1 and 2	Antibacterial and antifungal action	Bacteria, candida	[176]
	Igs	IgG	Immune hematologic disorders	Fc receptors	[177, 178]
		IgA	[a]	[a]	[179]
		IgM	[a]	[a]	[180]
	Granule membrane-specific proteins	CD63	[a]	Integrins	[181]
		GMP33	[a]	[a]	[182]
	Chemokines	CCL3 (MIP-1α)	Leukocyte activation/recruitment	Monocytes/macrophages, eosinophils, basophils, NK cells, DCs	[183, 184]
		CCL5 (RANTES)	Monocyte recruitment	Monocytes, eosinophils, basophils, NK cells, T cells, DCs, platelets	[81, 99, 184, 185]
		CXCL1 (GROα)	Monocyte recruitment	Neutrophils/CXCR2	[183, 186]
		CXCL4 (PF4)	Monocyte recruitment and differentiation, anti-angiogenic, T-cell modulation	Monocytes, neutrophils, T cells, platelets	[187–190]
		CXCL5 (ENA78)	Modulation of chemokine scavenging, neutrophil chemotaxis	Neutrophils	[191, 192]
		CXCL7 (NAP2, β-thromboglobulin)	(Most abundant platelet chemokine, several variants) neutrophil recruitment and activation, EPC homing	Neutrophils, EPCs/CXCR1,2	[44, 193]
Dense granules	Amines	Serotonin (5-HT)	Neutrophil recruitment, leukocyte modulation	Neutrophils, monocytes, lymphocytes, NK cells, platelets/5-HT receptors	[50, 194]
		Histamine (?)	[a]	Endothelial cells, monocytes, neutrophils, NK cells, T cells, B cells, eosinophils	[49]

(continued)

Table 4.1 (continued)

	Superfamily	Molecule	Associated with	Effector cell/ligand	Ref.
	Cations	Calcium	Signal transduction	Effector proteins	[12, 195]
		Magnesium	Platelet–neutrophil interactions	Neutrophils	[196]
	Nucleotides	ATP	Th2 cell sensitization following DC activation	DCs/P2X, P2Y receptors	[139]
		ADP	Second-wave platelet activation	Platelets/P2Y receptors	[197]
Lysosomes	Proteases	Carboxypeptidases A, B	Eosinophil chemotaxis	[a]	[74, 198]
		Cathepsin D, E	Chemokine cleavage	[a]	[199]
		Acid phosphatase	[a]	[a]	[200]
		Collagenase	Facilitation of cell recruitment	[a]	[201]
	Glycohydrolases	Heparinase	Facilitation of cell recruitment	Endothelial glycocylix	[202]
		β-N-acetyl-glucosaminidase β-glucuronidase β-glycerophosphatase β-galactosidase α-D-glucosidase α-L-fucosidase β-D-fucosidase	Facilitation of cell recruitment	Extracellular matrix	[203]
Granule-independent soluble mediators	Cytoplasmic or membrane components	CCL7 (MCP3)	Monocyte chemotaxis	Monocytes, basophils, NK cells, DCs/CCR1-3	[183]
		IL-1β	T-cell activation, atherosclerosis, obesity	Monocytes, macrophages, DCs, T cells	[204]
		HMGB1	Systemic sclerosis	[a]	[205, 206]
		β-defensin 1, 2, 3	Bacteria clustering	Bacteria	[75]
		Thromboxane A2	Vasoconstriction, leukocyte modulation	Platelets, macrophages, T cells	[207]
		PAF	Autoimmunity/anaphylaxis, inflammation, cancer	PAFR	[208]
		sCD40L	Acute coronary syndrome	CD40	[87]

Surface adhesion molecules	Integrins	α5β1 (VLA-5)	Adhesion	Fibrinogen, CD40L	[209]
		α6β1 (VLA-6)	Adhesion	Laminin	[210]
		α2β1	Adhesion	Collagen	[211]
		α2bβ3 (GPIIb/IIIa)	Aggregation, bacterial trapping	Fibrinogen, fibronectin, vitronectin, VWF, thrombospondin	[154]
	Adhesion receptors	GPIbα	Adhesion, "immunothrombosis," bacterial infections	VWF, P-selectin, Mac-1	[36, 128, 212, 213]
		ICAM-2	Leukocyte recruitment	Neutrophils, DCs, T cells, monocytes/LFA-1, DC-SIGN	[214, 215]
		GPVI	Vascular integrity, rheumatoid arthritis, atherogenesis	collagen	[85, 216]
		CLEC2	Vascular integrity	Podoplanin	[85]
Immune receptors	TLRs	TLR1	[a]	Heterodimerizes with TLR2	[106]
		TLR2	Antimicrobial	Gram-positive bacteria	[104]
		TLR4	Sepsis	Lipopolysaccharide, gram-negative bacteria	[104]
		TLR6	[a]	Heterodimerizes with TLR2	[106]
		TLR7	PNC formation	Viral genome	[114]
		TLR9	Antiatherosclerotic action	DNA rich in CpG motifs	[116]
	Costimulatory proteins (TNF and TNFR superfamily)	CD40	Atherosclerosis, I/R injury	T cells/CD40L	[91]
		CD40L (CD154)	Atherosclerosis	CD40 on monocytes and endothelial cells/B cells, DCs, monocytes, macrophages, platelets	[90, 92, 93, 96]

(continued)

Table 4.1 (continued)

	Superfamily	Molecule	Associated with	Effector cell/ligand	Ref.
	Ig superfamily	TREM-1 ligand	Sepsis	PMNs, DCs, macrophages, monocytes	[217, 218]
	Sheddase	TACE (ADAM17)	[a]	GPIb, TNFα	[122]
	Ig receptors	FcγRIIA (CD32)	Bactericidal action	IgG	[42, 219]
		FcεRI	Allergy, parasite defense	IgE (high affinity)	[81]
		FcεRII (CD23)	Allergy, parasite defense	IgE (low affinity)	[220]
		FcαRI	IL-1β production	IgA	[221]
	Complement components	gC1qR	Antimicrobial	Bacterial protein A	[222, 223]
		C5b-9	Dense granule release	anaphylatoxins	[224, 225]

[a]Unknown or not applicable

Ref. exemplary references, *PSGL-1* P-selectin glycoprotein ligand 1, *GP* glycoprotein, *VWF* von Willebrand factor, *PECAM-1* platelet-endothelial cell adhesion molecule 1, *n/a* not applicable, *PDGF* platelet-derived growth factor, *TGF-β* transforming growth factor β, *EGF* epidermal growth factor, *VEGF* vascular endothelial growth factor, *NK cells* natural killer cells, *DCs* dendritic cells, *ENA78* epithelial-derived neutrophil-activating peptide 78, *GROα* growth-related oncogene α, *NAP-2* neutrophil-activating protein 2, *PF4* platelet factor 4, *PAI-1* plasminogen activator inhibitor 1, *IgG* immunoglobulin G, *GMP* α-granule membrane protein, *CCL* chemokine (C-C motif) ligand, *RANTES* regulated on activation, normal T cell expressed and secreted, *CXCL* chemokine (C-X-C motif) ligand, *MIP-1α*: macrophage inflammatory protein-1α, *EPC* endothelial progenitor cell, *5-HT* 5-hydroxytyptamine, *ATP* adenosine triphosphate, *ADP* adenosine diphosphate, *MCP* monocyte chemotactic protein, *IL* interleukin, *HMGB1* high mobility group protein 1, *PAF* platelet-activating factor, *CD40L* cluster of differentiation 40 ligand, *sCD40L* soluble CD40L, *Mac-1* macrophage-1 antigen (=αmβ2), *LFA-1* lymphocyte function-associated antigen 1 (=αLβ2), *DC-SIGN* dendritic cell-specific, ICAM-grabbing non-integrin, *TLR* toll-like receptor, *PNC* platelet–neutrophil complex, *TNF* tumor necrosis factor, *TREM-1* triggering receptor expressed on myeloid cells 1, *TACE* tumor necrosis factor α converting enzyme, *ADAM17* a disintegrin and metalloproteinase 17, *FcR* Ig Fc region receptor, *gC1qR* binding protein for the globular head domains of complement component C1q

4.5.1 α-granule Factors

α-granules are the most abundant granules in platelets and contain a variety of hemostatic and inflammatory mediators, including a number of adhesive proteins [38]. Platelet aggregation and (micro-)thrombus formation during immunothrombosis are promoted by fibrinogen, VWF, fibronectin, and vitronectin and serve not only as thrombus matrix [32, 39, 40] but also as immobilizing matrix for pathogen capture (reviewed in [38, 41, 42]). This limits pathogen growth and multiplication in the vasculature and facilitates exposure of these captured pathogens to neutralizing leukocytes. Although this pathophysiologic sequence has not been deciphered directly in mechanistic studies, observations in septic patients (and mice) suggest that platelet aggregation is not only a complication (in disseminated intravascular coagulation—DIC) but rather a feature of primary host defense [43].

Platelet factor 4 (PF4, CXCL4) and the β-thromboglobulin neutrophil-activating protein 2 (NAP2, CXCL7) regulate neutrophil and monocyte functions and promote their recruitment [44]. PF4 furthermore suppresses neutrophil apoptosis, which was demonstrated in a platelet depletion study of murine limb ischemia [45].

Several α-granule-derived chemokines have been studied extensively, especially in atherogenesis, and are considered important messengers of immune functions (reviewed in [46]). Chemokine functions include chemotaxis and modulation of different leukocyte functions. Platelets are able to take up immunoglobulins from plasma to store them in α-granules and to release this cargo on-site following inflammatory stimulation (reviewed in [41, 47]).

4.5.2 Dense Granule Factors

Dense granules store serotonin, calcium, magnesium, ATP, and ADP (whether they also contain histamine is controversial [48, 49]) and secrete these factors upon activation (reviewed in [16, 48]). At the site of acute inflammation, platelets release serotonin at micromolar concentrations, boosting the recruitment of neutrophils into the inflamed tissue (e.g., during murine pneumonia, peritonitis, and skin wounds) [50]. In mice, this translates into improved sepsis outcome when platelet serotonin stores were depleted. The observation that antidepressants inhibiting the uptake of serotonin modulate the release of several cytokines suggests that platelet serotonin may also be important in human inflammation (reviewed in [51]). In fact, numerous immunomodulatory functions of peripheral serotonin have been characterized, including differential effects on chemokine/cytokine secretion by immune cells. Serotonin is one of several soluble factors of platelet immunomodulation [50, 52–73].

4.5.3 Lysosomes

Lysosomes release glycosidases, proteases, and bactericidal enzymes such as β-glucuronidase, elastase, and collagenase (reviewed in [74]). The lysosome releasate facilitates pathogen clearance and breakdown of extracellular matrix, but studies are rare and clinical implications have not been addressed systematically.

4.5.4 Defensins

Other secretable factors are not associated with any of the known granules. They are derived from cytoplasmic stores, are newly synthesized proteins, or are components of a yet unknown type of granule. β-defensin is an example of granule-independent mediators with anti-bactericidal activity [75] and belongs to the group of antimicrobial peptides. Human platelets express β-defensins 1, 2, and 3 [75–77]. Platelets release β-defensin 1 from cytoplasm in response to *S. aureus* α-toxin to induce neutrophil extracellular trap (NET) formation and limit bacterial growth [75].

4.6 Platelets Express Immune Receptors

Different immune receptors operate on the platelet surface (and in some cases intracellularly) (reviewed in [78]). Toll-like receptors (TLRs) recognize pathogen- and danger-associated molecular patterns (DAMPs and PAMPS, respectively) (reviewed in [79]), complement receptors mediate complement activation at sites of platelet accumulation [80], and Fc receptors recognizing immunoglobulins (FcR, notably the Fcγ receptor FcγRIIA, but also Fcα and Fcε receptors) provide a link to the adaptive immune system [81, 82]. In a murine model of Arthus reaction, platelets facilitate the immune complex-induced recruitment of neutrophils in microvessels [83]. P-selectin on activated platelets appears to participate in complement activation, and platelet-associated

immune complexes mediate autoimmune diseases such as immune thrombocytopenia or systemic lupus erythematosus [81, 84]. C-type lectin-like receptor 2 (CLEC-2) is involved in the regulation of vascular integrity in acute inflammation [85, 86].

4.6.1 CD40/CD40 Ligand

CD40 and CD40L are not only expressed by platelets but also by endothelial cells, smooth muscle cells, and several leukocyte subtypes [87–98] (reviewed in [88–90, 92, 95]). Their surface expression levels are differentially regulated by the degree of cell activation. The inflammatory CD40/CD40L axis mediates a variety of cell–cell interactions. CD40/CD40L-mediated interactions have been well characterized in atherosclerosis but seem to be also involved in many other immune reactions. Platelets release CD40L and RANTES upon stimulation with IgG complexes without showing signs of general activation [99]. Platelet CD40L triggers inflammatory activation of endothelium: it induces upregulation of E-selectin, ICAM-1, and VCAM-1 and provokes chemokine secretion by endothelial cells [100].

4.6.2 Toll-like Receptors

Platelets express functional TLRs and respond to ligand binding and activation [101–103] (reviewed in [104]). TLR2 and TLR4 and the predominately intracellular TLR9 [105] are the most extensively studied platelet TLRs [105–108]. Platelets also contain the adapter molecules MYD88 and TRIF required for specific downstream signaling [109, 110]. The TLR2/1-specific agonist Pam3CSK4 has been utilized by different groups to stimulate platelets. It induces a dose-dependent response involving different intracellular signaling pathways, which may be part of defense mechanisms against gram-positive bacteria [111, 112]. Stimulation of platelet TLR4 has been linked to NET formation and subsequent capture of gram-negative bacteria in the bloodstream [113]. Platelet TLR7 binds viral RNA triggering PNC formation in mice, improving the animals' survival [114]. Platelet TLR9 recognizes viral and bacterial DNA and promotes platelet reactivity [105, 115]. Finally, TLR9 mediates protection from atherosclerosis in mice by suppressing the influx of CD4+ T cells into plaques [116]. Platelet TLRs are an interesting target for future therapeutic studies because pharmacological manipulation is uncomplicated.

4.7 Membrane Receptor Shedding in Inflammation

4.7.1 Soluble CD40 Ligand

Platelets are an important source of soluble CD40 ligand (sCD40L) (reviewed in [88, 89]). Platelet-derived sCD40L induces reactive oxygen species (ROS) production, neutrophil adhesion receptor upregulation, macrophage activation, and cytotoxic T-cell and B-cell stimulation (reviewed in [42]). CD40L can also be carried by platelet microparticles, regulating antigen-specific IgG production (reviewed in [42, 88]). Whether platelet-derived sCD40L is accessible to pharmacological intervention remains to be shown.

4.7.2 TACE

TNFα converting enzyme (TACE, ADAM17) is a sheddase expressed not only by neutrophils (where it regulates shedding of L-selectin and pro-TNFα) but also by platelets [117, 118]. Numerous signals can activate TACE, including atherosclerotic plaque components [119, 120]. Although specific immune functions of platelet surface TACE have not yet been deciphered, it is intriguing to note that oxidative stress and serotonin receptor activation induce GPIbα and GPV shedding by TACE [117, 121–124].

4.8 Acute Inflammation

4.8.1 Infections

Thrombocytopenia is a common feature of severe bacterial or viral infection and a marker of poor outcome (reviewed in [125]). This clinical observation suggests that platelets actively participate in the struggle against pathogens. Interestingly, recovery is often associated with reactive thrombocytosis [126]. Platelet consumption during DIC is well known but does not explain these frequent findings in milder clinical courses. Direct antimicrobial activity of platelets has been discovered in animal models of infectious diseases. In murine malaria, platelets eradicate intra-erythrocytic parasites (improving survival of the host), an effect that could be reproduced in ex vivo models with human blood cells [127]. When platelet binding to hepatic Kupffer cells via GPIbα was imaged in intravital microscopy of mice, challenging these mice with bacteria resulted in firm platelet immobilization via GPIIb and encapsulating of bacteria [128].

Human data supporting a role of platelets in immune defense are rare. Patients with chronic thrombocytopenia or patients with GPIb deficiency in Bernard–Soulier syndrome are not known to suffer from immune defects. Of note however, a recent tragic case report suggested an association between GPIIb/IIIa deficiency in Glanzmann thrombasthenia and HIV susceptibility [129].

4.8.2 Sepsis

In human sepsis, the number of circulating platelet–neutrophil complexes (PNCs) and platelet–monocyte complexes (PMCs) increases dramatically [43], correlating with the severity of multi-organ failure [130]. A differential release of growth factors from platelets was observed in septic patients [131]. The hemostatic functions were attenuated in relation to the severity of sepsis, but the release of VEGF was enhanced. Moreover, the transcriptome is altered in platelets from septic patients, facilitating differential release of proteins such as tissue factor [132]. Microthrombotic complications are provoked by disseminated platelet activation and platelet–leukocyte interactions [42, 43, 133].

4.8.3 Autoimmune Disease

Immune complex formation and complement activation by platelets were found in patients with immune thrombocytopenia and systemic lupus erythematosus (reviewed in [84]). Platelets release serotonin into acutely inflamed joints in murine rheumatoid arthritis, increasing synovial permeability [73]. Both mechanisms may represent attractive targets for therapeutic intervention.

4.8.4 Asthma

Platelet activation and granule secretion enhance bronchoconstriction and bronchial obstruction during airway inflammation and asthmatic attacks in mice and humans [134–138]. ATP and serotonin are released from dense granules sustaining these attacks [52, 139]. Bronchoalveolar lavage fluid after segmental allergen challenge of asthmatic patients contains high levels of platelet-derived serotonin, which enhances leukocyte infiltration, Th2-priming capacity of dendritic cells, and ultimately all cardinal features of allergic airway inflammation [52].

4.8.5 Ischemia/Reperfusion Injury

Ischemia/reperfusion (IR) injury contributes to the final infarct size in myocardial infarction. PNC infiltration correlates with myocardial reperfusion damage [140] and PMCs form rapidly after myocardial infarction in animal studies (reviewed in [42]). It has been suggested that the $P2Y_{12}$ inhibitor ticagrelor may limit IR injury [141, 142]. IR injury of the liver is also mediated by platelet–neutrophil interactions in mice [143], and liver regeneration depends on platelet-derived serotonin [144]. Whether specific antiplatelet intervention could limit IR injury in patients with myocardial infarction remains to be shown. This is of particular interest, because although reperfusion is often rapidly ensured by percutaneous coronary intervention, the subsequent inflammatory sequelae still dictate the final extent of the myocardial scar.

4.9 Chronic Inflammation

4.9.1 Metabolic Syndrome

Data from the Framingham heart study suggest that inflammatory platelet activation correlates with obesity and cardiovascular risk [145, 146]. Inflammatory gene transcripts derived from isolated platelets such as tumor necrosis factor (TNF), TLR2, and TLR4 were associated with increased body mass index [146]. Platelets therefore likely promote the inflammatory phenotype of metabolic syndrome (but direct data are not yet conclusive) [147].

4.9.2 Atherosclerosis

Platelets not only drive the atherothrombotic occlusion of a coronary artery in acute myocardial infarction, but they also mediate the chronic progression of vessel wall inflammation in atherosclerosis [148–150] (reviewed in [148, 149]). The various aspects ranging from cytokine release to monocyte recruitment have been examined in several in-depth animal studies and complemented by human ex vivo data (reviewed in [151]).

4.10 Conclusion

Platelets regulate thrombosis and inflammation using a unique set of different tools summarized in Table 4.1. The adhesion receptors GPIbα and GPIIb for example are not only required for initiation and growth of stable thrombi, but they also interact with leukocytes during host defense. In the microcirculation, immunothrombosis is a complex interplay of platelets, endothelial cells, and leukocytes with plasma proteins to promote pathogen capture, neutralization, and adaptive immune responses. These mechanisms are not yet completely understood but may enable future therapeutic interventions.

Compliance with Ethical Standards

Conflict of Interest: Daniel Duerschmied and Steffen Massberg declare that they have no conflict of interest.

Ethical Approval: This article does not contain any studies with human participants or animals performed by any of the authors.

References

1. Chu SG, Becker RC, Berger PB, et al. Mean platelet volume as a predictor of cardiovascular risk: a systematic review and meta-analysis. J Thromb Haemost. 2010;8:148–56.
2. Zhang L, Orban M, Lorenz M, et al. A novel role of sphingosine 1-phosphate receptor S1pr1 in mouse thrombopoiesis. J Exp Med. 2012;209:2165–81.
3. Junt T, Schulze H, Chen Z, et al. Dynamic visualization of thrombopoiesis within bone marrow. Science. 2007;317:1767–70.
4. Denis CV, Wagner DD. Platelet adhesion receptors and their ligands in mouse models of thrombosis. Arterioscler Thromb Vasc Biol. 2007;27:728–39.
5. Savage B, Saldivar E, Ruggeri ZM. Initiation of platelet adhesion by arrest onto fibrinogen or translocation on von Willebrand factor. Cell. 1996;84:289–97.
6. Reininger AJ, Bernlochner I, Penz SM, et al. A 2-step mechanism of arterial thrombus formation induced by human atherosclerotic plaques. J Am Coll Cardiol. 2010;55:1147–58.
7. Bergmeier W, Piffath CL, Goerge T, et al. The role of platelet adhesion receptor GPIbalpha far exceeds that of its main ligand, von Willebrand factor, in arterial thrombosis. Proc Natl Acad Sci USA. 2006;103:16900–5.

8. Nieswandt B, Brakebusch C, Bergmeier W, et al. Glycoprotein VI but not alpha2beta1 integrin is essential for platelet interaction with collagen. EMBO J. 2001;20:2120–30.
9. Nesbitt WS, Westein E, Tovar-Lopez FJ, et al. A shear gradient-dependent platelet aggregation mechanism drives thrombus formation. Nat Med. 2009;15:665–73.
10. Ruggeri ZM, Orje JN, Habermann R, Federici AB, Reininger AJ. Activation-independent platelet adhesion and aggregation under elevated shear stress. Blood. 2006;108:1903–10.
11. Gurbel PA, Tantry US. Combination antithrombotic therapies. Circulation. 2010;121:569–83.
12. Stefanini L, Roden RC, Bergmeier W. CalDAG-GEFI is at the nexus of calcium-dependent platelet activation. Blood. 2009;114:2506–14.
13. Cifuni SM, Wagner DD, Bergmeier W. CalDAG-GEFI and protein kinase C represent alternative pathways leading to activation of integrin alphaIIbbeta3 in platelets. Blood. 2008;112:1696–703.
14. Crittenden JR, Bergmeier W, Zhang Y, et al. CalDAG-GEFI integrates signaling for platelet aggregation and thrombus formation. Nat Med. 2004;10:982–6.
15. Chrzanowska-Wodnicka M, Smyth SS, Schoenwaelder SM, Fischer TH, White GC. Rap1b is required for normal platelet function and hemostasis in mice. J Clin Invest. 2005;115:680–7.
16. McNicol A, Israels SJ. Platelet dense granules: structure, function and implications for haemostasis. Thromb Res. 1999;95:1–18.
17. Davi G, Patrono C. Platelet activation and atherothrombosis. N Engl J Med. 2007;357:2482–94.
18. Gachet C, Leon C, Hechler B. The platelet P2 receptors in arterial thrombosis. Blood Cells Mol Dis. 2006;36:223–7.
19. Jin J, Daniel JL, Kunapuli SP. Molecular basis for ADP-induced platelet activation. II. The P2Y1 receptor mediates ADP-induced intracellular calcium mobilization and shape change in platelets. J Biol Chem. 1998;273:2030–4.
20. Leon C, Hechler B, Freund M, et al. Defective platelet aggregation and increased resistance to thrombosis in purinergic P2Y(1) receptor-null mice. J Clin Invest. 1999;104:1731–7.
21. Kauffenstein G, Bergmeier W, Eckly A, et al. The P2Y(12) receptor induces platelet aggregation through weak activation of the alpha(IIb)beta(3) integrin—a phosphoinositide 3-kinase-dependent mechanism. FEBS Lett. 2001;505:281–90.
22. Collet JP, Montalescot G. P2Y12 inhibitors: thienopyridines and direct oral inhibitors. Hamostaseologie. 2009;29:339–48.
23. Coughlin SR. Thrombin signalling and protease-activated receptors. Nature. 2000;407:258–64.
24. Vu TK, Hung DT, Wheaton VI, Coughlin SR. Molecular cloning of a functional thrombin receptor reveals a novel proteolytic mechanism of receptor activation. Cell. 1991;64:1057–68.
25. Sambrano GR, Weiss EJ, Zheng YW, Huang W, Coughlin SR. Role of thrombin signalling in platelets in haemostasis and thrombosis. Nature. 2001;413:74–8.
26. Andersen H, Greenberg DL, Fujikawa K, Xu W, Chung DW, Davie EW. Protease-activated receptor 1 is the primary mediator of thrombin-stimulated platelet procoagulant activity. Proc Natl Acad Sci USA. 1999;96:11189–93.
27. Shah R. Protease-activated receptors in cardiovascular health and diseases. Am Heart J. 2009;157:253–62.
28. Kahn ML, Zheng YW, Huang W, et al. A dual thrombin receptor system for platelet activation. Nature. 1998;394:690–4.
29. The Task Force for the Diagnosis and Management, Konstantinides SV, Torbicki A, et al. 2014 ESC guidelines on the diagnosis and management of acute pulmonary embolism: The Task Force for the Diagnosis and Management of Acute Pulmonary Embolism of the European Society of Cardiology (ESC) Endorsed by the European Respiratory Society (ERS). Eur Heart J. 2014;35(43):3033–69.
30. Becattini C, Agnelli G, Schenone A, et al. Aspirin for preventing the recurrence of venous thromboembolism. N Engl J Med. 2012;366:1959–67.
31. Brighton TA, Eikelboom JW, Mann K, et al. Low-dose aspirin for preventing recurrent venous thromboembolism. N Engl J Med. 2012;367:1979–87.

32. von Bruhl ML, Stark K, Steinhart A, et al. Monocytes, neutrophils, and platelets cooperate to initiate and propagate venous thrombosis in mice in vivo. J Exp Med. 2012;209:819–35.
33. Brill A, Fuchs TA, Chauhan AK, et al. von Willebrand factor-mediated platelet adhesion is critical for deep vein thrombosis in mouse models. Blood. 2011;117:1400–7.
34. Clemetson KJ. Platelets and primary haemostasis. Thromb Res. 2012;129:220–4.
35. Guenther F, Herr N, Mauler M, et al. Contrast ultrasound for the quantification of deep vein thrombosis in living mice: effects of enoxaparin and P2Y12 receptor inhibition. J Thromb Haemost. 2013;11:1154–62.
36. Engelmann B, Massberg S. Thrombosis as an intravascular effector of innate immunity. Nat Rev Immunol. 2013;13:34–45.
37. Denis MM, Tolley ND, Bunting M, et al. Escaping the nuclear confines: signal-dependent pre-mRNA splicing in anucleate platelets. Cell. 2005;122:379–91.
38. Zarbock A, Polanowska-Grabowska RK, Ley K. Platelet-neutrophil-interactions: linking hemostasis and inflammation. Blood Rev. 2007;21:99–111.
39. Riegger J, Byrne RA, Joner M, et al. Histopathological evaluation of thrombus in patients presenting with stent thrombosis. A multicenter European study: a report of the prevention of late stent thrombosis by an interdisciplinary global European effort consortiumdagger. Eur Heart J. 2016 May 14;37(19):1538–49.
40. Silvain J, Collet JP, Nagaswami C, et al. Composition of coronary thrombus in acute myocardial infarction. J Am Coll Cardiol. 2011;57:1359–67.
41. Jenne CN, Urrutia R, Kubes P. Platelets: bridging hemostasis, inflammation, and immunity. Int J Lab Hematol. 2013;35:254–61.
42. Rondina MT, Weyrich AS, Zimmerman GA. Platelets as cellular effectors of inflammation in vascular diseases. Circ Res. 2013;112:1506–19.
43. Gawaz M, Fateh-Moghadam S, Pilz G, Gurland HJ, Werdan K. Platelet activation and interaction with leucocytes in patients with sepsis or multiple organ failure. Eur J Clin Investig. 1995;25:843–51.
44. Brandt E, Petersen F, Ludwig A, Ehlert JE, Bock L, Flad HD. The beta-thromboglobulins and platelet factor 4: blood platelet-derived CXC chemokines with divergent roles in early neutrophil regulation. J Leukoc Biol. 2000;67:471–8.
45. Hartwig H, Drechsler M, Lievens D, et al. Platelet-derived PF4 reduces neutrophil apoptosis following arterial occlusion. Thromb Haemost. 2013;111:562–4.
46. Gleissner CA, von Hundelshausen P, Ley K. Platelet chemokines in vascular disease. Arterioscler Thromb Vasc Biol. 2008;28:1920–7.
47. Verheul HM, Lolkema MP, Qian DZ, et al. Platelets take up the monoclonal antibody bevacizumab. Clin Cancer Res. 2007;13:5341–7.
48. de Jong JS, Dekker LR. Platelets and cardiac arrhythmia. Front Physiol. 2010;1:166.
49. Mannaioni PF, Di Bello MG, Raspanti S, et al. Storage and release of histamine in human platelets. Inflamm Res. 1995;44(Suppl 1):S16–7.
50. Duerschmied D, Suidan GL, Demers M, et al. Platelet serotonin promotes the recruitment of neutrophils to sites of acute inflammation in mice. Blood. 2013;121:1008–15.
51. Kenis G, Maes M. Effects of antidepressants on the production of cytokines. Int J Neuropsychopharmacol. 2002;5:401–12.
52. Durk T, Duerschmied D, Muller T, et al. Production of serotonin by tryptophan hydroxylase 1 and release via platelets contribute to allergic airway inflammation. Am J Respir Crit Care Med. 2013;187:476–85.
53. Durk T, Panther E, Muller T, et al. 5-Hydroxytryptamine modulates cytokine and chemokine production in LPS-primed human monocytes via stimulation of different 5-HTR subtypes. Int Immunol. 2005;17:599–606.
54. Walther DJ, Peter J-U, Bashammakh S, et al. Synthesis of serotonin by a second tryptophan hydroxylase isoform. Science. 2003;299:76.

55. Iken K, Chheng S, Fargin A, Goulet AC, Kouassi E. Serotonin upregulates mitogen-stimulated B lymphocyte proliferation through 5-HT1A receptors. Cell Immunol. 1995;163:1–9.
56. Ito T, Ikeda U, Shimpo M, Yamamoto K, Shimada K. Serotonin increases interleukin-6 synthesis in human vascular smooth muscle cells. Circulation. 2000;102:2522–7.
57. Yu B, Becnel J, Zerfaoui M, Rohatgi R, Boulares AH, Nichols CD. Serotonin 5-hydroxytryptamine(2A) receptor activation suppresses tumor necrosis factor-alpha-induced inflammation with extraordinary potency. J Pharmacol Exp Ther. 2008;327:316–23.
58. Muller T, Durk T, Blumenthal B, et al. 5-hydroxytryptamine modulates migration, cytokine and chemokine release and T-cell priming capacity of dendritic cells in vitro and in vivo. PLoS One. 2009;4:e6453.
59. Walther A, Petri E, Peter C, Czabanka M, Martin E. Selective serotonin-receptor antagonism and microcirculatory alterations during experimental endotoxemia. J Surg Res. 2007;143:216–23.
60. Schuff-Werner P, Splettstoesser W. Antioxidative properties of serotonin and the bactericidal function of polymorphonuclear phagocytes. Adv Exp Med Biol. 1999;467:321–5.
61. Ciz M, Komrskova D, Pracharova L, et al. Serotonin modulates the oxidative burst of human phagocytes via various mechanisms. Platelets. 2007;18:583–90.
62. Pracharova L, Okenkova K, Lojek A, Ciz M. Serotonin and its 5-HT(2) receptor agonist DOI hydrochloride inhibit the oxidative burst in total leukocytes but not in isolated neutrophils. Life Sci. 2010;86:518–23.
63. Bondesson L, Nordlind K, Liden S, Sundstrom E. Inhibiting effects of serotonin and serotonin antagonists on the migration of mononuclear leucocytes. Immunopharmacol Immunotoxicol. 1993;15:243–50.
64. Northover BJ. The effect of histamine and 5-hydroxytryptamine on phagocytosis of staphylococci in vitro by polymorphs and macrophages. J Pathol Bacteriol. 1961;82:355–61.
65. Nordlind K, Sundstrom E, Bondesson L. Inhibiting effects of serotonin antagonists on the proliferation of mercuric chloride stimulated human peripheral blood T lymphocytes. Int Arch Allergy Immunol. 1992;97:105–8.
66. Sternberg EM, Trial J, Parker CW. Effect of serotonin on murine macrophages: suppression of Ia expression by serotonin and its reversal by 5-HT2 serotonergic receptor antagonists. J Immunol. 1986;137:276–82.
67. Hellstrand K, Czerkinsky C, Ricksten A, et al. Role of serotonin in the regulation of interferon-gamma production by human natural killer cells. J Interf Res. 1993;13:33–8.
68. Young MR, Matthews JP. Serotonin regulation of T-cell subpopulations and of macrophage accessory function. Immunology. 1995;84:148–52.
69. Kut JL, Young MR, Crayton JW, Wright MA, Young ME. Regulation of murine T-lymphocyte function by spleen cell-derived and exogenous serotonin. Immunopharmacol Immunotoxicol. 1992;14:783–96.
70. Arzt E, Costas M, Finkielman S, Nahmod VE. Serotonin inhibition of tumor necrosis factor-alpha synthesis by human monocytes. Life Sci. 1991;48:2557–62.
71. Cloez-Tayarani I, Petit-Bertron AF, Venters HD, Cavaillon JM. Differential effect of serotonin on cytokine production in lipopolysaccharide-stimulated human peripheral blood mononuclear cells: involvement of 5-hydroxytryptamine2A receptors. Int Immunol. 2003;15:233–40.
72. Kubera M, Maes M, Kenis G, Kim YK, Lason W. Effects of serotonin and serotonergic agonists and antagonists on the production of tumor necrosis factor alpha and interleukin-6. Psychiatry Res. 2005;134:251–8.
73. Cloutier N, Pare A, Farndale RW, et al. Platelets can enhance vascular permeability. Blood. 2012;120:1334–43.
74. Rendu F, Brohard-Bohn B. The platelet release reaction: granules' constituents, secretion and functions. Platelets. 2001;12:261–73.
75. Kraemer BF, Campbell RA, Schwertz H, et al. Novel anti-bacterial activities of beta-defensin 1 in human platelets: suppression of pathogen growth and signaling of neutrophil extracellular trap formation. PLoS Pathog. 2011;7:e1002355.
76. Tohidnezhad M, Varoga D, Wruck CJ, et al. Platelets display potent antimicrobial activity and release human beta-defensin 2. Platelets. 2012;23:217–23.

77. Tohidnezhad M, Varoga D, Podschun R, et al. Thrombocytes are effectors of the innate immune system releasing human beta defensin-3. Injury. 2011;42:682–6.
78. Kasirer-Friede A, Kahn ML, Shattil SJ. Platelet integrins and immunoreceptors. Immunol Rev. 2007;218:247–64.
79. Semple JW, Italiano JE Jr, Freedman J. Platelets and the immune continuum. Nat Rev Immunol. 2011;11:264–74.
80. Peerschke EI, Yin W, Ghebrehiwet B. Platelet mediated complement activation. Adv Exp Med Biol. 2008;632:81–91.
81. Hasegawa S, Tashiro N, Matsubara T, Furukawa S, Ra C. A comparison of FcepsilonRI-mediated RANTES release from human platelets between allergic patients and healthy individuals. Int Arch Allergy Immunol. 2001;125(Suppl 1):42–7.
82. Ginsberg MH, Henson PM. Enhancement of platelet response to immune complexes and IgG aggregates by lipid A-rich bacterial lipopolysaccharides. J Exp Med. 1978;147:207–17.
83. Lister KJ, James WG, Hickey MJ. Immune complexes mediate rapid alterations in microvascular permeability: roles for neutrophils, complement, and platelets. Microcirculation. 2007;14:709–22.
84. Puram V, Giuliani D, Morse BS. Circulating immune complexes and platelet IgG in various diseases. Clin Exp Immunol. 1984;58:672–6.
85. Boulaftali Y, Hess PR, Getz TM, et al. Platelet ITAM signaling is critical for vascular integrity in inflammation. J Clin Invest. 2013;123:908–16.
86. Suzuki-Inoue K. Essential in vivo roles of the platelet activation receptor CLEC-2 in tumour metastasis, lymphangiogenesis and thrombus formation. J Biochem. 2011;150:127–32.
87. Aukrust P, Muller F, Ueland T, et al. Enhanced levels of soluble and membrane-bound CD40 ligand in patients with unstable angina. Possible reflection of T lymphocyte and platelet involvement in the pathogenesis of acute coronary syndromes. Circulation. 1999;100:614–20.
88. Elzey BD, Ratliff TL, Sowa JM, Crist SA. Platelet CD40L at the interface of adaptive immunity. Thromb Res. 2011;127:180–3.
89. Ferroni P, Santilli F, Guadagni F, Basili S, Davi G. Contribution of platelet-derived CD40 ligand to inflammation, thrombosis and neoangiogenesis. Curr Med Chem. 2007;14:2170–80.
90. Gerdes N, Zirlik A. Co-stimulatory molecules in and beyond co-stimulation – tipping the balance in atherosclerosis? Thromb Haemost. 2011;106:804–13.
91. Lapchak PH, Ioannou A, Kannan L, Rani P, Dalle Lucca JJ, Tsokos GC. Platelet-associated CD40/CD154 mediates remote tissue damage after mesenteric ischemia/reperfusion injury. PLoS One. 2012;7:e32260.
92. Lievens D, Eijgelaar WJ, Biessen EA, Daemen MJ, Lutgens E. The multi-functionality of CD40L and its receptor CD40 in atherosclerosis. Thromb Haemost. 2009;102:206–14.
93. Lievens D, Zernecke A, Seijkens T, et al. Platelet CD40L mediates thrombotic and inflammatory processes in atherosclerosis. Blood. 2010;116:4317–27.
94. Lutgens E, Lievens D, Beckers L, et al. Deficient CD40-TRAF6 signaling in leukocytes prevents atherosclerosis by skewing the immune response toward an antiinflammatory profile. J Exp Med. 2010;207:391–404.
95. Nurden AT. Platelets, inflammation and tissue regeneration. Thromb Haemost. 2011;105 (Suppl 1):S13–33.
96. Wolf D, Hohmann JD, Wiedemann A, et al. Binding of CD40L to Mac-1's I-domain involves the EQLKKSKTL motif and mediates leukocyte recruitment and atherosclerosis—but does not affect immunity and thrombosis in mice. Circ Res. 2011;109:1269–79.
97. Zirlik A, Bavendiek U, Libby P, et al. TRAF-1, -2, -3, -5, and -6 are induced in atherosclerotic plaques and differentially mediate proinflammatory functions of CD40L in endothelial cells. Arterioscler Thromb Vasc Biol. 2007;27:1101–7.
98. Lutgens E, Poggi M, Weber C. CD40L CD40 fuel ignites obesity. Thromb Haemost. 2010;103:694–5.

99. Antczak AJ, Singh N, Gay SR, Worth RG. IgG-complex stimulated platelets: a source of sCD40L and RANTES in initiation of inflammatory cascade. Cell Immunol. 2010;263:129–33.
100. Henn V, Slupsky JR, Grafe M, et al. CD40 ligand on activated platelets triggers an inflammatory reaction of endothelial cells. Nature. 1998;391:591–4.
101. Andonegui G, Kerfoot SM, McNagny K, Ebbert KV, Patel KD, Kubes P. Platelets express functional Toll-like receptor-4. Blood. 2005;106:2417–23.
102. Rex S, Beaulieu LM, Perlman DH, et al. Immune versus thrombotic stimulation of platelets differentially regulates signalling pathways, intracellular protein-protein interactions, and alpha-granule release. Thromb Haemost. 2009;102:97–110.
103. Blair P, Rex S, Vitseva O, et al. Stimulation of Toll-like receptor 2 in human platelets induces a thromboinflammatory response through activation of phosphoinositide 3-kinase. Circ Res. 2009;104:346–54.
104. Beaulieu LM, Freedman JE. The role of inflammation in regulating platelet production and function: Toll-like receptors in platelets and megakaryocytes. Thromb Res. 2010;125:205–9.
105. Thon JN, Peters CG, Machlus KR, et al. T granules in human platelets function in TLR9 organization and signaling. J Cell Biol. 2012;198:561–74.
106. Shiraki R, Inoue N, Kawasaki S, et al. Expression of Toll-like receptors on human platelets. Thromb Res. 2004;113:379–85.
107. Cognasse F, Hamzeh H, Chavarin P, Acquart S, Genin C, Garraud O. Evidence of Toll-like receptor molecules on human platelets. Immunol Cell Biol. 2005;83:196–8.
108. Aslam R, Speck ER, Kim M, et al. Platelet Toll-like receptor expression modulates lipopolysaccharide-induced thrombocytopenia and tumor necrosis factor-alpha production in vivo. Blood. 2006;107:637–41.
109. Berthet J, Damien P, Hamzeh-Cognasse H, et al. Human platelets can discriminate between various bacterial LPS isoforms via TLR4 signaling and differential cytokine secretion. Clin Immunol. 2012;145:189–200.
110. Zhang G, Han J, Welch EJ, et al. Lipopolysaccharide stimulates platelet secretion and potentiates platelet aggregation via TLR4/MyD88 and the cGMP-dependent protein kinase pathway. J Immunol. 2009;182:7997–8004.
111. Falker K, Klarstrom-Engstrom K, Bengtsson T, Lindahl TL, Grenegard M. The toll-like receptor 2/1 (TLR2/1) complex initiates human platelet activation via the src/Syk/LAT/PLCgamma2 signalling cascade. Cell Signal. 2014;26:279–86.
112. Rivadeneyra L, Carestia A, Etulain J, et al. Regulation of platelet responses triggered by Toll-like receptor 2 and 4 ligands is another non-genomic role of nuclear factor-kappaB. Thromb Res. 2014;133:235–43.
113. Clark SR, Ma AC, Tavener SA, et al. Platelet TLR4 activates neutrophil extracellular traps to ensnare bacteria in septic blood. Nat Med. 2007;13:463–9.
114. Koupenova M, Vitseva O, Mackay CR, et al. Platelet-TLR7 mediates host survival and platelet count during viral infection in the absence of platelet-dependent thrombosis. Blood. 2014;124(5):791–802.
115. Panigrahi S, Ma Y, Hong L, et al. Engagement of platelet toll-like receptor 9 by novel endogenous ligands promotes platelet hyperreactivity and thrombosis. Circ Res. 2013;112:103–12.
116. Koulis C, Chen YC, Hausding C, et al. Protective role for Toll-like receptor-9 in the development of atherosclerosis in apolipoprotein E-deficient mice. Arterioscler Thromb Vasc Biol. 2014;34(3):516–25.
117. Li Y, Brazzell J, Herrera A, Walcheck B. ADAM17 deficiency by mature neutrophils has differential effects on L-selectin shedding. Blood. 2006;108:2275–9.
118. Black RA, Rauch CT, Kozlosky CJ, et al. A metalloproteinase disintegrin that releases tumour-necrosis factor-alpha from cells. Nature. 1997;385:729–33.
119. Canault M, Leroyer AS, Peiretti F, et al. Microparticles of human atherosclerotic plaques enhance the shedding of the tumor necrosis factor-alpha converting enzyme/ADAM17 substrates, tumor necrosis factor and tumor necrosis factor receptor-1. Am J Pathol. 2007;171:1713–23.

120. Canault M, Peiretti F, Kopp F, et al. The TNF alpha converting enzyme (TACE/ADAM17) is expressed in the atherosclerotic lesions of apolipoprotein E-deficient mice: possible contribution to elevated plasma levels of soluble TNF alpha receptors. Atherosclerosis. 2006;187:82–91.
121. Duerschmied D, Canault M, Lievens D, et al. Serotonin stimulates platelet receptor shedding by tumor necrosis factor-alpha-converting enzyme (ADAM17). J Thromb Haemost. 2009;7:1163–71.
122. Bergmeier W, Piffath CL, Cheng G, et al. Tumor necrosis factor-alpha-converting enzyme (ADAM17) mediates GPIbalpha shedding from platelets in vitro and in vivo. Circ Res. 2004;95:677–83.
123. Peschon JJ, Slack JL, Reddy P, et al. An essential role for ectodomain shedding in mammalian development. Science. 1998;282:1281–4.
124. Brill A, Chauhan AK, Canault M, Walsh MT, Bergmeier W, Wagner DD. Oxidative stress activates ADAM17/TACE and induces its target receptor shedding in platelets in a p38-dependent fashion. Cardiovasc Res. 2009;84:137–44.
125. Hui P, Cook DJ, Lim W, Fraser GA, Arnold DM. The frequency and clinical significance of thrombocytopenia complicating critical illness: a systematic review. Chest. 2011;139:271–8.
126. Forehand CC, Cribb J, May JR. Examination of the relationship between antimicrobials and thrombocytosis. Ann Pharmacother. 2012;46:1425–9.
127. McMorran BJ, Marshall VM, de Graaf C, et al. Platelets kill intraerythrocytic malarial parasites and mediate survival to infection. Science. 2009;323:797–800.
128. Wong CH, Jenne CN, Petri B, Chrobok NL, Kubes P. Nucleation of platelets with blood-borne pathogens on Kupffer cells precedes other innate immunity and contributes to bacterial clearance. Nat Immunol. 2013;14:785–92.
129. Manne RK, Natarajan K, Patil R, Prathi VS, Beeraka SS, Kolaparthi VS. Glanzmann thrombasthenia associated with human immunodeficiency virus-positive patient. Int J Prev Med. 2014;5:500–4.
130. Russwurm S, Vickers J, Meier-Hellmann A, et al. Platelet and leukocyte activation correlate with the severity of septic organ dysfunction. Shock. 2002;17:263–8.
131. Yaguchi A, Lobo FL, Vincent JL, Pradier O. Platelet function in sepsis. J Thromb Haemost. 2004;2:2096–102.
132. Rondina MT, Schwertz H, Harris ES, et al. The septic milieu triggers expression of spliced tissue factor mRNA in human platelets. J Thromb Haemost. 2011;9:748–58.
133. Grabarek J, Timmons S, Hawiger J. Modulation of human platelet protein kinase C by endotoxic lipid A. J Clin Invest. 1988;82:964–71.
134. Gresele P, Dottorini M, Selli ML, et al. Altered platelet function associated with the bronchial hyperresponsiveness accompanying nocturnal asthma. J Allergy Clin Immunol. 1993;91:894–902.
135. Kowal K, Pampuch A, Kowal-Bielecka O, DuBuske LM, Bodzenta-Lukaszyk A. Platelet activation in allergic asthma patients during allergen challenge with Dermatophagoides pteronyssinus. Clin Exp Allergy. 2006;36:426–32.
136. Johansson MW, Han ST, Gunderson KA, Busse WW, Jarjour NN, Mosher DF. Platelet activation, P-selectin, and eosinophil beta1-integrin activation in asthma. Am J Respir Crit Care Med. 2012;185:498–507.
137. Pitchford SC, Momi S, Baglioni S, et al. Allergen induces the migration of platelets to lung tissue in allergic asthma. Am J Respir Crit Care Med. 2008;177:604–12.
138. Pitchford SC, Yano H, Lever R, et al. Platelets are essential for leukocyte recruitment in allergic inflammation. J Allergy Clin Immunol. 2003;112:109–18.
139. Idzko M, Hammad H, van Nimwegen M, et al. Extracellular ATP triggers and maintains asthmatic airway inflammation by activating dendritic cells. Nat Med. 2007;13:913–9.
140. Kohler D, Straub A, Weissmuller T, et al. Phosphorylation of vasodilator-stimulated phosphoprotein prevents platelet-neutrophil complex formation and dampens myocardial ischemia-reperfusion injury. Circulation. 2011;123:2579–90.

141. Wang K, Zhou X, Huang Y, et al. Adjunctive treatment with ticagrelor, but not clopidogrel, added to tPA enables sustained coronary artery recanalisation with recovery of myocardium perfusion in a canine coronary thrombosis model. Thromb Haemost. 2010;104:609–17.
142. de Groot H, Rauen U. Ischemia-reperfusion injury: processes in pathogenetic networks: a review. Transplant Proc. 2007;39:481–4.
143. Pak S, Kondo T, Nakano Y, et al. Platelet adhesion in the sinusoid caused hepatic injury by neutrophils after hepatic ischemia reperfusion. Platelets. 2010;21:282–8.
144. Lesurtel M, Graf R, Aleil B, et al. Platelet-derived serotonin mediates liver regeneration. Science. 2006;312:104–7.
145. McManus DD, Beaulieu LM, Mick E, et al. Relationship among circulating inflammatory proteins, platelet gene expression, and cardiovascular risk. Arterioscler Thromb Vasc Biol. 2013;33:2666–73.
146. Freedman JE, Larson MG, Tanriverdi K, et al. Relation of platelet and leukocyte inflammatory transcripts to body mass index in the Framingham heart study. Circulation. 2010;122:119–29.
147. Wolf D, Jehle F, Anto Michel N, et al. Co-inhibitory suppression of T cell activation by CD40 protects from obesity and adipose tissue inflammation in mice. Circulation. 2014;129(23):2414–25.
148. Lievens D, von Hundelshausen P. Platelets in atherosclerosis. Thromb Haemost. 2011;106:827–38.
149. Gawaz M, Langer H, May AE. Platelets in inflammation and atherogenesis. J Clin Invest. 2005;115:3378–84.
150. Koenen RR, von Hundelshausen P, Nesmelova IV, et al. Disrupting functional interactions between platelet chemokines inhibits atherosclerosis in hyperlipidemic mice. Nat Med. 2009;15:97–103.
151. von Hundelshausen P, Weber C. Platelets as immune cells: bridging inflammation and cardiovascular disease. Circ Res. 2007;100:27–40.
152. Wagner DD, Frenette PS. The vessel wall and its interactions. Blood. 2008;111:5271–81.
153. Moore KL, Stults NL, Diaz S, et al. Identification of a specific glycoprotein ligand for P-selectin (CD62) on myeloid cells. J Cell Biol. 1992;118:445–56.
154. Ruggeri ZM, Mendolicchio GL. Adhesion mechanisms in platelet function. Circ Res. 2007;100:1673–85.
155. Coller BS, Peerschke EI, Scudder LE, Sullivan CA. A murine monoclonal antibody that completely blocks the binding of fibrinogen to platelets produces a thrombasthenic-like state in normal platelets and binds to glycoproteins IIb and/or IIIa. J Clin Invest. 1983;72:325–38.
156. Savage B, Almus-Jacobs F, Ruggeri ZM. Specific synergy of multiple substrate-receptor interactions in platelet thrombus formation under flow. Cell. 1998;94:657–66.
157. Ginsberg MH, Forsyth J, Lightsey A, Chediak J, Plow EF. Reduced surface expression and binding of fibronectin by thrombin-stimulated thrombasthenic platelets. J Clin Invest. 1983;71:619–24.
158. Asch E, Podack E. Vitronectin binds to activated human platelets and plays a role in platelet aggregation. J Clin Invest. 1990;85:1372–8.
159. Wencel-Drake JD, Painter RG, Zimmerman TS, Ginsberg MH. Ultrastructural localization of human platelet thrombospondin, fibrinogen, fibronectin, and von Willebrand factor in frozen thin section. Blood. 1985;65:929–38.
160. Varga-Szabo D, Pleines I, Nieswandt B. Cell adhesion mechanisms in platelets. Arterioscler Thromb Vasc Biol. 2008;28:403–12.
161. Newman PJ, Newman DK. Signal transduction pathways mediated by PECAM-1: new roles for an old molecule in platelet and vascular cell biology. Arterioscler Thromb Vasc Biol. 2003;23:953–64.
162. Hayward CP, Furmaniak-Kazmierczak E, Cieutat AM, et al. Factor V is complexed with multimerin in resting platelet lysates and colocalizes with multimerin in platelet alpha-granules. J Biol Chem. 1995;270:19217–24.

163. Schwarz HP, Heeb MJ, Wencel-Drake JD, Griffin JH. Identification and quantitation of protein S in human platelets. Blood. 1985;66:1452–5.
164. Hu CJ, Baglia FA, Mills DC, Konkle BA, Walsh PN. Tissue-specific expression of functional platelet factor XI is independent of plasma factor XI expression. Blood. 1998;91:3800–7.
165. Marx G, Korner G, Mou X, Gorodetsky R. Packaging zinc, fibrinogen, and factor XIII in platelet alpha-granules. J Cell Physiol. 1993;156:437–42.
166. Barrientos S, Stojadinovic O, Golinko MS, Brem H, Tomic-Canic M. Growth factors and cytokines in wound healing. Wound Repair Regen. 2008;16:585–601.
167. Linder BL, Chernoff A, Kaplan KL, Goodman DS. Release of platelet-derived growth factor from human platelets by arachidonic acid. Proc Natl Acad Sci USA. 1979;76:4107–11.
168. Lo Re S, Lecocq M, Uwambayinema F, et al. Platelet-derived growth factor-producing CD4+ Foxp3+ regulatory T lymphocytes promote lung fibrosis. Am J Respir Crit Care Med. 2011;184:1270–81.
169. Fava RA, Casey TT, Wilcox J, Pelton RW, Moses HL, Nanney LB. Synthesis of transforming growth factor-beta 1 by megakaryocytes and its localization to megakaryocyte and platelet alpha-granules. Blood. 1990;76:1946–55.
170. Pinzani M, Gesualdo L, Sabbah GM, Abboud HE. Effects of platelet-derived growth factor and other polypeptide mitogens on DNA synthesis and growth of cultured rat liver fat-storing cells. J Clin Invest. 1989;84:1786–93.
171. Salgado R, Benoy I, Bogers J, et al. Platelets and vascular endothelial growth factor (VEGF): a morphological and functional study. Angiogenesis. 2001;4:37–43.
172. Schmaier AH, Amenta S, Xiong T, Heda GD, Gewirtz AM. Expression of platelet C1 inhibitor. Blood. 1993;82:465–74.
173. Shieh BH, Travis J. The reactive site of human alpha 2-antiplasmin. J Biol Chem. 1987;262:6055–9.
174. Nylander M, Osman A, Ramstrom S, Aklint E, Larsson A, Lindahl TL. The role of thrombin receptors PAR1 and PAR4 for PAI-1 storage, synthesis and secretion by human platelets. Thromb Res. 2012;129:e51–8.
175. Brogren H, Karlsson L, Andersson M, Wang L, Erlinge D, Jern S. Platelets synthesize large amounts of active plasminogen activator inhibitor 1. Blood. 2004;104:3943–8.
176. Kwakman PH, Krijgsveld J, de Boer L, et al. Native thrombocidin-1 and unfolded thrombocidin-1 exert antimicrobial activity via distinct structural elements. J Biol Chem. 2011;286:43506–14.
177. George JN, Saucerman S, Levine SP, Knieriem LK, Bainton DF. Immunoglobulin G is a platelet alpha granule-secreted protein. J Clin Invest. 1985;76:2020–5.
178. George JN. Platelet IgG: measurement, interpretation, and clinical significance. Prog Hemost Thromb. 1991;10:97–126.
179. Falet H, Marchetti MP, Hoffmeister KM, Massaad MJ, Geha RS, Hartwig JH. Platelet-associated IgAs and impaired GPVI responses in platelets lacking WIP. Blood. 2009;114:4729–37.
180. George JN, Saucerman S. Platelet IgG, IgA, IgM, and albumin: correlation of platelet and plasma concentrations in normal subjects and in patients with ITP or dysproteinemia. Blood. 1988;72:362–5.
181. Berditchevski F, Bazzoni G, Hemler ME. Specific association of CD63 with the VLA-3 and VLA-6 integrins. J Biol Chem. 1995;270:17784–90.
182. Damas C, Vink T, Nieuwenhuis HK, Sixma JJ. The 33-kDa platelet alpha-granule membrane protein (GMP-33) is an N-terminal proteolytic fragment of thrombospondin. Thromb Haemost. 2001;86:887–93.
183. Gear AR, Camerini D. Platelet chemokines and chemokine receptors: linking hemostasis, inflammation, and host defense. Microcirculation. 2003;10:335–50.
184. Al-Bannawi A, Al-Wesebai K, Taha S, Bakhiet M. Chlamydia pneumoniae induces chemokine expression by platelets in patients with atherosclerosis. Med Princ Pract. 2011;20:438–43.

185. von Hundelshausen P, Weber KS, Huo Y, et al. RANTES deposition by platelets triggers monocyte arrest on inflamed and atherosclerotic endothelium. Circulation. 2001;103:1772–7.
186. Smith DF, Galkina E, Ley K, Huo Y. GRO family chemokines are specialized for monocyte arrest from flow. Am J Physiol Heart Circ Physiol. 2005;289:H1976–84.
187. Scheuerer B, Ernst M, Durrbaum-Landmann I, et al. The CXC-chemokine platelet factor 4 promotes monocyte survival and induces monocyte differentiation into macrophages. Blood. 2000;95:1158–66.
188. von Hundelshausen P, Koenen RR, Sack M, et al. Heterophilic interactions of platelet factor 4 and RANTES promote monocyte arrest on endothelium. Blood. 2005;105:924–30.
189. Struyf S, Burdick MD, Proost P, Van Damme J, Strieter RM. Platelets release CXCL4L1, a nonallelic variant of the chemokine platelet factor-4/CXCL4 and potent inhibitor of angiogenesis. Circ Res. 2004;95:855–7.
190. Fleischer J, Grage-Griebenow E, Kasper B, et al. Platelet factor 4 inhibits proliferation and cytokine release of activated human T cells. J Immunol. 2002;169:770–7.
191. Mei J, Liu Y, Dai N, et al. CXCL5 regulates chemokine scavenging and pulmonary host defense to bacterial infection. Immunity. 2010;33:106–17.
192. Power CA, Furness RB, Brawand C, Wells TN. Cloning of a full-length cDNA encoding the neutrophil-activating peptide ENA-78 from human platelets. Gene. 1994;151:333–4.
193. Hristov M, Zernecke A, Bidzhekov K, et al. Importance of CXC chemokine receptor 2 in the homing of human peripheral blood endothelial progenitor cells to sites of arterial injury. Circ Res. 2007;100:590–7.
194. Berger M, Gray JA, Roth BL. The expanded biology of serotonin. Annu Rev Med. 2009;60:355–66.
195. King SM, McNamee RA, Houng AK, Patel R, Brands M, Reed GL. Platelet dense-granule secretion plays a critical role in thrombosis and subsequent vascular remodeling in atherosclerotic mice. Circulation. 2009;120:785–91.
196. Evangelista V, Manarini S, Rotondo S, et al. Platelet/polymorphonuclear leukocyte interaction in dynamic conditions: evidence of adhesion cascade and cross talk between P-selectin and the beta 2 integrin CD11b/CD18. Blood. 1996;88:4183–94.
197. Frelinger AL 3rd, Jakubowski JA, Li Y, et al. The active metabolite of prasugrel inhibits ADP-stimulated thrombo-inflammatory markers of platelet activation: Influence of other blood cells, calcium, and aspirin. Thromb Haemost. 2007;98:192–200.
198. Weissmann G. The role of lysosomes in inflammation and disease. Annu Rev Med. 1967;18:97–112.
199. Sixma JJ, van den Berg A, Hasilik A, von Figura K, Geuze HJ. Immuno-electron microscopical demonstration of lysosomes in human blood platelets and megakaryocytes using anti-cathepsin D. Blood. 1985;65:1287–91.
200. Radzun HJ, Parwaresch MR, Kulenkampff C, Stein H. Lysosomal acid phosphatase: activity and isoenzymes in separated normal human blood cells. Clin Chim Acta. 1980;102:227–35.
201. Chesney CM, Harper E, Colman RW. Human platelet collagenase. J Clin Invest. 1974;53:1647–54.
202. Chappell D, Jacob M, Rehm M, et al. Heparinase selectively sheds heparan sulphate from the endothelial glycocalyx. Biol Chem. 2008;389:79–82.
203. Dangelmaier CA, Holmsen H. Determination of acid hydrolases in human platelets. Anal Biochem. 1980;104:182–91.
204. Beaulieu LM, Lin E, Mick E, et al. Interleukin 1 receptor 1 and interleukin 1beta regulate megakaryocyte maturation, platelet activation, and transcript profile during inflammation in mice and humans. Arterioscler Thromb Vasc Biol. 2014;34(3):552–64.
205. Maugeri N, Franchini S, Campana L, et al. Circulating platelets as a source of the damage-associated molecular pattern HMGB1 in patients with systemic sclerosis. Autoimmunity. 2012;45:584–7.

206. Rouhiainen A, Imai S, Rauvala H, Parkkinen J. Occurrence of amphoterin (HMG1) as an endogenous protein of human platelets that is exported to the cell surface upon platelet activation. Thromb Haemost. 2000;84:1087–94.
207. Prancan A, Simon D, Pope L. Platelet thromboxane production during endotoxin shock. Agents Actions. 1981;11:648–50.
208. Edwards LJ, Constantinescu CS. Platelet activating factor/platelet activating factor receptor pathway as a potential therapeutic target in autoimmune diseases. Inflamm Allergy Drug Targets. 2009;8:182–90.
209. Leveille C, Bouillon M, Guo W, et al. CD40 ligand binds to alpha5beta1 integrin and triggers cell signaling. J Biol Chem. 2007;282:5143–51.
210. Schaff M, Tang C, Maurer E, et al. Integrin alpha6beta1 is the main receptor for vascular laminins and plays a role in platelet adhesion, activation, and arterial thrombosis. Circulation. 2013;128:541–52.
211. Bix G, Iozzo RA, Woodall B, et al. Endorepellin, the C-terminal angiostatic module of perlecan, enhances collagen-platelet responses via the alpha2beta1-integrin receptor. Blood. 2007;109:3745–8.
212. Yin H, Stojanovic-Terpo A, Xu W, et al. Role for platelet glycoprotein Ib-IX and effects of its inhibition in endotoxemia-induced thrombosis, thrombocytopenia, and mortality. Arterioscler Thromb Vasc Biol. 2013;33:2529–37.
213. Verschoor A, Neuenhahn M, Navarini AA, et al. A platelet-mediated system for shuttling blood-borne bacteria to CD8alpha+ dendritic cells depends on glycoprotein GPIb and complement C3. Nat Immunol. 2011;12:1194–201.
214. Diacovo TG, deFougerolles AR, Bainton DF, Springer TA. A functional integrin ligand on the surface of platelets: intercellular adhesion molecule-2. J Clin Invest. 1994;94:1243–51.
215. Weber KS, Alon R, Klickstein LB. Sialylation of ICAM-2 on platelets impairs adhesion of leukocytes via LFA-1 and DC-SIGN. Inflammation. 2004;28:177–88.
216. Schulz C, Penz S, Hoffmann C, et al. Platelet GPVI binds to collagenous structures in the core region of human atheromatous plaque and is critical for atheroprogression in vivo. Basic Res Cardiol. 2008;103:356–67.
217. Haselmayer P, Grosse-Hovest L, von Landenberg P, Schild H, Radsak MP. TREM-1 ligand expression on platelets enhances neutrophil activation. Blood. 2007;110:1029–35.
218. Derive M, Bouazza Y, Sennoun N, et al. Soluble TREM-like transcript-1 regulates leukocyte activation and controls microbial sepsis. J Immunol. 2012;188:5585–92.
219. Riaz AH, Tasma BE, Woodman ME, Wooten RM, Worth RG. Human platelets efficiently kill IgG-opsonized E coli. FEMS Immunol Med Microbiol. 2012;65:78–83.
220. Rogala B, Gumprecht J, Gawlik R, Strojek K. Platelet aggregation in IgE-mediated allergy with elevated soluble Fc epsilon RII/CD23 level. J Investig Allergol Clin Immunol. 1995;5:161–5.
221. Qian K, Xie F, Gibson AW, Edberg JC, Kimberly RP, Wu J. Functional expression of IgA receptor FcalphaRI on human platelets. J Leukoc Biol. 2008;84:1492–500.
222. Peerschke EI, Reid KB, Ghebrehiwet B. Platelet activation by C1q results in the induction of alpha IIb/beta 3 integrins (GPIIb-IIIa) and the expression of P-selectin and procoagulant activity. J Exp Med. 1993;178:579–87.
223. Nguyen T, Ghebrehiwet B, Peerschke EI. Staphylococcus aureus protein A recognizes platelet gC1qR/p33: a novel mechanism for staphylococcal interactions with platelets. Infect Immun. 2000;68:2061–8.
224. Ando B, Wiedmer T, Sims PJ. The secretory release reaction initiated by complement proteins C5b-9 occurs without platelet aggregation through glycoprotein IIb-IIIa. Blood. 1989;73:462–7.
225. Martel C, Cointe S, Maurice P, et al. Requirements for membrane attack complex formation and anaphylatoxins binding to collagen-activated platelets. PLoS One. 2011;6:e18812.

Diversity of Inflammatory Cells in Vascular Degenerative Disease

5

Ingo Hilgendorf and Filip K. Swirski

Abstract

A variety of leukocytes reside and function in the vascular wall in health and disease. Although inflammatory cells are meant to protect from diseases and injuries in general, in the context of atherosclerosis, chronic inflammation mounted by some cell types is actually harmfull. This chapter reviews the multifaceted contribution of inflammatory cells to vascular degenerative disease.

Contents

Atherosclerosis is the most common form of vascular degenerative diseases and shares many mechanistic cues with other pathologies such as aortic aneurysm, vasculitis, and restenosis. For this reason, and because atherosclerosis is the

I. Hilgendorf, MD (✉)
Faculty of Medicine, Department of Cardiology and Angiology I, University Heart Center Freiburg-Bad Krozingen, University of Freiburg, Freiburg, Germany
e-mail: ingo.hilgendorf@uniklinik-freiburg.de

F.K. Swirski, PhD
Massachusetts General Hospital Center for Systems Biology, Harvard Medical School, Boston, MA, USA

A. Zirlik et al. (eds.), *Platelets, Haemostasis and Inflammation*,
Cardiac and Vascular Biology 5, https://doi.org/10.1007/978-3-319-66224-4_5

underlying pathology of the leading causes of death worldwide, we will mainly focus on atherosclerosis in this article.

The term atherosclerosis derives from the Greek. "Athera" means gruel and describes the yellowish flaky lipid core of plaques, and "scleros" means hard and describes the dense fibrous cap and calcifications. Atherosclerotic plaques contain varying ratios of macrophage foam cells and other leukocytes, extracellular lipid depositions and debris, smooth muscle cells, and collagen fibers that determine their stability. Early lesions, fatty streaks, contain layers of macrophage foam cells and lipid droplet-filled smooth muscle cells in the intima and are found in the aortas of nearly all humans aged 15–35 years [1]. The processes that drive their progression into advanced and rupture-prone lesions with clinical relevance are not fully understood but involve traditional cardiovascular risk factors such as hypertension, smoking, diabetes, hypercholesterolemia, and familial disposition. These conditions result in vessel wall inflammation [2]. Shear stress, radicals, and glycated mediators activate circulating leukocytes and the endothelium, increasing its permeability and expression of adhesion molecules and chemokines that mediate leukocyte infiltration. Low-density lipoprotein particles are trapped in the subintimal space binding to proteoglycans and lipoprotein lipase which renders them more susceptible to chemical modification by reactive oxygen species. These modifications may occur in the circulation as well. Modified lipoproteins further stimulate the endothelium and vascular smooth muscle cells and therefore need to be cleared by macrophages in the nascent plaque. As stimuli persist, their capacity for clearing lipids and cell debris is overwhelmed, resulting in plaque progression and persistence of inflammation. Elevated serum levels of C-reactive protein (CRP) report on the dwelling inflammation in patients at risk for coronary heart disease and recurrent cardiovascular events.

In the following, we will discuss the many cell types that contribute to the inflammatory process underlying atherosclerosis and other vascular degenerative diseases.

5.1 Monocytes

Monocytes are a heterogenous population of leukocytes that are produced in the bone marrow and spleen and give rise to macrophages and dendritic cells under certain conditions. In the mouse, two subsets are distinguished by their level of Ly6C (Gr-1) expression. Ly6C^{high} monocytes constitute the majority of monocytes found in the bone marrow attesting to their relative immature nature. In fact, common monocyte progenitors, the most committed proliferating precursor of monocytes [3]), also express high levels of Ly6C. Ly6C^{high} monocytes express the chemokine receptor CCR2 and are mobilized from the bone marrow via CCL2 (monocyte chemoattractant protein-1, MCP-1). Ly6C^{high} monocytes give rise to the Ly6C^{low} subset in the blood and bone marrow [4, 5]. The generation of Ly6C^{low} monocytes depends on the nuclear orphan receptor Nr4a1 (Nur77) and was suggested to occur independent of the Ly6C^{high} subset as well [6].

Hypercholesterolemia drives medullary and extramedullary myelopoiesis in atherosclerotic mice leading to Ly6C^{high} monocytosis [7]. Insufficient cholesterol efflux from stem cell progenitors via ABC transporters results in the overexpression of GM-CSF and IL3 receptors rendering them more susceptible to myeloid growth factor stimulation and proliferation [8, 9].

The murine Ly6C^{high} monocyte subset corresponds to classical CD14^{high} CD16^{-} and intermediate CD14^{high} CD16^{+} monocytes in humans while Ly6C^{low} monocytes resemble nonclassical CD14^{dim} CD16^{high} monocytes. Elevated numbers in classical and intermediate monocytes indicate an increased risk for first and recurrent cardiovascular events and associate with adverse cardiac remodeling after myocardial infarction [10].

Ly6C^{high} monocytes preferentially infiltrate sites of vascular inflammation where they may undergo differentiation into macrophages or inflammatory dendritic cells. They may also exit via lymph vessels for MHCII-dependent antigen presentation or die locally.

Deficiency or inhibition of the CCR2/CCL2 interaction limits the number of Ly6C^{high} monocytes in circulation and their invasion into inflamed tissues and results in protection from early atherosclerosis, aortic aneurysm, and neointima formation [11, 12]. Many other chemokine receptors, e.g., CCR1, CCR5, CCR6, CXCR2, and CX3CR1, are also involved in monocyte recruitment to plaque lesions [13, 14].

Ly6C^{low} monocytes patrol the vasculature and scavenge microparticles. When endothelial cells are damaged, Ly6C^{low} monocytes sense danger signals via TLR7 and attract neutrophils to eliminate the damaged endothelial cell (Carlin. Cell. 2013). Nur77 deficiency and lack of Ly6C^{low} monocytes aggravate atherogenesis and impair cardiac remodeling after myocardial infarction in some experimental models. These effects may partly result from the absence of Ly6C^{low} monocytes but also from Nur77 limiting Ly6C^{high} monocyte and macrophage inflammation and recruitment [5, 15, 16].

5.2 Macrophages

Macrophages reside in all tissues and function as phagocytes and antigen-presenting cells. For many decades, macrophages were thought to derive exclusively from monocytes. However, nowadays we realize that macrophages are among the first immune cells to seed tissues during embryogenesis at a time point when bone marrow hematopoiesis has not even begun. Microglia are unique in that they seem to solely derive from yolk sac macrophages in the steady state. Most other tissues harbor macrophages that mainly derive from fetal liver monocytes except the small intestine. The healthy aorta contains macrophages primarily in the adventitia. A recent study revealed that both yolk sac macrophages and fetal liver monocytes seed the aorta before birth. Within the first two weeks after birth, monocytes from the bone marrow contribute to the adventitial macrophage mix. Thereafter, adventitial macrophages sustain into adulthood through proliferation with minimal contribution from the blood [17]. During atherogenesis, however, Ly6C^{high} monocytes infiltrate the intima and differentiate into lesional macrophages. It is currently unknown whether adventitial macrophages contribute to the intimal macrophage pool. Limiting macrophage accumulation protects from atherosclerosis, vascular aneurysm, and neointima formation [18–20]. Macrophages are equipped with a set of scavenger receptors that allow for the receptor-mediated uptake of modified lipoproteins and apoptotic bodies in addition to macropinocytosis. Cholesterol loading of macrophages in vitro inhibits cholesterol biosynthesis leading to the accumulation of desmosterol that activates liver X receptor transcription factors, drives cholesterol efflux, and inhibits pro-inflammatory NFκB pathways. Pro-inflammatory arachidonic acid derivatives activate 12/15-lipoxygenase and lipoxin production in macrophages initiating a negative feedback loop that limits inflammation and promotes its resolution [21]. When cholesterol is packed in modified lipoproteins, however, uptake via scavenger receptors and toll-like receptor co-signaling limits cholesterol efflux and rather stimulates inflammation. Cholesterol is esterified intracellularly and stored in lipid droplets. If the capacities for cholesterol esterification, storage, or efflux are overwhelmed, free cholesterol can crystallize resulting in inflammasome activation and IL1β secretion [22, 23]. Free cholesterol can also integrate into the endoplasmic reticulum (ER) membrane and interfere with the protein folding machinery which leads to an ER stress response and, if unmet, to apoptosis [24, 25]. Macrophages adopt different phenotypes in response to external stimuli. The original classification into IFNγ-stimulated M1 and IL-4- and IL-10-stimulated M2 macrophages has meanwhile been expanded by Mhem, Mox, and M4 macrophages with different inflammatory and phagocytic properties. Even human plaques contain macrophages that share some features of these subsets although there is considerable phenotypic overlap and plasticity [26].

If foam cells are largely pro-inflammatory in the context of atherosclerosis, would not their cell death be beneficial? By deleting pro-apoptotic bcl2 in macrophages of ApoE$^{-/-}$ mice, macrophages survived longer and accumulated in higher numbers in early atherosclerotic lesions. Paradoxically, with disease progression, less macrophages were found in those lesions where macrophages survived longer [27]. This finding underscores the importance of clearing apoptotic cells, a process called efferocytosis. When apoptotic cells are not cleared by macrophages, they undergo secondary necrosis and propagate inflammation and lesion progression [28].

Net macrophage accumulation in plaque lesions is a function of monocyte entry and differentiation, macrophage proliferation, apoptosis, and egress. In the nascent plaque, the majority of macrophages directly derive from infiltrating Ly6C^{high} monocytes. With disease progression, however, macrophages renew primarily through local proliferation of macrophages in the plaque [29]. Transdifferentiation of vascular smooth muscle cells into macrophages [30, 31] was proposed as yet another monocyte independent source of lesional macrophages besides a minor contribution from aortic progenitor cells [32, 33]. For a long time it has been known that smooth muscle cells in atherosclerotic lesions accumulate lipid droplets like macrophage foam cells. Now, two different tamoxifen-inducible Cre-reporter mice were used to track the fate of vascular smooth muscle cells in atherosclerosis [30, 31]. Cells of smooth muscle cell origin were found in intimal lesions that expressed the macrophage marker galectin-3 (Mac 2). By flow cytometry, they also expressed the integrin CD11b but mostly not the leukocyte marker PTPRC. It is therefore fair to conclude that some vascular smooth muscle cells obtain some features of macrophages in atherosclerotic lesions. Taken together, macrophages remain the prototypical proatherogenic culprits in atherogenesis with clinically relevant correlations between lesional macrophage content and plaque inflammation and stability [34]. At the same time, given their plasticity, they may serve as the ideal candidate for immunomodulation and induction of plaque regression.

5.3 Neutrophils

Neutrophils are often the first immune cells to infiltrate sites of inflammation. Although they are found in low numbers in atherosclerotic lesions, they still function importantly in disease onset. Antibody-mediated neutrophil depletion reduced early lesion formation but not plaque progression in $ApoE^{-/-}$ mice [35]. Neutrophils are recruited to plaque lesions via CCR1, CCR2, CCR5, and CXCR2. When entering the intima, neutrophils secrete cathelicidins (e.g., CRAMP in mice) which are transported to the endothelium and attract $Ly6C^{high}$ monocytes. Consequently, CRAMP deficiency reduced the macrophage burden in atherosclerotic lesions [36]. In the context of neointima formation, however, neutrophil-derived CRAMP was rather protective by promoting reendothelialization [37], while myeloperoxidase (MPO) aggravated both atherosclerosis and neointima formation [38]. Neutrophil-derived lipocalin may aid in matrix metallopeptidase-9 activation [39]. Depleting neutrophils are protected from experimental aortic aneurysm formation [40]. When neutrophils die, they may release chromatin fibers containing histones and intracellular proteins forming so-called neutrophil extracellular traps (NET). NETs induce apoptosis in endothelial cells, promote thrombus formation, and stimulate dendritic cells and macrophages [41, 42]. Cholesterol crystals trigger NETosis in atherosclerotic plaques with NETs stimulating pro-IL1β expression in plaque macrophages. Their cholesterol crystals stimulate the NLRP3 inflammasome resulting in caspase-1 activation and secretion of cleaved IL1β [43]. Neutrophils propagate atherogenesis on multiple levels, and in humans neutrophilia associates with increased risk for first and recurrent cardiovascular events [44].

5.4 Mast Cells

Mast cells are tissue-resident inflammatory cells containing basophilic granules filled with proteases, histamine, growth factors, chemokines, and cytokines [45]. They are found in the aortic adventitia and in plaque lesions in humans and mice [46–48]. Chymase and tryptase secretions catalyze the activation of matrix metalloproteases and cathepsins by vascular cells and may trigger their apoptosis. Heparin binds LDL promoting macrophage foam cell formation, and histamine increases perivascular leakage [49]. Mast cells may be activated by toll-like, IgE, and complement receptors. $LDLR^{-/-}$ mice deficient in mast cells develop smaller lesions. Adoptive transfer studies revealed that mast cell-derived IL-6 and IFNγ but not TNFα promoted atherogenesis partly by extracellular matrix degradation [48]. Similarly, mast cells promote aortic aneurysm formation [50–52]. Therapeutically, the mast cell stabilizer cromolyn reduced the development of atherosclerosis in the brachiocephalic artery of $ApoE^{-/-}$ mice [47].

5.5 Dendritic Cells

Dendritic cells (DCs) are a heterogenous group of antigen-presenting cells best suited to prime an adaptive T cell response. DC precursors distinct from monocytes give rise to plasmacytoid DC and classical DCs. Classical DCs are $MHCII^{high}$ $CD11c^{high}$ *Zbtb46*$^{+}$ and can be divided into a Flt3-dependent $CD8\alpha^{+}$/$CD103^{+}$ and a $CD11b^{+}$ subset. Monocytes may supplement the $CD11b^{+}$ DC population in an M-CSF-dependent manner [53] underscoring the difficulty of distinguishing DC and macrophages in sites of inflammation. Functionally DCs are more efficient stimulators of T cell proliferation but less phagocytic compared to macrophages [53]. The healthy aorta contains cDC both in the adventitia and the intima at sites prone to develop atherosclerosis where cDC accumulate lipids in the nascent plaque [54]. In advanced disease stages, tertiary lymphoid organs form in the adventitia that contains DCs, B cells, and T cells [55]. Flt3 deficiency results in the loss of $CD103^{+}$ DC in atherosclerotic aortas sparing M-CSF-dependent monocyte-derived $CD11b^{+}$ DC. This leads to a reduction in protective regulatory T cells (Treg) and increased atherosclerosis [53]. In addition, a CCL17-producing subset of $CD11b^{+}$ DC was shown to suppress Treg differentiation and promote atherosclerosis [56]. Plasmacytoid DCs (pDCs) circulate through blood, enter into tissues, and produce type I interferons. $LDLR^{-/-}$ mice devoid of pDC develop reduced atherosclerosis. Mechanistically, pDCs induce the production of $IFN\gamma^{+}$ $CD4^{+}$ T cells directed against ApoB100 [57]. Antibody-mediated depletion of pDCs in atherosclerotic mice yielded conflicting results [58, 59].

Typically DCs migrate from tissues to lymphoid organs to present antigens to naive T cells. It is unclear whether this kind of trafficking is functionally important to the development and progression of atherosclerosis or whether circulating antigens are primarily taken up by DCs residing in lymphoid organs. Similarly, the site where T cell priming occurs in atherosclerosis, lymphoid organs, and/or plaque lesions is a matter of debate [60]. Both DCs isolated from lymphoid organs of atherosclerotic mice and DCs inside atherosclerotic lesions were shown to stimulate CD4 T cell proliferation and production of the type I T helper cell (TH1) cytokines IFNγ and TNFα [61, 62]. Notably, activated $CD44^{high}$ $CD62L^{low}$ T cells interact with antigen-presenting cells in the plaque more efficiently and produce more cytokines [62]. The stimulation and polarization of T cells require additional external cues that determine the inflammatory phenotype of the antigen-presenting DC. Deletion of the toll-like receptor adaptor Myd88 in $CD11c^+$ cells resulted in the pronounced loss of protective regulatory T cells over $CD4^+$ effector cells and aggravated atherosclerosis [63]. B cell-derived GM-CSF promotes the generation of IL-12-secreting DCs that stimulate the production of $IFN\gamma^+$ T cells in a model of atherosclerosis [64], whereas TGFβ suppresses the production of IL-12 and TNFα by DCs limiting the expansion of effector T cells [65]. The concept that DCs instruct both protective and atherogenic T cell responses fuels the idea of vaccinating against atherosclerosis. The transfer of bone marrow-derived DCs treated with anti-inflammatory IL-10 and pulsed with ApoB100 into atherosclerotic $LDLR^{-/-}$ mice transgenic for human ApoB100 limited the number of $IFN\gamma^+$ TH1 cells, increased the number of regulatory T cells, and reduced atherosclerosis [66]. Even direct vaccination with modified LDL plus adjuvant protects from atherogenesis [37]. Besides, DCs have been implicated in limiting experimental hypercholesterolemia although the underlying mechanism remains speculative [67]. DC may participate in lipoprotein clearance directly or alter the inflammatory status that in turn influences lipid metabolism.

5.6 T Cells

Several subsets of T cells infiltrate atherosclerotic lesions depending on CCR5, CXCR3, CXCR6, and L-selectin [68]. $CD4^+$ T helper cells are up to ten times more frequent in lesions than cytotoxic $CD8^+$ cells. Most of them produce the type I cytokine IFNγ and recognize, for example, epitopes of oxidized but also native LDL [69–71] and of heat shock proteins in mice and men [72, 73]. Transfer of $CD4^+$ T cells sensitized against modified LDL into lymphocyte-deficient $ApoE^{-/-}$ mice accelerated atherosclerosis [74], while CD4 deficiency reduced atherogenesis in another study [75]. Deficiencies in IFNγ, IFNγ-receptor, and the TH1transcription factor T-bet protected mice from atherosclerosis and confirmed the proatherogenic role of the dominating type I helper cell population [76–78]. Other major T helper cell populations include T_{H2} and T_{H17} cells that are identified by their prototypic cytokine and transcription factor profiles.

While IL-12 and IL-18 drive TH1 cell differentiation, IL-4 and IL-6 promote the generation of $GATA3^+$ T_{H2} cells that produce IL-4, IL-5, IL-10, and IL-13. IL-6 and TGFβ in concert activate the transcription factor RORγT in activated T cells and drive T_{H17} cell differentiation [79]. Controversies exist with regard to the respective roles of T_{H2} and T_{H17} cells in atherosclerosis most likely due to opposing cytokine effects and modes of cytokine silencing. IL-4 deficiency, for example, are protected from early atherosclerosis in some but not all studies, suggesting a neutral or weakly proatherogenic role of T_{H2} cells [80–82]. On the other hand, deficiencies in IL-5 and IL-13, cytokines likewise produced by T_{H2} cells, aggravated plaque formation in mice [83, 84]. Mechanistically, IL-5 was shown to promote protective natural IgM production by B1a cells [83]. Antibody blockade as well as genetic deletion of IL17A and its receptor in mouse models of atherosclerosis yielded conflicting results. One study reported accelerated unstable early plaque formation in ApoE$^{-/-}$ IL17A$^{-/-}$ mice [85], while another demonstrated reduced atherosclerosis in both ApoE$^{-/-}$ IL17A$^{-/-}$ and ApoE$^{-/-}$ IL17RA$^{-/-}$ mice [86]. Supplementation of IL17A attenuated atherogenesis in ApoE$^{-/-}$ mice [87] but increased it in LDLR$^{-/-}$ mice [88]. Evaluating the role of T cell subsets by targeting cytokines systemically or by bone marrow transplantation is further complicated by the fact that none of the abovementioned cytokines is exclusively produced by T cells.

CD4^{+} regulatory T cells (T reg) are naturally occurring suppressive cells that mature in the thymus and escape negative selection against self-antigens, or that are induced by dendritic cells and TGFβ stimulation. Both regulatory and inducible T reg express the transcription factor Foxp3 and function via secretion of anti-inflammatory cytokines IL-10 and TGFβ, inhibition of antigen presentation and co-stimulation, and suppression of IL-2-dependent T effector cell proliferation [89, 90]. T reg depletion with anti-CD25 antibodies, diphtheria toxin in transgenic mice, and vaccination against Foxp3 unequivocally increased atherosclerosis, while adoptive transfer of T reg protected from atherosclerosis and aneurysm formation [91–94].

The role of cytotoxic CD8^{+} T cells in atherosclerosis is less well understood. Anti-CD8-mediated cell depletion in ApoE$^{-/-}$ mice attenuated lesion formation, while CD8^{+} T cell transfer to lymphocyte-deficient ApoE$^{-/-}$ mice increased atherosclerosis. Mechanistically granzyme B, perforin, and TNFα promoted vascular cell apoptosis and inflammation, respectively. IFNγ seemed dispensable in this study but relevant in another one in LDLR$^{-/-}$ mice where it stimulated monocytosis [95, 96]. Regulatory CD8+ CD25^{+} T cells attenuate atherosclerosis [97]. Other lymphocytes include natural killer (NK) cells, NKT cells, and innate lymphocytes also engage in atherosclerosis [98].

5.7 B Cells

B cells are unique immune cells in that they produce antibodies, but they also function through cytokine, chemokine, growth factor secretion, and antigen presentation. Historically they were regarded as atheroprotective. Splenectomy of ApoE$^{-/-}$ mice aggravated atherosclerotic lesion formation, while adoptive transfer of B cells rescued the phenotype [99]. B cell-deficient LDLR$^{-/-}$ mice developed larger lesions [100]. It was therefore surprising to learn that anti-CD20-mediated B cell depletion is protected from atherosclerosis in mice [101]. Depletion predominantly affected IgG-producing B2 cells and spared B1 cells that produce natural IgM and inhibited T_{H1} cell proliferation while expanding T_{H17} cells. Also, BAFF receptor deficiency resulted in the selective loss of B2 over B1 cells and attenuated atherosclerosis [102, 103]. B1 cells are the first to arise during fetal development and seed serosal membranes and the spleen. B2 cells derive from and are continuously replaced by bone marrow precursors divide into different subsets including follicular and marginal zone B cells and produce different classes of antigen-specific immunoglobulins (Ig). Antibodies directed against modified lipoproteins modulate atherogenesis. Deficiency in activating FcγRI and FcγRIII protected against atherosclerosis [104], while deficiency in inhibitory FcγRIIb aggravated atherosclerosis [105]. Natural IgM are secreted by B1a and B1b cells and are thought to protect from atherosclerosis by scavenging modified lipoproteins thus attenuating vascular cell inflammation [106, 107].

In atherosclerosis, the number of GM-CSF producing innate response activator (IRA) B cells expanded in lymphoid organs and promoted the generation of IL-12 producing DC that in turn mounted an antigen-specific T_{H1} response and increased lesion formation [64].

IRA B cells may also stimulate myelopoiesis in $ApoE^{-/-}$ mice [108], and B2 cells aid in monocyte mobilization from the bone marrow during myocardial infarction [109]. Regulatory B cells are a heterogenous group of cells that suppress inflammation primarily by secreting IL10. B cell-restricted IL-10 deficiency had no effect on atherosclerosis [110]. On the other hand, transfer of B cells isolated from subiliac and para-aortic lymph nodes but not from the spleen of atherosclerotic $ApoE^{-/-}$ mice is protected from carotid artery plaque formation after perivascular collar injury [111]. Protection partially depended on B cell-derived IL-10 suggesting that site-specific education and cell composition (CD21high CD23high CD24high subset) may be important. In fact, B cells in and around the adventitia may protect from atherosclerosis and abdominal aneurysm formation [112, 113]. B cell homing to the aorta depends on inhibitor of differentiation-3 (Id3), CCR6, and L-selectin, and loss of Id3 aggravates atherosclerosis in mice [112, 114, 115]. Mechanistically confounding, however, Id3 deficiency also influences VCAM1 expression in vascular smooth muscle cells and B1a cell homeostasis with protective IgM production [115, 116]. Adventitial B cells in concert with T cells and DC form artery tertiary lymphoid organs (TLO) in advanced atherosclerosis instructed by vascular smooth muscle cells (VSMC) and immune cells [55]. Lymphotoxin β stimulates the production of the lymphorganogenic chemokines CXCL13 and CCL21 by VSMC, and VSMC-specific deletion of the lymphotoxin β receptor corrupted ATLO formation and accelerated atherosclerosis in $ApoE^{-/-}$ mice [117]. These data are the first to indicate that ATLO form in advanced disease stages to dampen inflammation and plaque progression.

In conclusion, multiple innate and adaptive immune cells engage in vascular degenerative disease. Disease progression results from an immunologic imbalance that favors pro-inflammatory over anti-inflammatory responses. Innovative therapies directed against inflammation in cardiovascular disease are currently evaluated in clinics and may hopefully improve morbidity and mortality in the future [118].

5.8 Translational Perspective

Experimental studies and clinical research suggested that inflammation propagates atherosclerosis and its complications. Recent clinical trials identified inflammation as a potent therapeutic target. Colchicine disrupts microtubule formation, thereby interfering with the assembly of the inflammasome and migration of inflammatory cells. Low-dose colchicine prevented cardiovascular events in patients with stable coronary artery disease [119]. Following myocardial infarction, at least one-third of patients may carry a residual inflammatory risk based on an elevated hsCRP [120]. In CANTOS, this high-risk patient population was protected from future cardiovascular events by Canakinumab, an IL-1ß blocking antibody [121]. Immunomodulation may represent a novel therapeutic strategy in vascular disease.

Compliance with Ethical Standards

Conflict of Interest: Ingo Hilgendorf and Filip K. Swirski declares that they have no conflict of interest.

Ethical Approval: This article does not contain any studies with human participants or animals performed by any of the authors.

References

1. Strong JP, Malcom GT, McMahan CA, Tracy RE, Newman WP, Herderick EE, Cornhill JF. Prevalence and extent of atherosclerosis in adolescents and young adults: implications for prevention from the Pathobiological Determinants of Atherosclerosis in Youth Study. JAMA. 1999;281:727–35.
2. Libby P. Inflammation in atherosclerosis. Arterioscler Thromb Vasc Biol. 2012;32:2045–51.
3. Hettinger J, Richards DM, Hansson J, Barra MM, Joschko AC, Krijgsveld J, Feuerer M. Origin of monocytes and macrophages in a committed progenitor. Nat Immunol. 2013;14:821–30.
4. Yona S, Kim KW, Wolf Y, Mildner A, Varol D, Breker M, Strauss-Ayali D, Viukov S, Guilliams M, Misharin A, Hume DA, Perlman H, Malissen B, Zelzer E, Jung S. Fate mapping reveals origins and dynamics of monocytes and tissue macrophages under homeostasis. Immunity. 2013;38:79–91.
5. Hilgendorf I, Gerhardt L, Tan TC, Winter C, Holderried TA, Chousterman BG, Iwamoto Y, Liao R, Zirlik A, Scherrer-Crosbie M, Hedrick CC, Libby P, Nahrendorf M, Weissleder R, Swirski FK. Ly-6Chigh monocytes depend on Nr4a1 to balance both inflammatory and reparative phases in the infarcted myocardium. Circ Res. 2014;114(10):1611–22.
6. Hanna RN, Carlin LM, Hubbeling HG, Nackiewicz D, Green AM, Punt JA, Geissmann F, Hedrick CC. The transcription factor NR4A1 (Nur77) controls bone marrow differentiation and the survival of Ly6C- monocytes. Nat Immunol. 2011;12:778–85.
7. Swirski FK, Libby P, Aikawa E, Alcaide P, Luscinskas FW, Weissleder R, Pittet MJ. Ly-6Chi monocytes dominate hypercholesterolemia-associated monocytosis and give rise to macrophages in atheromata. J Clin Invest. 2007;117:195–205.
8. Yvan-Charvet L, Pagler T, Gautier EL, Avagyan S, Siry RL, Han S, Welch CL, Wang N, Randolph GJ, Snoeck HW, Tall AR. ATP-binding cassette transporters and HDL suppress hematopoietic stem cell proliferation. Science. 2010;328:1689–93.

9. Murphy AJ, Akhtari M, Tolani S, Pagler T, Bijl N, Kuo CL, Wang M, Sanson M, Abramowicz S, Welch C, Bochem AE, Kuivenhoven JA, Yvan-Charvet L, Tall AR. ApoE regulates hematopoietic stem cell proliferation, monocytosis, and monocyte accumulation in atherosclerotic lesions in mice. J Clin Invest. 2011;121:4138–49.
10. Hilgendorf I, Swirski FK, Robbins CS. Monocyte fate in atherosclerosis. Arterioscler Thromb Vasc Biol. 2015;35:272–9.
11. Boring L, Gosling J, Cleary M, Charo IF. Decreased lesion formation in CCR2-/- mice reveals a role for chemokines in the initiation of atherosclerosis. Nature. 1998;394:894–7.
12. Jerath MR, Liu P, Struthers M, Demartino JA, Peng R, Peterson LB, Cumiskey AM, Yang L, Rojas M, Patel DD, Fong AM. Dual targeting of CCR2 and CX3CR1 in an arterial injury model of vascular inflammation. Thromb J. 2010;8:14.
13. Soehnlein O, Drechsler M, Doring Y, Lievens D, Hartwig H, Kemmerich K, Ortega-Gomez-A, Mandl M, Vijayan S, Projahn D, Garlichs CD, Koenen RR, Hristov M, Lutgens E, Zernecke A, Weber C. Distinct functions of chemokine receptor axes in the atherogenic mobilization and recruitment of classical monocytes. EMBO Mol Med. 2013;5:471–81.
14. Zernecke A, Weber C. Chemokines in atherosclerosis: proceedings resumed. Arterioscler Thromb Vasc Biol. 2014;34:742–50.
15. Hanna RN, Shaked I, Hubbeling HG, Punt JA, Wu R, Herrley E, Zaugg C, Pei H, Geissmann F, Ley K, Hedrick CC. NR4A1 (Nur77) deletion polarizes macrophages toward an inflammatory phenotype and increases atherosclerosis. Circ Res. 2012;110:416–27.
16. Thomas G, Tacke R, Hedrick CC, Hanna RN. Nonclassical patrolling monocyte function in the vasculature. Arterioscler Thromb Vasc Biol. 2015;35:1306–16.
17. Ensan S, Li A, Besla R, Degousee N, Cosme J, Roufaiel M, Shikatani EA, El-Maklizi M, Williams JW, Robins L, Li C, Lewis B, Yun TJ, Lee JS, Wieghofer P, Khattar R, Farrokhi K, Byrne J, Ouzounian M, Zavitz CC, Levy GA, Bauer CM, Libby P, Husain M, Swirski FK, Cheong C, Prinz M, Hilgendorf I, Randolph GJ, Epelman S, Gramolini AO, Cybulsky MI, Rubin BB, Robbins CS. Self-renewing resident arterial macrophages arise from embryonic CX3CR1 precursors and circulating monocytes immediately after birth. Nat Immunol. 2015;17(2):159–68.
18. Stoneman V, Braganza D, Figg N, Mercer J, Lang R, Goddard M, Bennett M. Monocyte/macrophage suppression in CD11b diphtheria toxin receptor transgenic mice differentially affects atherogenesis and established plaques. Circ Res. 2007;100:884–93.
19. Danenberg HD, Fishbein I, Gao J, Monkkonen J, Reich R, Gati I, Moerman E, Golomb G. Macrophage depletion by clodronate-containing liposomes reduces neointimal formation after balloon injury in rats and rabbits. Circulation. 2002;106:599–605.
20. Kanematsu Y, Kanematsu M, Kurihara C, Tada Y, Tsou TL, van Rooijen N, Lawton MT, Young WL, Liang EI, Nuki Y, Hashimoto T. Critical roles of macrophages in the formation of intracranial aneurysm. Stroke. 2011;42:173–8.
21. Merched AJ, Ko K, Gotlinger KH, Serhan CN, Chan L. Atherosclerosis: evidence for impairment of resolution of vascular inflammation governed by specific lipid mediators. FASEB J. 2008;22:3595–606.
22. Duewell P, Kono H, Rayner KJ, Sirois CM, Vladimer G, Bauernfeind FG, Abela GS, Franchi L, Nunez G, Schnurr M, Espevik T, Lien E, Fitzgerald KA, Rock KL, Moore KJ, Wright SD, Hornung V, Latz E. NLRP3 inflammasomes are required for atherogenesis and activated by cholesterol crystals. Nature. 2010;464:1357–61.
23. Sheedy FJ, Grebe A, Rayner KJ, Kalantari P, Ramkhelawon B, Carpenter SB, Becker CE, Ediriweera HN, Mullick AE, Golenbock DT, Stuart LM, Latz E, Fitzgerald KA, Moore KJ. CD36 coordinates NLRP3 inflammasome activation by facilitating intracellular nucleation of soluble ligands into particulate ligands in sterile inflammation. Nat Immunol. 2013;14:812–20.
24. Li Y, Schwabe RF, DeVries-Seimon T, Yao PM, Gerbod-Giannone MC, Tall AR, Davis RJ, Flavell R, Brenner DA, Tabas I. Free cholesterol-loaded macrophages are an abundant source

of tumor necrosis factor-alpha and interleukin-6: model of NF-kappaB- and map kinase-dependent inflammation in advanced atherosclerosis. J Biol Chem. 2005;280:21763–72.
25. Moore KJ, Sheedy FJ, Fisher EA. Macrophages in atherosclerosis: a dynamic balance. Nat Rev Immunol. 2013;13:709–21.
26. Chinetti-Gbaguidi G, Colin S, Staels B. Macrophage subsets in atherosclerosis. Nat Rev Cardiol. 2015;12:10–7.
27. Gautier EL, Huby T, Witztum JL, Ouzilleau B, Miller ER, Saint-Charles F, Aucouturier P, Chapman MJ, Lesnik P. Macrophage apoptosis exerts divergent effects on atherogenesis as a function of lesion stage. Circulation. 2009;119:1795–804.
28. Thorp E, Cui D, Schrijvers DM, Kuriakose G, Tabas I. Mertk receptor mutation reduces efferocytosis efficiency and promotes apoptotic cell accumulation and plaque necrosis in atherosclerotic lesions of apoe-/- mice. Arterioscler Thromb Vasc Biol. 2008;28:1421–8.
29. Robbins CS, Hilgendorf I, Weber GF, Theurl I, Iwamoto Y, Figueiredo JL, Gorbatov R, Sukhova GK, Gerhardt LM, Smyth D, Zavitz CC, Shikatani EA, Parsons M, Rooijen NV, Lin HY, Husain M, Libby P, Nahrendorf M, Weissleder R, Swirski FK. Local proliferation dominates lesional macrophage accumulation in atherosclerosis. Nat Med. 2013;19:1166–72.
30. Feil S, Fehrenbacher B, Lukowski R, Essmann F, Schulze-Osthoff K, Schaller M, Feil R. Transdifferentiation of vascular smooth muscle cells to macrophage-like cells during atherogenesis. Circ Res. 2014;115:662–7.
31. Shankman LS, Gomez D, Cherepanova OA, Salmon M, Alencar GF, Haskins RM, Swiatlowska P, Newman AA, Greene ES, Straub AC, Isakson B, Randolph GJ, Owens GK. KLF4-dependent phenotypic modulation of smooth muscle cells has a key role in atherosclerotic plaque pathogenesis. Nat Med. 2015;21:628–37.
32. Psaltis PJ, Harbuzariu A, Delacroix S, Witt TA, Holroyd EW, Spoon DB, Hoffman SJ, Pan S, Kleppe LS, Mueske CS, Gulati R, Sandhu GS, Simari RD. Identification of a monocyte-predisposed hierarchy of hematopoietic progenitor cells in the adventitia of postnatal murine aorta. Circulation. 2012;125:592–603.
33. Psaltis PJ, Puranik AS, Spoon DB, Chue CD, Hoffman SJ, Witt TA, Delacroix S, Kleppe LS, Mueske CS, Pan S, Gulati R, Simari RD. Characterization of a resident population of adventitial macrophage progenitor cells in postnatal vasculature. Circ Res. 2014;115:364–75.
34. Fuster V, Moreno PR, Fayad ZA, Corti R, Badimon JJ. Atherothrombosis and highrisk plaque: part I: evolving concepts. J Am Coll Cardiol. 2005;46:937–54.
35. Drechsler M, Megens RT, van Zandvoort M, Weber C, Soehnlein O. Hyperlipidemia-triggered neutrophilia promotes early atherosclerosis. Circulation. 2010;122:1837–45.
36. Doring Y, Drechsler M, Wantha S, Kemmerich K, Lievens D, Vijayan S, Gallo RL, Weber C, Soehnlein O. Lack of neutrophil-derived CRAMP reduces atherosclerosis in mice. Circ Res. 2012;110:1052–6.
37. Soehnlein O, Wantha S, Simsekyilmaz S, Doring Y, Megens RT, Mause SF, Drechsler M, Smeets R, Weinandy S, Schreiber F, Gries T, Jockenhoevel S, Moller M, Vijayan S, van Zandvoort MA, Agerberth B, Pham CT, Gallo RL, Hackeng TM, Liehn EA, Zernecke A, Klee D, Weber C. Neutrophil-derived cathelicidin protects from neointimal hyperplasia. Sci Transl Med. 2011;3:103ra98.
38. Tiyerili V, Camara B, Becher MU, Schrickel JW, Lutjohann D, Mollenhauer M, Baldus S, Nickenig G, Andrie RP. Neutrophil-derived myeloperoxidase promotes atherogenesis and neointima formation in mice. Int J Cardiol. 2015;204:29–36.
39. Hemdahl AL, Gabrielsen A, Zhu C, Eriksson P, Hedin U, Kastrup J, Thoren P, Hansson GK. Expression of neutrophil gelatinase-associated lipocalin in atherosclerosis and myocardial infarction. Arterioscler Thromb Vasc Biol. 2006;26:136–42.
40. Eliason JL, Hannawa KK, Ailawadi G, Sinha I, Ford JW, Deogracias MP, Roelofs KJ, Woodrum DT, Ennis TL, Henke PK, Stanley JC, Thompson RW, Upchurch GRJ. Neutrophil

depletion inhibits experimental abdominal aortic aneurysm formation. Circulation. 2005;112:232–40.
41. Doring Y, Soehnlein O, Weber C. Neutrophils cast NETs in atherosclerosis: employing peptidylarginine deiminase as a therapeutic target. Circ Res. 2014;114:931–4.
42. Grayson PC, Kaplan MJ. At the bench: neutrophil extracellular traps (NETs) highlight novel aspects of innate immune system involvement in autoimmune diseases. J Leukoc Biol. 2015;99:253–64.
43. Warnatsch A, Ioannou M, Wang Q, Papayannopoulos V. Inflammation. Neutrophil extracellular traps license macrophages for cytokine production in atherosclerosis. Science. 2015;349:316–20.
44. Madjid M, Awan I, Willerson JT, Casscells SW. Leukocyte count and coronary heart disease: implications for risk assessment. J Am Coll Cardiol. 2004;44:1945–56.
45. Wernersson S, Pejler G. Mast cell secretory granules: armed for battle. Nat Rev Immunol. 2014;14:478–94.
46. Kaartinen M, Penttila A, Kovanen PT. Accumulation of activated mast cells in the shoulder region of human coronary atheroma, the predilection site of atheromatous rupture. Circulation. 1994;90:1669–78.
47. Bot I, de Jager SC, Zernecke A, Lindstedt KA, van Berkel TJ, Weber C, Biessen EA. Perivascular mast cells promote atherogenesis and induce plaque destabilization in apolipoprotein E-deficient mice. Circulation. 2007;115:2516–25.
48. Sun J, Sukhova GK, Wolters PJ, Yang M, Kitamoto S, Libby P, MacFarlane LA, Mallen-St Clair J, Shi GP. Mast cells promote atherosclerosis by releasing proinflammatory cytokines. Nat Med. 2007;13:719–24.
49. Bot I, Biessen EA. Mast cells in atherosclerosis. Thromb Haemost. 2011;106:820–6.
50. Sun J, Sukhova GK, Yang M, Wolters PJ, MacFarlane LA, Libby P, Sun C, Zhang Y, Liu J, Ennis TL, Knispel R, Xiong W, Thompson RW, Baxter BT, Shi GP. Mast cells modulate the pathogenesis of elastase-induced abdominal aortic aneurysms in mice. J Clin Invest. 2007;117:3359–68.
51. Sun J, Zhang J, Lindholt JS, Sukhova GK, Liu J, He A, Abrink M, Pejler G, Stevens RL, Thompson RW, Ennis TL, Gurish MF, Libby P, Shi GP. Critical role of mast cell chymase in mouse abdominal aortic aneurysm formation. Circulation. 2009;120:973–82.
52. Zhang J, Sun J, Lindholt JS, Sukhova GK, Sinnamon M, Stevens RL, Adachi R, Libby P, Thompson RW, Shi GP. Mast cell tryptase deficiency attenuates mouse abdominal aortic aneurysm formation. Circ Res. 2011;108:1316–27.
53. Choi JH, Cheong C, Dandamudi DB, Park CG, Rodriguez A, Mehandru S, Velinzon K, Jung IH, Yoo JY, Oh GT, Steinman RM. Flt3 signaling-dependent dendritic cells protect against atherosclerosis. Immunity. 2011;35:819–31.
54. Paulson KE, Zhu SN, Chen M, Nurmohamed S, Jongstra-Bilen J, Cybulsky MI. Resident intimal dendritic cells accumulate lipid and contribute to the initiation of atherosclerosis. Circ Res. 2010;106:383–90.
55. Mohanta SK, Yin C, Peng L, Srikakulapu P, Bontha V, Hu D, Weih F, Weber C, Gerdes N, Habenicht AJ. Artery tertiary lymphoid organs contribute to innate and adaptive immune responses in advanced mouse atherosclerosis. Circ Res. 2014;114:1772–87.
56. Weber C, Meiler S, Doring Y, Koch M, Drechsler M, Megens RT, Rowinska Z, Bidzhekov K, Fecher C, Ribechini E, van Zandvoort MA, Binder CJ, Jelinek I, Hristov M, Boon L, Jung S, Korn T, Lutz MB, Forster I, Zenke M, Hieronymus T, Junt T, Zernecke A. CCL17-expressing dendritic cells drive atherosclerosis by restraining regulatory T cell homeostasis in mice. J Clin Invest. 2011;121:2898–910.
57. Sage AP, Murphy D, Maffia P, Masters LM, Sabir SR, Baker LL, Cambrook H, Finigan AJ, Ait-Oufella H, Grassia G, Harrison JE, Ludewig B, Reith W, Hansson GK, Reizis B, Hugues S, Mallat Z. MHC Class II-restricted antigen presentation by plasmacytoid dendritic cells drives proatherogenic T cell immunity. Circulation. 2014;130:1363–73.

58. Daissormont IT, Christ A, Temmerman L, Sampedro Millares S, Seijkens T, Manca M, Rousch M, Poggi M, Boon L, van der Loos C, Daemen M, Lutgens E, Halvorsen B, Aukrust P, Janssen E, Biessen EA. Plasmacytoid dendritic cells protect against atherosclerosis by tuning T-cell proliferation and activity. Circ Res. 2011;109:1387–95.
59. Macritchie N, Grassia G, Sabir SR, Maddaluno M, Welsh P, Sattar N, Ialenti A, Kurowska-Stolarska M, McInnes IB, Brewer JM, Garside P, Maffia P. Plasmacytoid dendritic cells play a key role in promoting atherosclerosis in apolipoprotein Edeficient mice. Arterioscler Thromb Vasc Biol. 2012;32:2569–79.
60. Ley K. The second touch hypothesis: T cell activation, homing and polarization. F1000Res. 2014;3:37.
61. Packard RR, Maganto-Garcia E, Gotsman I, Tabas I, Libby P, Lichtman AH. CD11c(+) dendritic cells maintain antigen processing, presentation capabilities, and CD4(+) T-cell priming efficacy under hypercholesterolemic conditions associated with atherosclerosis. Circ Res. 2008;103:965–73.
62. Koltsova EK, Garcia Z, Chodaczek G, Landau M, McArdle S, Scott SR, von Vietinghoff S, Galkina E, Miller YI, Acton ST, Ley K. Dynamic T cell-APC interactions sustain chronic inflammation in atherosclerosis. J Clin Invest. 2012;122:3114–26.
63. Subramanian M, Thorp E, Hansson GK, Tabas I. Treg-mediated suppression of atherosclerosis requires MYD88 signaling in DCs. J Clin Invest. 2013;123:179–88.
64. Hilgendorf I, Theurl I, Gerhardt LM, Robbins CS, Weber GF, Gonen A, Iwamoto Y, Degousee N, Holderried TA, Winter C, Zirlik A, Lin HY, Sukhova GK, Butany J, Rubin BB, Witztum JL, Libby P, Nahrendorf M, Weissleder R, Swirski FK. Innate response activator B cells aggravate atherosclerosis by stimulating TH1 adaptive immunity. Circulation. 2014;129(16):1677–87.
65. Lievens D, Habets KL, Robertson AK, Laouar Y, Winkels H, Rademakers T, Beckers L, Wijnands E, Boon L, Mosaheb M, Ait-Oufella H, Mallat Z, Flavell RA, Rudling M, Binder CJ, Gerdes N, Biessen EA, Weber C, Daemen MJ, Kuiper J, Lutgens E. Abrogated transforming growth factor beta receptor II (TGFbetaRII) signalling in dendritic cells promotes immune reactivity of T cells resulting in enhanced atherosclerosis. Eur Heart J. 2013;34:3717–27.
66. Hermansson A, Johansson DK, Ketelhuth DF, Andersson J, Zhou X, Hansson GK. Immunotherapy with tolerogenic apolipoprotein B-100-loaded dendritic cells attenuates atherosclerosis in hypercholesterolemic mice. Circulation. 2011;123:1083–91.
67. Gautier EL, Huby T, Saint-Charles F, Ouzilleau B, Pirault J, Deswaerte V, Ginhoux F, Miller ER, Witztum JL, Chapman MJ, Lesnik P. Conventional dendritic cells at the crossroads between immunity and cholesterol homeostasis in atherosclerosis. Circulation. 2009;119:2367–75.
68. Li J, Ley K. Lymphocyte migration into atherosclerotic plaque. Arterioscler Thromb Vasc Biol. 2015;35:40–9.
69. de Boer OJ, van der Wal AC, Verhagen CE, Becker AE. Cytokine secretion profiles of cloned T cells from human aortic atherosclerotic plaques. J Pathol. 1999;188:174–9.
70. Stemme S, Faber B, Holm J, Wiklund O, Witztum JL, Hansson GK. T lymphocytes from human atherosclerotic plaques recognize oxidized low density lipoprotein. Proc Natl Acad Sci USA. 1995;92:3893–7.
71. Hermansson A, Ketelhuth DF, Strodthoff D, Wurm M, Hansson EM, Nicoletti A, Paulsson-Berne G, Hansson GK. Inhibition of T cell response to native low-density lipoprotein reduces atherosclerosis. J Exp Med. 2010;207:1081–93.
72. Buono C, Pang H, Uchida Y, Libby P, Sharpe AH, Lichtman AH. B7-1/B7-2 costimulation regulates plaque antigen-specific T-cell responses and atherogenesis in low-density lipoprotein receptor-deficient mice. Circulation. 2004;109:2009–15.
73. Almanzar G, Ollinger R, Leuenberger J, Onestingel E, Rantner B, Zehm S, Cardini B, van der Zee R, Grundtman C, Wick G. Autoreactive HSP60 epitope-specific Tcells in early human atherosclerotic lesions. J Autoimmun. 2012;39:441–50.

74. Zhou X, Robertson AK, Hjerpe C, Hansson GK. Adoptive transfer of CD4+ T cells reactive to modified low-density lipoprotein aggravates atherosclerosis. Arterioscler Thromb Vasc Biol. 2006;26:864–70.
75. Zhou X, Robertson AK, Rudling M, Parini P, Hansson GK. Lesion development and response to immunization reveal a complex role for CD4 in atherosclerosis. Circ Res. 2005;96:427–34.
76. Gupta S, Pablo AM, Jiang X, Wang N, Tall AR, Schindler C. IFN-gamma potentiates atherosclerosis in ApoE knock-out mice. J Clin Invest. 1997;99:2752–61.
77. Buono C, Come CE, Stavrakis G, Maguire GF, Connelly PW, Lichtman AH. Influence of interferon-gamma on the extent and phenotype of diet-induced atherosclerosis in the LDLR-deficient mouse. Arterioscler Thromb Vasc Biol. 2003;23:454–60.
78. Buono C, Binder CJ, Stavrakis G, Witztum JL, Glimcher LH, Lichtman AH. T-bet deficiency reduces atherosclerosis and alters plaque antigen-specific immune responses. Proc Natl Acad Sci USA. 2005;102:1596–601.
79. Ait-Oufella H, Sage AP, Mallat Z, Tedgui A. Adaptive (T and B cells) immunity and control by dendritic cells in atherosclerosis. Circ Res. 2014;114:1640–60.
80. King VL, Szilvassy SJ, Daugherty A. Interleukin-4 deficiency decreases atherosclerotic lesion formation in a site-specific manner in female LDL receptor-/- mice. Arterioscler Thromb Vasc Biol. 2002;22:456–61.
81. King VL, Cassis LA, Daugherty A. Interleukin-4 does not influence development of hypercholesterolemia or angiotensin II-induced atherosclerotic lesions in mice. Am J Pathol. 2007;171:2040–7.
82. Davenport P, Tipping PG. The role of interleukin-4 and interleukin-12 in the progression of atherosclerosis in apolipoprotein E-deficient mice. Am J Pathol. 2003;163:1117–25.
83. Binder CJ, Hartvigsen K, Chang MK, Miller M, Broide D, Palinski W, Curtiss LK, Corr M, Witztum JL. IL-5 links adaptive and natural immunity specific for epitopes of oxidized LDL and protects from atherosclerosis. J Clin Invest. 2004;114:427–37.
84. Cardilo-Reis L, Gruber S, Schreier SM, Drechsler M, Papac-Milicevic N, Weber C, Wagner O, Stangl H, Soehnlein O, Binder CJ. Interleukin-13 protects from atherosclerosis and modulates plaque composition by skewing the macrophage phenotype. EMBO Mol Med. 2012;4:1072–86.
85. Danzaki K, Matsui Y, Ikesue M, Ohta D, Ito K, Kanayama M, Kurotaki D, Morimoto J, Iwakura Y, Yagita H, Tsutsui H, Uede T. Interleukin-17A deficiency accelerates unstable atherosclerotic plaque formation in apolipoprotein E-deficient mice. Arterioscler Thromb Vasc Biol. 2012;32:273–80.
86. Butcher MJ, Gjurich BN, Phillips T, Galkina EV. The IL-17A/IL-17RA axis plays a proatherogenic role via the regulation of aortic myeloid cell recruitment. Circ Res. 2012;110:675–87.
87. Gao Q, Jiang Y, Ma T, Zhu F, Gao F, Zhang P, Guo C, Wang Q, Wang X, Ma C, Zhang Y, Chen W, Zhang L. A critical function of Th17 proinflammatory cells in the development of atherosclerotic plaque in mice. J Immunol. 2010;185:5820–7.
88. Taleb S, Romain M, Ramkhelawon B, Uyttenhove C, Pasterkamp G, Herbin O, Esposito B, Perez N, Yasukawa H, Van Snick J, Yoshimura A, Tedgui A, Mallat Z. Loss of SOCS3 expression in T cells reveals a regulatory role for interleukin-17 in atherosclerosis. J Exp Med. 2009;206:2067–77.
89. Sakaguchi S, Yamaguchi T, Nomura T, Ono M. Regulatory T cells and immune tolerance. Cell. 2008;133:775–87.
90. Foks AC, Lichtman AH, Kuiper J. Treating atherosclerosis with regulatory T cells. Arterioscler Thromb Vasc Biol. 2015;35:280–7.
91. Ait-Oufella H, Salomon BL, Potteaux S, Robertson AK, Gourdy P, Zoll J, Merval R, Esposito B, Cohen JL, Fisson S, Flavell RA, Hansson GK, Klatzmann D, Tedgui A, Mallat Z. Natural regulatory T cells control the development of atherosclerosis in mice. Nat Med. 2006;12:178–80.

92. van Es T, van Puijvelde GH, Foks AC, Habets KL, Bot I, Gilboa E, Van Berkel TJ, Kuiper J. Vaccination against Foxp3(+) regulatory T cells aggravates atherosclerosis. Atherosclerosis. 2010;209:74–80.
93. Klingenberg R, Gerdes N, Badeau RM, Gistera A, Strodthoff D, Ketelhuth DF, Lundberg AM, Rudling M, Nilsson SK, Olivecrona G, Zoller S, Lohmann C, Luscher TF, Jauhiainen M, Sparwasser T, Hansson GK. Depletion of FOXP3+ regulatory T cells promotes hypercholesterolemia and atherosclerosis. J Clin Invest. 2013;123:1323–34.
94. Ait-Oufella H, Wang Y, Herbin O, Bourcier S, Potteaux S, Joffre J, Loyer X, Ponnuswamy P, Esposito B, Dalloz M, Laurans L, Tedgui A, Mallat Z. Natural regulatory T cells limit angiotensin II-induced aneurysm formation and rupture in mice. Arterioscler Thromb Vasc Biol. 2013;33:2374–9.
95. Kyaw T, Winship A, Tay C, Kanellakis P, Hosseini H, Cao A, Li P, Tipping P, Bobik A, Toh BH. Cytotoxic and proinflammatory CD8+ T lymphocytes promote development of vulnerable atherosclerotic plaques in apoE-deficient mice. Circulation. 2013;127:1028–39.
96. Cochain C, Koch M, Chaudhari SM, Busch M, Pelisek J, Boon L, Zernecke A. CD8+ T cells regulate monopoiesis and circulating Ly6C-high monocyte levels in atherosclerosis in mice. Circ Res. 2015;117:244–53.
97. Zhou J, Dimayuga PC, Zhao X, Yano J, Lio WM, Trinidad P, Honjo T, Cercek B, Shah PK, Chyu KY. CD8(+)CD25(+) T cells reduce atherosclerosis in apoE(-/-) mice. Biochem Biophys Res Commun. 2014;443:864–70.
98. Hedrick CC. Lymphocytes in atherosclerosis. Arterioscler Thromb Vasc Biol. 2015;35:253–7.
99. Caligiuri G, Nicoletti A, Poirier B, Hansson GK. Protective immunity against atherosclerosis carried by B cells of hypercholesterolemic mice. J Clin Invest. 2002;109:745–53.
100. Major AS, Fazio S, Linton MF. B-lymphocyte deficiency increases atherosclerosis in LDL receptor-null mice. Arterioscler Thromb Vasc Biol. 2002;22:1892–8.
101. Ait-Oufella H, Herbin O, Bouaziz JD, Binder CJ, Uyttenhove C, Laurans L, Taleb S, Van Vre E, Esposito B, Vilar J, Sirvent J, Van Snick J, Tedgui A, Tedder TF, Mallat Z. B cell depletion reduces the development of atherosclerosis in mice. J Exp Med. 2010;207:1579–87.
102. Kyaw T, Tay C, Hosseini H, Kanellakis P, Gadowski T, MacKay F, Tipping P, Bobik A, Toh BH. Depletion of B2 but not B1a B cells in BAFF receptor-deficient ApoE mice attenuates atherosclerosis by potently ameliorating arterial inflammation. PLoS One. 2012;7:e29371.
103. Sage AP, Tsiantoulas D, Baker L, Harrison J, Masters L, Murphy D, Loinard C, Binder CJ, Mallat Z. BAFF receptor deficiency reduces the development of atherosclerosis in mice—brief report. Arterioscler Thromb Vasc Biol. 2012;32:1573–6.
104. Hernandez-Vargas P, Ortiz-Munoz G, Lopez-Franco O, Suzuki Y, Gallego-Delgado J, Sanjuan G, Lazaro A, Lopez-Parra V, Ortega L, Egido J, Gomez-Guerrero C. Fcgamma receptor deficiency confers protection against atherosclerosis in apolipoprotein E knockout mice. Circ Res. 2006;99:1188–96.
105. Schneider MP, Leusen JH, Herrmann M, Garlichs CD, Amann K, John S, Schmieder RE. The Fcgamma receptor IIA R131H gene polymorphism is associated with endothelial function in patients with hypercholesterolaemia. Atherosclerosis. 2011;218:411–5.
106. Kyaw T, Tay C, Krishnamurthi S, Kanellakis P, Agrotis A, Tipping P, Bobik A, Toh BH. B1a B lymphocytes are atheroprotective by secreting natural IgM that increases IgM deposits and reduces necrotic cores in atherosclerotic lesions. Circ Res. 2011;109:830–40.
107. Rosenfeld SM, Perry HM, Gonen A, Prohaska TA, Srikakulapu P, Grewal S, Das D, McSkimming C, Taylor AM, Tsimikas S, Bender TP, Witztum JL, McNamara CA. B-1b cells secrete atheroprotective IgM and attenuate atherosclerosis. Circ Res. 2015;117:e28–39.
108. Wang M, Subramanian M, Abramowicz S, Murphy AJ, Gonen A, Witztum J, Welch C, Tabas I, Westerterp M, Tall AR. Interleukin-3/granulocyte macrophage colonystimulating factor receptor promotes stem cell expansion, monocytosis, and atheroma macrophage burden in mice with hematopoietic ApoE deficiency. Arterioscler Thromb Vasc Biol. 2014;34:976–84.

109. Zouggari Y, Ait-Oufella H, Bonnin P, Simon T, Sage AP, Guerin C, Vilar J, Caligiuri G, Tsiantoulas D, Laurans L, Dumeau E, Kotti S, Bruneval P, Charo IF, Binder CJ, Danchin N, Tedgui A, Tedder TF, Silvestre JS, Mallat Z. B lymphocytes trigger monocyte mobilization and impair heart function after acute myocardial infarction. Nat Med. 2013;19:1273–80.
110. Sage AP, Nus M, Baker LL, Finigan AJ, Masters LM, Mallat Z. Regulatory B cellspecific interleukin-10 is dispensable for atherosclerosis development in mice. Arterioscler Thromb Vasc Biol. 2015;35:1770–3.
111. Strom AC, Cross AJ, Cole JE, Blair PA, Leib C, Goddard ME, Rosser EC, Park I, Hultgardh Nilsson A, Nilsson J, Mauri C, Monaco C. B regulatory cells are increased in hypercholesterolaemic mice and protect from lesion development via IL-10. Thromb Haemost. 2015;114:835–47.
112. Doran AC, Lipinski MJ, Oldham SN, Garmey JC, Campbell KA, Skaflen MD, Cutchins A, Lee DJ, Glover DK, Kelly KA, Galkina EV, Ley K, Witztum JL, Tsimikas S, Bender TP, McNamara CA. B-cell aortic homing and atheroprotection depend on Id3. Circ Res. 2012;110:e1–12.
113. Zhang L, Wang Y. B lymphocytes in abdominal aortic aneurysms. Atherosclerosis. 2015;242:311–7.
114. Gjurich BN, Taghavie-Moghadam PL, Ley K, Galkina EV. L-selectin deficiency decreases aortic B1a and Breg subsets and promotes atherosclerosis. Thromb Haemost. 2014;112:803–11.
115. Lipinski MJ, Campbell KA, Duong SQ, Welch TJ, Garmey JC, Doran AC, Skaflen MD, Oldham SN, Kelly KA, McNamara CA. Loss of Id3 increases VCAM-1 expression, macrophage accumulation, and atherogenesis in Ldlr-/- mice. Arterioscler Thromb Vasc Biol. 2012;32:2855–61.
116. Perry HM, Oldham SN, Fahl SP, Que X, Gonen A, Harmon DB, Tsimikas S, Witztum JL, Bender TP, McNamara CA. Helix-loop-helix factor inhibitor of differentiation 3 regulates interleukin-5 expression and B-1a B cell proliferation. Arterioscler Thromb Vasc Biol. 2013;33:2771–9.
117. Hu D, Mohanta SK, Yin C, Peng L, Ma Z, Srikakulapu P, Grassia G, MacRitchie N, Dever G, Gordon P, Burton FL, Ialenti A, Sabir SR, McInnes IB, Brewer JM, Garside P, Weber C, Lehmann T, Teupser D, Habenicht L, Beer M, Grabner R, Maffia P, Weih F, Habenicht AJ. Artery tertiary lymphoid organs control aorta immunity and protect against atherosclerosis via vascular smooth muscle cell lymphotoxin beta receptors. Immunity. 2015;42:1100–15.
118. Ridker PM, Luscher TF. Anti-inflammatory therapies for cardiovascular disease. Eur Heart J. 2014;35:1782–91.
119. Nidorf SM, Eikelboom JW, Budgeon CA, Thompson PL. Low-dose colchicine for secondary prevention of cardiovascular disease. J Am Coll Cardiol. 2013;61:404–10.
120. Ridker, PM. Circ Res. 2017; 120:617–619.
121. Ridker PM, Everett BM, Thuren T, MacFadyen JG, Chang WH, Ballantyne C, Fonseca F, Nicolau J, Koenig W, Anker SD, Kastelein JJP, Cornel JH, Pais P, Pella D, Genest J, Cifkova R, Lorenzatti A, Forster T, Kobalava Z, Vida-Simiti L, Flather M, Shimokawa H, Ogawa H, Dellborg M, Rossi PRF, Troquay RPT, Libby P, Glynn RJ, CANTOS Trial Group. N Engl J Med. 2017;377:1119–31.

Platelet Inhibition as a Therapeutic Approach in Intravascular Intervention

6

Ingo Ahrens and Hector Bueno

Abstract

Platelets are the primary mediators of vascular haemostasis. They express numerous adhesion receptors on their surface enabling a direct interaction with the endothelial cells of the blood vessel wall and the underlying extracellular matrix. In addition they interact with leucocytes and fulfil functions of innate immunity. Furthermore they contain a plethora of stored proteins in intracellular granula, including cytokines, which are released upon activation. Platelets undergo a rapid and extreme change of their surface membrane once they are activated. Platelet activation can occur via soluble platelet activators (e.g. thrombin and ADP) or by direct interaction of platelet adhesion receptors with components of the vessel wall.

Besides naturally occurring platelet activators, medical devices used for intravascular interventions do represent surfaces that may directly or indirectly lead to platelet activation and subsequent platelet aggregation ultimately causing intravascular thrombosis and thereby clinical adverse events. For many decades, aspirin was the mainstay of platelet inhibition. Within the last two decades, a rapidly evolving era of extensive research on platelets and atherothrombosis led to the clinical development of several novel antiplatelet agents that

I. Ahrens (✉)
Cardiology and Angiology I, Heart Center, University of Freiburg, Freiburg, Germany

Departement of Cardiology and Medical Intensive Care, Augustinerinnen Hospital, Cologne, Academic Teaching Hospital University of Cologne, Jakobstr. 27-31, 50678 Cologne, Germany
e-mail: iahrens@severinskloesterchen.de

H. Bueno
Centro Nacional de Investigaciones Cardiovasculares (CNIC), Madrid, Spain

Departamento de Cardiología, Instituto de Investigación I+12, Hospital Universitario 12 de Octubre, Madrid, Spain

Universidad Complutense de Madrid, Madrid, Spain

A. Zirlik et al. (eds.), *Platelets, Haemostasis and Inflammation*,
Cardiac and Vascular Biology 5, https://doi.org/10.1007/978-3-319-66224-4_6

successfully entered daily clinical practice in intravascular interventions. In addition to aspirin, currently clinically approved antiplatelet agents used in intravascular interventions target the platelet P2Y12 receptor, the platelet PAR1 receptor and the platelet fibrinogen receptor (GPIIb/IIIa).

Contents

6.1 Platelets in ACS

Platelets are anucleate discoid cells with a lifespan of approximately 10 days. Their predominant purpose is vascular haemostasis (e.g. sealing of damaged vessel walls, thereby preventing blood loss). However, beside their haemostatic functions, platelets are also important mediators and effector cells of innate immunity and do express several functionally active receptors of the innate immune system [1, 2].

Atherosclerosis is a chronic inflammatory disease of the vessel wall [3] and platelets are critically involved in the initiation of this process [4]. The acute coronary syndrome (ACS), a clinical sequelae of coronary atherosclerosis, is characterized by ischaemic symptoms caused by total or subtotal occlusion of a coronary artery [5]. Upon rupture or superficial erosion of a coronary artery plaque, circulating platelets adhere and become activated, subsequently recruiting further circulating platelets from the bloodstream and cross-linking them via fibrinogen, a process known as platelet aggregation [6]. This local platelet aggregation at the site of the ruptured atherosclerotic plaque ultimately leads to local thrombus formation and total or subtotal vessel closure of the coronary artery. The activation of the platelet fibrinogen receptor GPIIb/IIIa is a prerequisite of this process and also an established therapeutic target [7]. Novel therapeutic strategies in preclinical development aim at targeting only the activated GPIIb/IIIa receptor, thereby allowing a targeted therapy at the site of the forming thrombus with a lower systemic bleeding risk [8, 9].

6.2 Platelets as Therapeutic Targets

The clinical development of the concept of dual antiplatelet therapy (DAPT) established percutaneous coronary artery interventions (PCI) with the deployment of stents as the standard therapy in patients with ACS and cardiac ischaemia. The discovery of the platelet P2Y12 receptor and the development of drugs that allow the selective blockade of this receptor were the prerequisites for the establishment of DAPT [6, 10]. Newer concepts of platelet inhibition include the blockade of the protease-activated receptor 1 (PAR1), the platelet thrombin receptor (Fig. 6.1). Vorapaxar is the first drug with clinical approval targeting PAR1, thereby expanding the possible combinations of aspirin in DAPT and also enabling a triple antiplatelet therapy. However, the concept of the more the platelet inhibition the better the outcome in patients with coronary artery disease and ACS has come to an end as currently available antiplatelet drugs ultimately lead to increased risk of major bleeding once they are used beyond the concept of DAPT [11–13].

Among currently clinically approved and orally available antiplatelet agents are aspirin; the P2Y12 receptor antagonists ticlopidine, clopidogrel, prasugrel and ticagrelor; and the PAR1 receptor antagonist vorapaxar. In addition there are the parenteral GPIIb/IIIa antagonists abciximab, eptifibatide and tirofiban, which are used in patients with ACS undergoing PCI [7] but more and more restricted to patients with high intracoronary thrombus burden and bail-out situations, which is

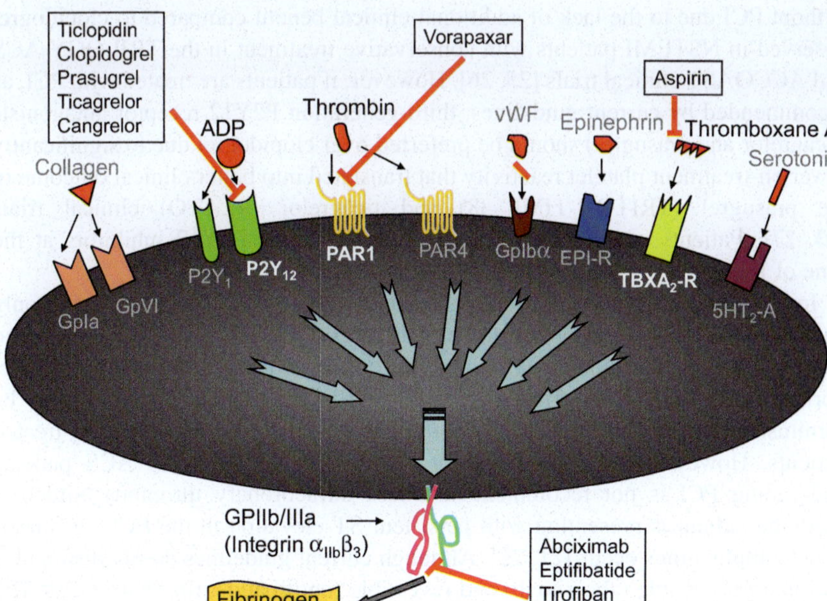

Fig. 6.1 Platelet receptor and antiplatelet agents with clinical approval. Modified with permission from Ahrens I. et al. *Curr Opin Investig Drugs*. 2009 Sep;10(9):902–11

very likely due to the clinical availability and potency of third-generation P2Y12 receptor antagonists (prasugrel and ticagrelor) [14–16]. Recently, cangrelor an intravenous P2Y12 receptor antagonist has also gained clinical approval (Fig. 6.1).

6.3 Platelet Inhibition in ACS State of the Art

Aspirin was and still is the antiplatelet agent of choice that should be administered immediately (if no contraindication) in patients with ACS. Despite appealing novel therapeutic strategies that exclude aspirin in the intermediate and long-term treatment of patients with ACS who also require oral anticoagulation due to atrial fibrillation [17], there is currently no pharmacological evidence supporting the concept that inhibition of thromboxane A2 synthesis in platelets may be substituted by P2Y12 receptor inhibition + addition of either a PAR1 receptor antagonist or a low-dose non-vitamin K antagonist oral anticoagulant [18, 19].

Current guidelines recommend oral or intravenous administration of aspirin at a dosage of 150–300 mg in patients presenting with ACS [20–22]. In addition patients should be treated with a P2Y12 receptor antagonist. Clopidogrel and ticagrelor are recommended regardless of the treatment strategy (interventional or noninterventional) based on the data of the CURE (clopidogrel) and the PLATO (ticagrelor) studies [23, 24]. Prasugrel is not recommended if patients are treated without PCI due to the lack of additional clinical benefit compared to clopidogrel observed in NSTEMI patients with conservative treatment in the TRILOGY ACS and ACCOAST clinical trials [25, 26]. However, if patients are treated with PCI, as recommended by current guidelines, third-generation P2Y12 receptor antagonists (ticagrelor and prasugrel) should be preferred over clopidogrel due to significantly lower on-treatment platelet reactivity that translated into better clinical outcome in the prasugrel (TRITON-TIMI 38) and ticagrelor (PLATO) clinical trials [23, 27]. Patients who cannot receive orally available P2Y12 inhibitors at the time of PCI may be treated with intravenous cangrelor [21].

In the acute phase of an ACS in patients undergoing PCI, there is currently only one more state-of-the-art option to extend platelet inhibition beyond aspirin and P2Y12 receptor inhibition (DAPT), and this option is the inhibition of the platelet fibrinogen receptor GPIIb/IIIa (Fig. 6.1). GPIIb/IIIa inhibitors should only be administered in patients with ACS undergoing PCI but not in conservatively treated patients. However, the general use of a GPIIb/IIIa inhibitor in ACS patients undergoing PCI is not recommended unless intracoronary thrombus burden is high, the patient is presenting with persistent ST-elevation in the ECG or thrombotic complications occur [21, 22]. Although current guidelines do not distinguish between irreversible (abciximab) and reversible (eptifibatide, tirofiban) GPIIb/IIIa inhibitors, it could be useful to consider short-acting and reversible GPIIb/IIIa inhibitors in patients that may require procedures with a high bleeding risk in the immediate period following PCI for an ACS [7].

6.4 Platelet Inhibition in Peripheral Interventions

Patients with peripheral artery disease (PAD) have an increased risk for myocardial infarction and stroke. Unlike in patients with coronary artery disease, the role of antiplatelet therapy is less well examined. The hypothesis that aspirin therapy is effective in all patients with PAD has been rejected after the results of the Aspirin for Asymptomatic Atherosclerosis Trial were published. This trial investigated 3350 patients living in central Scotland with PAD but free from clinical cardiovascular disease and randomized them to either aspirin 100 mg QD or placebo. The mean follow-up was 8.2 years. There was no difference in the primary endpoint of fatal or nonfatal coronary events, stroke or revascularization [28].

In contrast to asymptomatic patients with PAD, there is some evidence supporting single antiplatelet therapy in patients with symptomatic PAD and patients with prior revascularization [29–31]. The AHA guidelines currently have a class I (level of evidence A) recommendation for the use of antiplatelet monotherapy in these patients [32, 33]. Despite a lack of evidence and guideline recommendations, dual antiplatelet therapy (DAPT) is widely used after peripheral interventions and with variable durations from 1 to 6 months, which largely depends on the results of smaller trials examining a special device, stent or drug-eluting balloon in peripheral interventions [34]. The optimal duration of DAPT following peripheral interventions is currently examined in 400 patients enrolling in the Antiplatelet Strategy for Peripheral Arterial Interventions for Revascularization of Lower Extremities trial (clinicaltrials.gov, NCT02217501, accessed 02 February 2016).

6.5 Platelet Inhibition in Transcatheter Valvular Interventions

Transcatheter aortic valve implantation (TAVI) is by far the fastest-growing procedure in interventional cardiology worldwide. Antiplatelet therapy is a cornerstone of the prevention of thromboembolic complications following TAVI. However, despite common perception, there is a variety of recommendations regarding the intensity (mono vs. dual antiplatelet therapy) and duration of antiplatelet therapy following TAVI. In addition the availability of different valve systems that are either balloon expandable, self-expandable or without a metal frame may require a differential use of antiplatelet therapy based on the assumed time frame that is necessary for endothelialization of the implanted aortic valve and the expected blood flow velocities in the aortic bulb, more specific in the space between the side wall of the implanted valve and the aortic wall.

Current recommendations for antiplatelet therapy range from a general recommendation for dual antiplatelet therapy (DAPT) without specified duration [35] (European Society for Cardiology) to a recommendation of 1–3 months [36] (Canadian Cardiovascular Society) and 3–6 months of DAPT, respectively [37] (American College of Cardiology and Society for Cardiovascular Angiography and Interventions). In general there appears to be a widespread adoption of the concept of dual antiplatelet therapy in TAVI at least for a month, while 3 months of DAPT have been reported as the preferred strategy in the majority of TAVI centres in the Netherlands [38]. The currently running ARTE trial (NCT01559298) will help to clarify whether low-dose aspirin or aspirin and clopidogrel (DAPT) given for at least 6 months post TAVI are the preferred strategy in balloon-expandable valves in 155 patients undergoing TAVI. The strategy of aspirin monotherapy versus aspirin plus clopidogrel for 3 months post TAVI will also be assessed in cohort A of the POPular-TAVI study, a multicentre trial currently recruiting patients in the Netherlands (NCT02247128).

More recently, computed tomography angiography routinely obtained in patients before and after TAVI revealed a phenomenon that is currently perceived as possible subclinical leaflet thrombosis in implanted transcatheter valves which appears to be reversible with oral anticoagulation [39, 40]. Therefore in addition to the current debate on the intensity and duration of antiplatelet therapy, the novel concept of oral anticoagulation in TAVI patients that do not have an indication of OAC for other purposes does raise even more questions on the optimal antithrombotic therapy in TAVI. At least two randomized trials with rivaroxaban and apixaban compared to single or dual antiplatelet therapy are currently under way to address this issue. In the ATLANTIS trial, patients with no indication for oral anticoagulation will be randomized to either antiplatelet therapy (single or dual) or apixaban 5 mg bid post TAVI. The GALILEO trial will assess rivaroxaban 10 mg qd and aspirin for 3 months followed by rivaroxaban 10 mg qd thereafter compared to DAPT for 3 months followed by aspirin therapy thereafter in patients after successful TAVI [41].

6.6 ESC/ACCA Guidelines on Platelet Inhibition in ACS

STEMI—In patients with STEMI, ESC guidelines and the current ACCA consensus document recommend to initiate antiplatelet therapy with aspirin in the initial treatment setting (including prehospital treatment and before patients are transferred to a cath lab for primary PCI) [22, 42]. In addition oral platelet P2Y12 inhibitors may be administered although evidence for a clinical benefit with pretreatment (out of hospital or before coronary anatomy is known) remains limited [42]. The third-generation P2Y12 inhibitors prasugrel and ticagrelor should be preferred over clopidogrel if there is no contraindication for their preferred use [22].

In high-risk patients with STEMI and low bleeding risk at the same time, the use of upstream GPIIb/IIIa inhibitors may be considered; however the level of recommendation is low (level of recommendation IIb, B) [22, 42]. Recommendations on pre- and periprocedural antiplatelet therapy in patients with ACS scheduled to undergo percutaneous intravascular coronary intervention (PCI) are summarized in Table 6.1.

NSTE-ACS—In patients with NSTE-ACS, current ESC guidelines and the current ACCA consensus document recommend to initiate antiplatelet therapy with aspirin in the initial treatment setting (including prehospital treatment) [22, 42]. However, there is also a lack of clinical evidence guiding physicians in

Table 6.1 ESC/ACCA recommendations for platelet inhibition patients with STEMI

ESC/ACCA recommendations for platelet inhibition patients with STEMI	Class	Level
Aspirin		
Aspirin either intravenously or orally is recommended in all patients	I	B
P2Y12 inhibitors		
Oral prasugrel and ticagrelor are recommended in addition to aspirin if no contraindication exists	I	B
Oral clopidogrel is recommended preferably when prasugrel and ticagrelor are either not available or contraindicated	I	C
GPIIb/IIIa inhibitors		
GPIIb/IIIa inhibitors should be considered for bail-out therapy if there is angiographic evidence of massive thrombus, slow or no reflow or a thrombotic complication	IIa	C
Upstream use of a GPIIb/IIIa inhibitor (vs. in-lab use) may be considered in high-risk patients undergoing transfer for primary PCI	IIb	B
ESC/ACCA recommendations for platelet inhibition patients with NSTE-ACS	Class	Level
Aspirin		
Aspirin is recommended for all patients without contraindication as an initial oral or intravenous loading dose of 150–300 mg	I	A
P2Y12 inhibitors		
Oral P2Y12 inhibitors are recommended in addition to aspirin unless there are contraindications such as excessive bleeding risk	I	A
Oral prasugrel and ticagrelor should be preferred if no contraindication; otherwise clopidogrel is recommended for patients who cannot receive prasugrel or ticagrelor	I	B
Intravenous cangrelor may be considered for PCI in patients who have not received any other P2Y12 inhibitor previously	IIb	A
Oral prasugrel should not be administered prior to PCI in patients in whom coronary anatomy is not known	III	B
GPIIb/IIIa inhibitors		
GPIIb/IIIa inhibitors should be considered for bail-out situations and thrombotic complications in patients undergoing PCI	IIa	C

Modified from [21, 22, 42]

the best possible timing of aspirin administration in the treatment of NSTE-ACS [42]. The recommendations on the additional administration of P2Y12 inihibitors on top of initial aspirin treatment are more differentiated in patients with NSTE-ACS compared to patients with STEMI. Although currently a matter of scientific debate, pretreatment with clopidogrel and ticagrelor is recommended regardless of conservative or early invasive treatment strategies, but prasugrel pretreatment is discouraged and should be left to patients with known coronary anatomy (meaning oral administration during or directly after PCI) [21, 42]. In addition, the current ESC guidelines do already contain a recommendation for the use of the periprocedural use of the intravenous P2Y12 inhibitor cangrelor in patients who have not yet received an oral P2Y12 inhibitor [21]. This is especially of interest in unconscious patients or patients with known or anticipated delayed intestinal absorption of orally administered P2Y12 antagonists.

In contrast to patients with STEMI, in patients with NSTE-ACS and unknown coronary anatomy, it is strongly discouraged to administer GPIIb/IIIa inhibitors (level of recommendation III, A). However, GPIIb/IIIa inhibitors may be used in patients with known coronary anatomy, especially in bail-out scenarios or periprocedural thrombotic complications (level of recommendation IIa, C) [21]. Recommendations on pre- and periprocedural antiplatelet therapy in patients with ACS scheduled to undergo percutaneous intravascular coronary intervention (PCI) are summarized in Table 6.1.

Compliance with Ethical Standards

Conflict of Interest: Ingo Ahrens and Hector Bueno declares that they have no conflict of interest.

Ethical Approval: This article does not contain any studies with human participants or animals performed by any of the authors.

References

1. Ahrens I, Chen YC, Topcic D, Bode M, Haenel D, Hagemeyer CE, Seeba H, Duerschmied D, Bassler N, Jandeleit-Dahm KA, Sweet MJ, Agrotis A, Bobik A, Peter K. HMGB1 binds to activated platelets via the receptor for advanced glycation end products and is present in platelet rich human coronary artery thrombi. Thromb Haemost. 2015;114:994–1003. https://doi.org/10.1160/TH14-12-1073
2. Duerschmied D, Bode C, Ahrens I. Immune functions of platelets. Thromb Haemost. 2014;112:678–91. https://doi.org/10.1160/TH14-02-0146
3. Libby P, Hansson GK. Inflammation and immunity in diseases of the arterial tree: players and layers. Circ Res. 2015;116:307–11. https://doi.org/10.1161/CIRCRESAHA.116.301313
4. Huo Y, Schober A, Forlow SB, Smith DF, Hyman MC, Jung S, Littman DR, Weber C, Ley K. Circulating activated platelets exacerbate atherosclerosis in mice deficient in apolipoprotein E. Nat Med. 2003;9:61–7. https://doi.org/10.1038/nm810
5. Libby P. Mechanisms of acute coronary syndromes and their implications for therapy. N Engl J Med. 2013;368:2004–13. https://doi.org/10.1056/NEJMra1216063
6. Ahrens I, Bode C, Peter K. Inhibition of platelet activation and aggregation. Handb Exp Pharmacol. 2005;170:443–62.
7. Ahrens I, Peter K, Bode C. Use of GPIIb/IIIa inhibitors in cardiovascular medicine. Expert Rev Cardiovasc Ther. 2003;1:233–42. https://doi.org/10.1586/14779072.1.2.233
8. Schwarz M, Meade G, Stoll P, Ylanne J, Bassler N, Chen YC, Hagemeyer CE, Ahrens I, Moran N, Kenny D, Fitzgerald D, Bode C, Peter K. Conformation-specific blockade of the integrin GPIIb/IIIa: a novel antiplatelet strategy that selectively targets activated platelets. Circ Res. 2006;99:25–33. https://doi.org/10.1161/01.RES.0000232317.84122.0c
9. Wang X, Hagemeyer CE, Hohmann JD, Leitner E, Armstrong PC, Jia F, Olschewski M, Needles A, Peter K, Ahrens I. Novel single-chain antibody-targeted microbubbles for molecular ultrasound imaging of thrombosis: validation of a unique noninvasive method for rapid and sensitive detection of thrombi and monitoring of success or failure of thrombolysis in mice. Circulation. 2012;125:3117–26. https://doi.org/10.1161/CIRCULATIONAHA.111.030312

10. Gachet C. ADP receptors of platelets and their inhibition. Thromb Haemost. 2001;86:222–32.
11. Diehl P, Bode C, Duerschmied D. Clinical potential of vorapaxar in cardiovascular risk reduction in patients with atherosclerosis. Ther Clin Risk Manag. 2015;11:1133–8. https://doi.org/10.2147/TCRM.S55469
12. Eikelboom JW, Mehta SR, Anand SS, Xie C, Fox KA, Yusuf S. Adverse impact of bleeding on prognosis in patients with acute coronary syndromes. Circulation. 2006;114:774–82. https://doi.org/10.1161/CIRCULATIONAHA.106.612812
13. Stachon P, Ahrens I, Bode C, Zirlik A. Dual pathway therapy in acute coronary syndrome. J Thromb Thrombolysis. 2015. https://doi.org/10.1007/s11239-015-1306-3
14. Ahrens I, Bode C, Zirlik A. Anticoagulation during and after acute coronary syndrome. Hamostaseologie. 2014;34:72–7. https://doi.org/10.5482/HAMO-13-09-0048
15. Bosch X, Marrugat J, Sanchis J. Platelet glycoprotein IIb/IIIa blockers during percutaneous coronary intervention and as the initial medical treatment of non-ST segment elevation acute coronary syndromes. Cochrane Database Syst Rev. 2013;11:CD002130. https://doi.org/10.1002/14651858.CD002130.pub4
16. Bosch X, Marrugat J, Sanchis J. Platelet glycoprotein IIb/IIIa blockers during percutaneous coronary intervention and as the initial medical treatment of non-ST segment elevation acute coronary syndromes. Cochrane Database Syst Rev. 2013;10:CD002130. https://doi.org/10.1002/14651858.CD002130.pub3
17. Dewilde WJ, Oirbans T, Verheugt FW, Kelder JC, De Smet BJ, Herrman JP, Adriaenssens T, Vrolix M, Heestermans AA, Vis MM, Tijsen JG, van't Hof AW, ten Berg JM, Investigators WOEST Study. Use of clopidogrel with or without aspirin in patients taking oral anticoagulant therapy and undergoing percutaneous coronary intervention: an open-label, randomised, controlled trial. Lancet. 2013;381:1107–15. https://doi.org/10.1016/S0140-6736(12)62177-1
18. Hosokawa K, Ohnishi T, Miura N, Sameshima H, Koide T, Tanaka KA, Maruyama I. Antithrombotic effects of PAR1 and PAR4 antagonists evaluated under flow and static conditions. Thromb Res. 2014;133:66–72. https://doi.org/10.1016/j.thromres.2013.10.037
19. Scavone M, Femia EA, Caroppo V, Cattaneo M. Inhibition of the platelet P2Y12 receptor for adenosine diphosphate does not impair the capacity of platelet to synthesize thromboxane A2. Eur Heart J. 2015. https://doi.org/10.1093/eurheartj/ehv551
20. Risk of myocardial infarction and death during treatment with low dose aspirin and intravenous heparin in men with unstable coronary artery disease. The RISC Group. Lancet. 1990;336:827–30.
21. Ibanez B, James S, Agewall S, Antunes MJ, Bucciarelli-Ducci C, Bueno H, Caforio ALP, Crea F, Goudevenos JA, Halvorsen S, Hindricks G, Kastrati A, Lenzen MJ, Prescott E, Roffi M, Valgimigli M, Varenhorst C, Vranckx P, Widimský P. 2017 ESC guidelines for the management of acute myocardial infarction in patients presenting with ST-segment elevation. Eur Heart J. 2017.
22. Task Force on the Management of STseamiotESoC, Steg PG, James SK, Atar D, Badano LP, Blomstrom-Lundqvist C, Borger MA, Di Mario C, Dickstein K, Ducrocq G, Fernandez-Aviles F, Gershlick AH, Giannuzzi P, Halvorsen S, Huber K, Juni P, Kastrati A, Knuuti J, Lenzen MJ, Mahaffey KW, Valgimigli M, van't Hof A, Widimsky P, Zahger D. ESC Guidelines for the management of acute myocardial infarction in patients presenting with ST-segment elevation. Eur Heart J. 2012;33:2569–619. https://doi.org/10.1093/eurheartj/ehs215
23. Wallentin L, Becker RC, Budaj A, Cannon CP, Emanuelsson H, Held C, Horrow J, Husted S, James S, Katus H, Mahaffey KW, Scirica BM, Skene A, Steg PG, Storey RF, Harrington RA, Investigators P, Freij A, Thorsen M. Ticagrelor versus clopidogrel in patients with acute coronary syndromes. N Engl J Med. 2009;361:1045–57. https://doi.org/10.1056/NEJMoa0904327

24. Yusuf S, Zhao F, Mehta SR, Chrolavicius S, Tognoni G, Fox KK, Clopidogrel in Unstable Angina to Prevent Recurrent Events Trial I. Effects of clopidogrel in addition to aspirin in patients with acute coronary syndromes without ST-segment elevation. N Engl J Med. 2001;345:494–502. https://doi.org/10.1056/NEJMoa010746
25. Montalescot G, Bolognese L, Dudek D, Goldstein P, Hamm C, Tanguay JF, ten Berg JM, Miller DL, Costigan TM, Goedicke J, Silvain J, Angioli P, Legutko J, Niethammer M, Motovska Z, Jakubowski JA, Cayla G, Visconti LO, Vicaut E, Widimsky P, Investigators A. Pretreatment with prasugrel in non-ST-segment elevation acute coronary syndromes. N Engl J Med. 2013;369:999–1010. https://doi.org/10.1056/NEJMoa1308075
26. Roe MT, Armstrong PW, Fox KA, White HD, Prabhakaran D, Goodman SG, Cornel JH, Bhatt DL, Clemmensen P, Martinez F, Ardissino D, Nicolau JC, Boden WE, Gurbel PA, Ruzyllo W, Dalby AJ, McGuire DK, Leiva-Pons JL, Parkhomenko A, Gottlieb S, Topacio GO, Hamm C, Pavlides G, Goudev AR, Oto A, Tseng CD, Merkely B, Gasparovic V, Corbalan R, Cinteza M, McLendon RC, Winters KJ, Brown EB, Lokhnygina Y, Aylward PE, Huber K, Hochman JS, Ohman EM, Investigators TA. Prasugrel versus clopidogrel for acute coronary syndromes without revascularization. N Engl J Med. 2012;367:1297–309. https://doi.org/10.1056/NEJMoa1205512
27. Wiviott SD, Braunwald E, McCabe CH, Montalescot G, Ruzyllo W, Gottlieb S, Neumann FJ, Ardissino D, De Servi S, Murphy SA, Riesmeyer J, Weerakkody G, Gibson CM, Antman EM, Investigators T-T. Prasugrel versus clopidogrel in patients with acute coronary syndromes. N Engl J Med. 2007;357:2001–15. https://doi.org/10.1056/NEJMoa0706482
28. Fowkes FG, Price JF, Stewart MC, Butcher I, Leng GC, Pell AC, Sandercock PA, Fox KA, Lowe GD, Murray GD, Aspirin for Asymptomatic Atherosclerosis T. Aspirin for prevention of cardiovascular events in a general population screened for a low ankle brachial index: a randomized controlled trial. JAMA. 2010;303:841–8. https://doi.org/10.1001/jama.2010.221
29. Committee CS. A randomised, blinded, trial of clopidogrel versus aspirin in patients at risk of ischaemic events (CAPRIE). CAPRIE Steering Committee Lancet. 1996;348:1329–39.
30. Critical Leg Ischaemia Prevention Study G, Catalano M, Born G, Peto R. Prevention of serious vascular events by aspirin amongst patients with peripheral arterial disease: randomized, double-blind trial. J Intern Med. 2007;261:276–84. https://doi.org/10.1111/j.1365-2796.2006.01763.x
31. Spiliopoulos S, Pastromas G, Katsanos K, Kitrou P, Karnabatidis D, Siablis D. Platelet responsiveness to clopidogrel treatment after peripheral endovascular procedures: the PRECLOP study: clinical impact and optimal cutoff value of on-treatment high platelet reactivity. J Am Coll Cardiol. 2013;61:2428–34. https://doi.org/10.1016/j.jacc.2013.03.036
32. Rooke TW, Hirsch AT, Misra S, Sidawy AN, Beckman JA, Findeiss LK, Golzarian J, Gornik HL, Halperin JL, Jaff MR, Moneta GL, Olin JW, Stanley JC, White CJ, White JV, Zierler RE, Society for Cardiovascular A, Interventions, Society of Interventional R, Society for Vascular M, Society for Vascular S. 2011 ACCF/AHA focused update of the guideline for the management of patients with peripheral artery disease (updating the 2005 guideline): a report of the American College of Cardiology Foundation/American Heart Association Task Force on Practice Guidelines. J Am Coll Cardiol. 2011;58:2020–45. https://doi.org/10.1016/j.jacc.2011.08.023
33. Smith SC Jr, Benjamin EJ, Bonow RO, Braun LT, Creager MA, Franklin BA, Gibbons RJ, Grundy SM, Hiratzka LF, Jones DW, Lloyd-Jones DM, Minissian M, Mosca L, Peterson ED, Sacco RL, Spertus J, Stein JH, Taubert KA, World Heart F, The Preventive Cardiovascular Nurses A. AHA/ACCF secondary prevention and risk reduction therapy for patients with coronary and other atherosclerotic vascular disease: 2011 update: a guideline from the American Heart Association and American College of Cardiology Foundation. Circulation. 2011;124:2458–73. https://doi.org/10.1161/CIR.0b013e318235eb4d
34. Banerjee S, Sarode K, Vinas A, Banerjee A, Mohammad A, Brilakis ES. The role of antiplatelet therapy in patients with peripheral artery disease and lower extremity peripheral

artery revascularization. Curr Opin Cardiol. 2015;30:525–35. https://doi.org/10.1097/HCO.0000000000000208
35. Joint Task Force on the Management of Valvular Heart Disease of the European Society of C, European Association for Cardio-Thoracic S, Vahanian A, Alfieri O, Andreotti F, Antunes MJ, Baron-Esquivias G, Baumgartner H, Borger MA, Carrel TP, De Bonis M, Evangelista A, Falk V, Iung B, Lancellotti P, Pierard L, Price S, Schafers HJ, Schuler G, Stepinska J, Swedberg K, Takkenberg J, Von Oppell UO, Windecker S, Zamorano JL, Zembala M. Guidelines on the management of valvular heart disease (version 2012). Eur Heart J. 2012;33:2451–96. https://doi.org/10.1093/eurheartj/ehs109
36. Webb J, Rodes-Cabau J, Fremes S, Pibarot P, Ruel M, Ibrahim R, Welsh R, Feindel C, Lichtenstein S. Transcatheter aortic valve implantation: a Canadian Cardiovascular Society position statement. Can J Cardiol. 2012;28:520–8. https://doi.org/10.1016/j.cjca.2012.04.015
37. Holmes DR Jr, Mack MJ, Kaul S, Agnihotri A, Alexander KP, Bailey SR, Calhoon JH, Carabello BA, Desai MY, Edwards FH, Francis GS, Gardner TJ, Kappetein AP, Linderbaum JA, Mukherjee C, Mukherjee D, Otto CM, Ruiz CE, Sacco RL, Smith D, Thomas JD. 2012 ACCF/AATS/SCAI/STS expert consensus document on transcatheter aortic valve replacement. J Am Coll Cardiol. 2012;59:1200–54. https://doi.org/10.1016/j.jacc.2012.01.001
38. Nijenhuis VJ, Stella PR, Baan J, Brueren BR, de Jaegere PP, den Heijer P, Hofma SH, Kievit P, Slagboom T, van den Heuvel AF, van der Kley F, van Garsse L, van Houwelingen KG, Van't Hof AW, Ten Berg JM. Antithrombotic therapy in patients undergoing TAVI: an overview of Dutch hospitals. Neth Heart J. 2014;22:64–9. https://doi.org/10.1007/s12471-013-0496-6
39. Makkar RR, Fontana G, Jilaihawi H, Chakravarty T, Kofoed KF, de Backer O, Asch FM, Ruiz CE, Olsen NT, Trento A, Friedman J, Berman D, Cheng W, Kashif M, Jelnin V, Kliger CA, Guo H, Pichard AD, Weissman NJ, Kapadia S, Manasse E, Bhatt DL, Leon MB, Sondergaard L. Possible subclinical leaflet thrombosis in bioprosthetic aortic valves. N Engl J Med. 2015;373:2015–24. https://doi.org/10.1056/NEJMoa1509233
40. Pache G, Schoechlin S, Blanke P, Dorfs S, Jander N, Arepalli CD, Gick M, Buettner HJ, Leipsic J, Langer M, Neumann FJ, Ruile P. Early hypo-attenuated leaflet thickening in balloon-expandable transcatheter aortic heart valves. Eur Heart J. 2015. https://doi.org/10.1093/eurheartj/ehv526
41. Gargiulo G, Collet JP, Valgimigli M. Antithrombotic therapy in TAVI patients: changing concepts. EuroIntervention. 2015;11(Suppl W):W92–5. https://doi.org/10.4244/EIJV11SWA28
42. Beygui F, Castren M, Brunetti ND, Rosell-Ortiz F, Christ M, Zeymer U, Huber K, Folke F, Svensson L, Bueno H, Van't Hof A, Nikolaou N, Nibbe L, Charpentier S, Swahn E, Tubaro M, Goldstein P, care Asgop-h. Pre-hospital management of patients with chest pain and/or dyspnoea of cardiac origin. A position paper of the acute cardiovascular care association (ACCA) of the ESC. Eur. Heart J. Acute Card Care. 2015. https://doi.org/10.1177/2048872615604119

Diabetes, Thrombosis, and Cardiovascular Risks

7

Katharina Schuett and Nikolaus Marx

Abstract

Patients with diabetes are at an increased cardiovascular risk, and alterations of the coagulation system are pivotal in this context and reduce responsiveness to certain anticoagulants. Following plaque rupture, platelets are the first to be activated stabilizing the developing clot. In diabetes, hyperglycemia, oxidative stress, and endothelial dysfunction contribute to platelet dysfunction resulting in procoagulant hyperreactivity. Adherence of platelets is followed by the formation of a cross-linked fibrin clot. Subjects with diabetes exhibit a tight and rigid clot structure which is due to upregulation of coagulation factors and prolongation of clot lysis. Metabolic alterations and upregulation of inflammatory processes in diabetes are thought to be the main underlying causes. More recently, other factors such as erythrocytes, microparticles, and neutrophil extracellular traps have emerged as new players in this context directly influencing both platelet function and coagulation. This chapter provides an overview concerning the changes that lead to alterations of coagulation in diabetes.

Contents

K. Schuett • N. Marx (✉)
RWTH Aachen University, University Hospital Aachen, Pauwelsstraße 30, 52074 Aachen, Germany
e-mail: nmarx@ukaachen.de

A. Zirlik et al. (eds.), *Platelets, Haemostasis and Inflammation*,
Cardiac and Vascular Biology 5, https://doi.org/10.1007/978-3-319-66224-4_7

7.1 Diabetes and Cardiovascular Risk

The increasing prevalence of obesity and diabetes mellitus type 2 over the last two decades is one of the major healthcare problems in Western societies. Patients with diabetes exhibit an increased propensity to develop macrovascular complications such as myocardial infarction (MI) and stroke, leading to an increased risk for cardiovascular death. Early analyses from the Framingham cohort pointed to this association by showing a two- to threefold increased risk for atherosclerotic complications. In 1998, Haffner and colleagues published a 7-year follow-up of 1373 nondiabetic and 1059 diabetic patients showing that the incidence of myocardial infarction in subjects with diabetes was similar to the MI incidence in nondiabetic subjects after their first myocardial infarction, suggesting that diabetes may be a coronary heart disease equivalent. Interestingly, once patients with diabetes have experienced a myocardial infarction, there is an exponential increase in their risk for future events, and the study reveals similar results for stroke and cardiovascular death. Overall, the 7-year risk of patients with diabetes in this population was 22% to develop a myocardial infarction [1]. Studies in other populations confirmed this increased risk for patients with diabetes albeit some of the study suggested that the risk was not directly comparable to the risk of nondiabetic subjects post-myocardial infarction. Mechanistic data from a prospective registry analysis including patients with diabetes who underwent coronary intravascular ultrasound virtual histology (IVUS-VH) support the epidemiological results: subjects with a long duration of diabetes exhibited a higher overall plaque burden as well as a higher proportion of thin-cap fibroatheroma (TCFA) compared to individuals with a shorter duration of diabetes. However, the very early studies cited above enrolled patients with diabetes prior to the results of large cardiovascular outcome studies with statins and ACE inhibitors, raising the question whether current state-of-the-art therapy including these drugs influences the overall CV risk in diabetes. More recent data published from the European Prospective Investigation into Cancer and Nutrition (EPIC), a population-based cohort study from the UK, demonstrated that patients with diabetes exhibit an eightfold increased risk for cardiovascular mortality compared with patients with an HbA1c of less than 5%. Interestingly, this study revealed for the first time that HbA1c values ranging from 5% to >7% were associated with an increased risk for cardiovascular events and that even HbA1c levels in the upper range of the norm are associated with cardiovascular complications. These data were confirmed in a meta-analysis including more than 500,000 participants, suggesting that the presence of diabetes approximately doubles the risk for coronary artery disease, ischemic stroke, as well as death due to other vascular causes even after adjustment for potential confounders such as age, smoking, BMI, and blood pressure [2]. A very recent analysis from the Emerging Risk Factors Collaboration confirms a 1.9-fold risk for cardiovascular death in patients with diabetes and a 3.7-fold increased risk in patients with diabetes and MI compared to nondiabetic subjects. This translates into a significant loss of life years: compared to a subject without diabetes, the presence of diabetes without vascular disease in a 60-year-old patient leads to a loss of 6 years of life and a loss of 12 years in diabetic patients with a history of myocardial infarction. Overall, medical progress and improved patient care have led to a reduction of the incidence of myocardial infarction and stroke over the last 10–20 years, but the burden for diabetes-related cardiovascular complications remains high. Interestingly, many epidemiological studies revealed over the last decades that not only diabetes itself but also impaired fasting glucose (IFG) as well as impaired glucose tolerance (IGT) are also associated with an increased CV risk, suggesting that a prediabetic state also favors the development of vascular disease with its potentially deleterious sequelae.

Various factors contribute to the increased cardiovascular risk of patients with diabetes: among them are the associated risk factors like hypertension, dyslipidemia, obesity, as well as hypoglycemia itself. Moreover, altered vascular function, atherosclerotic manifestations in various vascular beds, as well as microvascular changes are only some of the pathophysiological mechanisms that are of importance in this context. In addition, subclinical inflammation as well as changes in platelet function and hypercoagulability seems to be crucially involved in the development of myocardial infarction.

7.2 Thrombosis in Diabetes

Alterations of the coagulation system in patients with diabetes are pivotal and contribute to the elevated cardiovascular risk and reduced responsiveness to certain anticoagulants. These variations include changes in platelet function, the coagulation system, altered erythrocyte function, as well as the frequency and composition of microparticles and neutrophil extracellular traps (NETs) (see Fig. 7.1). In the following section, the individual components of thrombosis in diabetes will be discussed in detail.

7.3 Platelets

Platelets are the first to respond following plaque rupture and exposure of thrombotic components to the blood. Platelet adhesion is followed by activation, further recruitment of platelets, and aggregation, thereby stabilizing the developing clot [3]. In diabetes, platelet function is disturbed leading to changes such as a more frequent

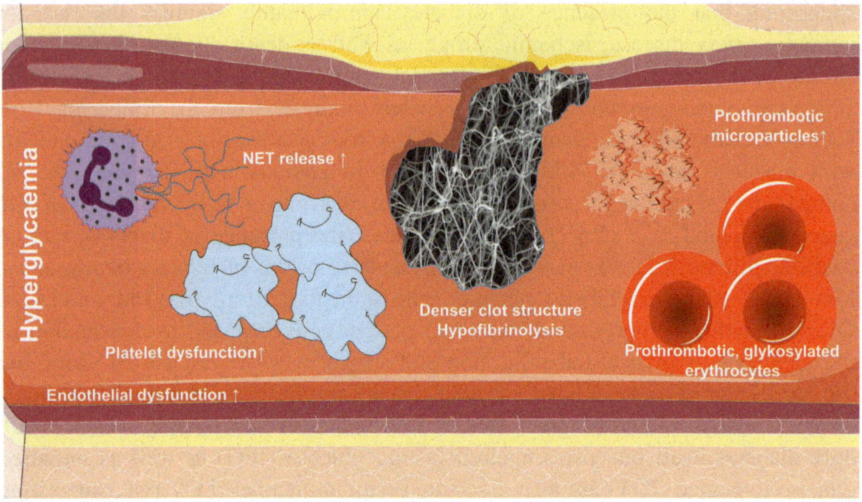

Fig. 7.1 Alterations of the coagulation in diabetes. In diabetes, several aspects of blood coagulation are altered. Platelet function is disturbed leading to changes such as a more frequent response to subthreshold stimuli, an increased turnover, and accelerated thrombopoiesis of hyperreactive platelets. Fibrin clots are more compact with impaired fibrinolysis. Furthermore, heavily glycosylated erythrocytes impact coagulation as well as increased neutrophil extracellular trap (NET) formation and microparticles do

response to subthreshold stimuli, an increased turnover, and accelerated thrombopoesis of hyperreactive platelets [4]. In particular, hyperglycemia, oxidative stress, and endothelial dysfunction are involved and will be described in more detail.

7.3.1 Hyperglycemia

Elevated blood glucose is known to impact platelet function by different mechanisms. Acute short-term hyperglycemia results in an increased activation of platelets exposed to high shear stress and an increased sensibility to agonists due to impairment of the Ca homeostasis, less NO production, activation of PKC, and augmented superoxide formation [5]. Platelets are known to retain an active insulin receptor [6] whereby insulin can regulate ADP- and thrombin-induced platelet functions. Following receptor binding, insulin activates the insulin-receptor substrate-1 (IRS-1) through tyrosine phosphorylation, which initiates association with $G_i\alpha$ subunit. This results in the inhibition of $G_i\alpha$ activity and impaired suppression of cyclic adenosine monophosphate (cAMP), thereby inhibiting $P2Y_{12}$ signaling leading to reduced platelet activity [7]. Short-term alterations in glucose metabolism, like hyperglycemia and hyperinsulinemia for 24 h, can downregulate insulin signaling [8]. In addition to acute mechanism, chronic alteration of glucose levels also impairs platelet function. In diabetes, platelets are less responsive to insulin contributing to increased adhesion, aggregation, and procoagulant activity [9]. Furthermore, platelet hyperreactivity in diabetes is associated with increased platelet production of thromboxane and tightly regulated by glucose control [10]. Calcium mobilization is augmented with an increase of intracellular Ca^{2+} resulting in enhanced platelet activation and aggregation [11]. Animal studies suggest that the downregulation of calsequestrin (CSQ), a Ca^{2+} storage protein, and insufficiency of the Na^{+}/Ca^{2+} exchange may play a role in this context [12].

Hyperglycemia further results in glycation of platelet membrane proteins which potentially causes changes in protein structure, thereby enhancing surface expression of P-selectin and GP receptors rendering platelets more prone to activation [13]. Studies examining the effect of glycemic control on platelet reactivity yielded conflicting data. Besides glycation of platelet membrane proteins, hyperglycemia can result in glycation of low-density lipoproteins (LDL), which are able to contribute to platelet dysfunction by increasing their intracellular Ca^{2+} concentration and NO production and decreasing the platelet membrane Na^{+}/K^{+}-adenosine triphosphatase activity [14]. Further to these posttranslational modifications, alterations of the lipid profile [increased LDL, triglycerides, and decreased high-density lipoproteins (HDL)] may affect platelet function by interacting with the intracellular system and decreasing membrane fluidity [5].

7.3.2 Oxidative Stress

Oxidative stress also modulates platelet function. Hyperglycemia directly induces reactive oxygen production (ROS) via glucose metabolism and auto-oxidation and indirectly due to the formation of advanced glycation end products (AGEs) and their receptor binding. ROS activate endothelial cell signaling including protein kinase C and nuclear factor-κB, thereby inducing the production of pro-inflammatory and prothrombotic molecules [15]. In addition, ROS lead to the formation of 8-iso-prostaglandin $F2_{\alpha}$, a nonenzymatic oxidation product of circulating LDL and arachidonic acid, which induces vasoconstriction and platelet hyperreactivity [16].

7.3.3 Endothelial Dysfunction

Endothelial cells are an important source for mediators regulating vasoconstriction (e.g., angiotensin II, thromboxane) and vasodilatation (e.g., NO, prostacyclin), thus impacting among others thrombotic processes. In diabetes, endothelial homeostasis is impaired with a reduction in vasodilatation [17]. Hyperglycemia induces endothelial dysfunction by activation of protein kinase C, increased activity of the polyol pathway, nonenzymatic glycation, and oxidative stress. Together this results in an elevated expression of pro-inflammatory cytokines and platelet adhesion molecules [18]. Moreover, platelets from patients with diabetes seem to have a direct effect on endothelial cells. A recent study in rats suggests that platelets from diabetic animals impair NO production via Akt/endothelial NO synthase signaling pathway and increase thromboxane synthesis [19].

7.4 Fibrin Networks

A compact fibrin clot structure and impaired fibrinolysis are associated with more severe cardiovascular disease [20]. Subjects with diabetes exhibit a prothrombotic clot structure characterized by small pores and resistance to fibrinolysis due to several mechanisms.

7.4.1 Hypercoagulability

Tissue factor (TF) is the key initiator of the coagulation cascade and produced by several cell types including endothelial cells, vascular smooth muscle cells, monocytes/macrophages, and platelets. Under basal conditions, TF expression is very low [21] but can be upregulated following stimulation with cytokines (e.g., TNFα, IL-1β) [22, 23] or mediators including thrombin, oxLDL, or histamine [24–26]. The major source of vascular TF expression is monocytes, while the role of platelets in this context is still a matter of debate. As some studies did not detect TF on activated platelets [27, 28], others demonstrated functionally active TF [29–31]. In obesity, a common feature of diabetes, adipose tissue functions as a site of TF production [32]. In diabetes, TF-procoagulatory activity, assessed by cell-bound and microparticle-associated TF obtained from whole blood, is elevated [33] and plasma TF activity correlates with fasting insulin, glucose, as well as free fatty acids [34]. In clamp studies, hyperglycemia leads to increased TF levels [35] even in healthy individuals [36]. In normal platelets, insulin inhibits TF synthesis; however, this inhibition is lost in diabetes resulting in a 1.6-fold higher tissue factor expression [37]. In addition, both ROS and AGEs are able to elevate TF level [21, 38].

TF binds and activates factor VII (FVII) leading to activation of the prothrombinase complex (FXa, FVa, and Ca^{2+}). Originally in 1986, the Northwick Park Heart study described an association between high levels of FVII coagulant activity and an elevated risk for ischemic heart disease [39]. Since then, this association was a matter of debate with studies supporting the original data [40] and others yielded opposite results [41–43]. Nevertheless, elevated levels of FVII are associated with insulin resistance [44] and factor VII coagulant activity (FVII:c) is elevated in metabolic syndrome [45]. In first-degree relatives of patients with diabetes, FVII:c levels are increased and cluster with risk factors for insulin resistance [46]. However, in one study including patients with diabetes and age- and obesity-matched controls, FVII:c and FVII antigen levels were lower in diabetes compared to healthy controls [47]. These data suggest that FVII may be more related to the metabolic syndrome and obesity than to diabetes [45].

A different way to activate factor X and the prothrombinase complex is via the contact activation pathway which requires the clotting factors XII, XI, IX, and VIII.

Already in 1982, Patrassi et al. found increased plasma levels of FXII and FXI in patients with diabetes [48]. This was confirmed more recently in a small cohort showing an increase of FXII, FXI, and FVIII in patients with diabetes [49]. In the circulation, FVIII is stabilized by von Willebrand factor (VWF), thereby increasing the half-life of FVIII. The expression of VWF is regulated by several agonists including thrombin and complement components [50]. Elevated levels of VWF and FVIII are independently associated with the presence of diabetes [51, 52] but not with cardiovascular disease [53].

By activation of the prothrombinase complex, prothrombin is converted to thrombin. This causes the activation of fibrinogen by cleaving fibrinopeptides A and B leading to polymerization and the formation of a fibrin clot. Fibrinogen is synthesized by the liver and is a heterodimer composed of three pairs of nonidentical polypeptide chains (Aα, Bβ, and γ) [54]. Elevated fibrinogen levels determine clot structure and are associated with an increased cardiovascular risk [20, 55–57]. In diabetes, fibrinogen plasma levels are elevated due to a variety of reasons. The low-grade inflammation in diabetes with elevation of, e.g., interleukin 6, interleukin 1, and tumor necrosis factor α induces the production of acute phase proteins including fibrinogen [58]. Independent of diabetes, a number of polymorphisms have been associated with increased fibrinogen levels [59]. Posttranslational modification such as glycation of fibrinogen further results in alterations of fibrin function with formation of a tight and rigid fibrin clot [60, 61], which is associated with an increased risk for myocardial infarction [20]. Improvement of glycemic control has been shown to improve these alterations of clot structure [62].

While forming the clot, FXIII is required for stabilization by cross-linking fibrin and incorporation of antifibrinolytic proteins, thereby protecting us from bleeding. FXIII is a tetrameric pro-transglutaminase that consists of two A- and B-subunits. The B-subunit serves as carrier protein for the active A-unit, which is exposed after stimulation by thrombin. Besides its protective effect, a role for FXIII in cardiovascular disease has been suggested since the FXIII-A Val34Leu polymorphism protects from myocardial infarction [63, 64]. In addition, a more recent study demonstrated unfavorable changes of clot structure, including thinner fibers and smaller pores in the presence of FXIII [65]. In diabetes, FXIII levels are increased with no difference in cross-linking abilities. The B-subunit correlates with features of the metabolic syndrome; the A-unit does not. This differential association can be explained by the diverse production site. While the A-unit is produced by hematopoietic cells, the B-subunit is synthesized by the liver [66].

7.4.2 Hypofibrinolysis

Fibrinolysis is important for homeostasis of coagulation. Following activation by tissue plasminogen activator (tpA) or urokinase, plasminogen is converted to plasmin which cleaves fibrin into its degradation products. In diabetes, fibrinolysis is prolonged. Glycation of fibrin(ogen) enhances resistance to fibrinolysis [50] and hyperinsulinemia has been shown to inhibit fibrinolysis irrespective of glucose levels [35]. Plasminogen activator inhibitor 1 (PAI-1) is the main inhibitor of fibrinolysis in diabetes [67]. PAI-1 rapidly forms inactive complexes with tpA and urokinase to prevent plasmin generation. In diabetes, PAI-1 levels are upregulated and associated with an increased cardiovascular risk [68]. In addition to PAI-1, more recent studies demonstrated complement C3 to be a substrate for factor XIII resulting in cross-linking of C3 to fibrin during clot formation [69, 70]. The high-affinity binding of C3 to fibrin(ogen) leads to a prothrombotic clot structure and prolongation of clot lysis in vitro with pronounced effects in diabetes [71–73].

7.5 Erythrocytes

Besides platelets and fibrinogen, in vivo clots contain red blood cells. In addition to their traditional role for oxygen transport, recent studies suggest an impact of erythrocytes on coagulation. They induce platelet aggregation and degranulation due to the release of ADP and ATP under low oxygen saturation, low pH, or mechanical deformation [74, 75]. Furthermore, they contribute to the activation of the coagulation cascade by losing their phospholipid asymmetry and serve as a procoagulant surface [75]. In addition, the incorporation of red blood cells influences clot structure leading to thicker fibrin fibers and alters the mechanical properties of the clot [76, 77]. In diabetes, the erythrocyte membrane becomes rigid and non-deformable due to a decrease of the cholesterol to phospholipid ratio of the cells [77]. Furthermore, several membrane proteins are heavily glycosylated compared with nondiabetic erythrocyte membranes [78], leading to a significant decrease in cell deformability. This elevates blood viscosity resulting in an increased shear stress on endothelial cells [79]. In addition, in diabetes electron microscopy studies revealed changes in erythrocyte morphology with elongated cells forming extended projections twisting around fibrin fibers [80].

7.6 Microparticles

Various studies demonstrated a role for microparticles (MPs) in coagulation. Following activation or apoptosis, MPs are released from membranes of various cell types including platelets, endothelial cells, red blood cells, and leukocytes. Depending on their origin, they vary in size (0.2–1 μm) and membrane composition including phospholipids and proteins. MPs can be detected in the circulation of healthy individuals and are elevated in diabetes mellitus [81]. MPs are directly able to modulate nitric oxide production from endothelial cells and induce cytokine release and prostacyclin production as well as adherence of monocytes to the endothelium [82]. The two major procoagulants found on the surface of MPs are phosphatidylserine and TF [83], thereby contributing to a prothrombotic state. Furthermore, MPs can harbor and transport microRNA, thereby impacting protein expression of target cells [84, 85].

7.7 Neutrophil Extracellular Traps

Neutrophil extracellular traps (NETs) are weblike structures of DNA released upon activation of neutrophils. NETs can exhibit procoagulatory properties including induction of platelet adhesion, aggregation, and fibrin deposition on their surface. Activated inflammatory cells such as neutrophils can further secrete histones, cationic proteins that are associated with DNA, thereby contributing to the formation of NETs. In this context, histones have been shown to promote platelet aggregation and thrombin formation through platelet-dependent mechanisms including platelet toll-like receptor (TLR)2 and TLR4 [86]. The impact of NET formation in cardiovascular disease was shown more recently in a study in acute ST-elevation myocardial infarction, demonstrating that the interaction of thrombin-activated platelets with polymorphonuclear neutrophils at the site of plaque rupture results in local NET formation and delivery of active TF [87]. In diabetes, isolated neutrophils from type 1 and type 2 diabetic humans and mice are primed to produce NETs [88]. Accordingly, ex vitro experiments demonstrated an increased release of NETs in a high glucose setting [89, 90].

7.8 Conclusion

Patients with diabetes mellitus are at an increased cardiovascular risk with alterations of the blood being of critical importance. Platelet dysfunction, clot structure, and prolongation of fibrinolysis result in an enhanced prothrombotic milieu which is associated with an increased cardiovascular risk. More recent studies revealed a role for erythrocytes, microparticles, and neutrophil extracellular traps in this context. More studies are necessary to investigate the interactions between those components to discover additional potential pharmacological targets. This is of importance to develop new treatment strategies for those high-risk patients. Overall, alterations of the coagulation should be taken into account when evaluating the cardiovascular risk of patients with diabetes mellitus.

Compliance with Ethical Standards

Conflict of Interest: Katharina Schuett and Nikolaus Marx declares that they have no conflict of interest.

Ethical Approval: This article does not contain any studies with human participants or animals performed by any of the authors.

References

1. Haffner SM, Lehto S, Rönnemaa T, Pyörälä K, Laakso M. Mortality from coronary heart disease in subjects with type 2 diabetes and in nondiabetic subjects with and without prior myocardial infarction. N Engl J Med. 1998;339:229–34.
2. Emerging Risk Factors Collaboration, Sarwar N, Gao P, Seshasai SR, Gobin R, Kaptoge S, Di Angelantonio E, Ingelsson E, Lawlor DA, Selvin E, Stampfer M, Stehouwer CD, Lewington S, Pennells L, Thompson A, Sattar N, White IR, Ray KK, Danesh J. Diabetes mellitus, fasting blood glucose concentration, and risk of vascular disease: a collaborative meta-analysis of 102 prospective studies. Lancet. 2010;375(9733):2215–22. https://doi.org/10.1016/S0140-6736(10)60484-9
3. Angiolillo DJ, Ueno M, Goto S. Basic principles of platelet biology and clinical implications. Circ J. 2010;74:597–607.
4. Watala C. Blood platelet reactivity and its pharmacological modulation in (people with) diabetes mellitus. Curr Pharm Des. 2005;11:2331–65.
5. Ferroni P, Basili S, Falco A, Davì G. Platelet activation in type 2 diabetes mellitus. J Thromb Haemost. 2004;2:1282–91.
6. Falcon C, Pfliegler G, Deckmyn H, Vermylen J. The platelet insulin receptor: detection, partial characterization, and search for a function. Biochem Biophys Res Commun. 1988;157:1190–6.
7. Ferreira IA, Eybrechts KL, Mocking AI, Kroner C, Akkerman JW. IRS-1 mediates inhibition of Ca2+ mobilization by insulin via the inhibitory G-protein Gi. J Biol Chem. 2004;279:3254–64.
8. Rao AK, Freishtat RJ, Jalagadugula G, Singh A, Mao G, Wiles A, Cheung P, Boden G. Alterations in insulin-signaling and coagulation pathways in platelets during hyperglycemia-hyperinsulinemia in healthy non-diabetic subject. Thromb Res. 2014;134(3):704–10. https://doi.org/10.1016/j.thromres.2014.06.029
9. Ferreira IA, Mocking AI, Feijge MA, Gorter G, van Haeften TW, Heemskerk JW, Akkerman JW. Platelet inhibition by insulin is absent in type 2 diabetes mellitus. Arterioscler Thromb Vasc Biol. 2006;26:417–22.
10. Davì G, Catalano I, Averna M, Notarbartolo A, Strano A, Ciabattoni G, Patrono C. Thromboxane biosynthesis and platelet function in type II diabetes mellitus. N Engl J Med. 1990;322:1769–74.
11. Li Y, Woo V, Bose R. Platelet hyperactivity and abnormal Ca(2+) homeostasis in diabetes mellitus. Am J Physiol Heart Circ Physiol. 2001;280:H1480–9.
12. Zheng Y, Wang L, Zhu Z, Yan X, Zhang L, Xu P, Luo D. Altered platelet calsequestrin abundance, Na+/Ca2+ exchange and Ca2+ signaling responses with the progression of diabetes mellitus. Thromb Res. 2014;134:674–81.
13. Angiolillo DJ, Suryadevara S. Aspirin and clopidogrel: efficacy and resistance in diabetes mellitus. Best Pract Res Clin Endocrinol Metab. 2009;23:375–88.
14. Ferretti G, Rabini RA, Bacchetti T, Vignini A, Salvolini E, Ravaglia F, Curatola G, Mazzanti L. Glycated low density lipoproteins modify platelet properties: a compositional and functional study. J Clin Endocrinol Metab. 2002;87:2180–4.
15. Ha H, Lee HB. Oxidative stress in diabetic nephropathy: basic and clinical information. Curr Diab Rep. 2001;1:282–7.
16. Davì G, Falco A, Patrono C. Determinants of F2-isoprostane biosynthesis and inhibition in man. Chem Phys Lipids. 2004;128:149–63.
17. Hess K. The vulnerable blood. Coagulation and clot structure in diabetes mellitus. Hamostaseologie. 2015;35(1):25–33. https://doi.org/10.5482/HAMO-14-09-0039
18. De Vriese AS, Verbeuren TJ, Van de Voorde J, Lameire NH, Vanhoutte PM. Endothelial dysfunction in diabetes. Br J Pharmacol. 2000;130:963–74.
19. Ishida K, Taguchi K, Matsumoto T, Kobayashi T. Activated platelets from diabetic rats cause endothelial dysfunction by decreasing Akt/endothelial NO synthase signaling pathway. PLoS One. 2014;9(7). doi:https://doi.org/10.1371/journal.pone.0102310

20. Collet JP, Allali Y, Lesty C, Tanguy ML, Silvain J, Ankri A, Blanchet B, Dumaine R, Gianetti J, Payot L, Weisel JW, Montalescot G. Altered fibrin architecture is associated with hypofibrinolysis and premature coronary atherothrombosis. Arterioscler Thromb Vasc Biol. 2006;26:2567–73.
21. Breitenstein A, Tanner FC, Lüscher TF. Tissue factor and cardiovascular disease. Circ J. 2010;74:3–12.
22. Napoleone E, Di Santo A, Lorenzet R. Monocytes upregulate endothelial cell expression of tissue factor: a role for cell-cell contact and cross-talk. Blood. 1997;89:541–9.
23. Steffel J, Hermann M, Greutert H, Gay S, Lüscher TF, Ruschitzka F, Tanner FC. Celecoxib decreases endothelial tissue factor expression through inhibition of c-Jun terminal NH2 kinase phosphorylation. Circulation. 2005;111:1685–9.
24. Drake TA, Hannani K, Fei HH, Lavi S, Berliner JA. Minimally oxidized low-density lipoprotein induces tissue factor expression in cultured human endothelial cells. Am J Pathol. 1991;138:601–7.
25. Eto M, Kozai T, Cosentino F, Joch H, Lüscher TF. Statin prevents tissue factor expression in human endothelial cells: role of Rho/Rho-kinase and Akt pathways. Circulation. 2002;105:1756–9.
26. Steffel J, Akhmedov A, Greutert H, Lüscher TF, Tanner FC. Histamine induces tissue factor expression: implications for acute coronary syndromes. Circulation. 2005;112:341–9.
27. Bouchard BA, Gissel MT, Whelihan MF, Mann KG, Butenas S. Platelets do not express the oxidized or reduced forms of tissue factor. Biochim Biophys Acta. 2014;1840:1188–93.
28. Østerud B, Olsen JO. Human platelets do not express tissue factor. Thromb Res. 2013;132:112–5.
29. Müller I, Klocke A, Alex M, Kotzsch M, Luther T, Morgenstern E, Zieseniss S, Zahler S, Preissner K, Engelmann B. Intravascular tissue factor initiates coagulation via circulating microvesicles and platelets. FASEB J. 2003;17:476–8.
30. Panes O, Matus V, Sáez CG, Quiroga T, Pereira J, Mezzano D. Human platelets synthesize and express functional tissue factor. Blood. 2007;109:5242–50.
31. Vignoli A, Giaccherini C, Marchetti M, Verzeroli C, Gargantini C, Da Prada L, Giussani B, Falanga A. Tissue factor expression on platelet surface during preparation and storage of platelet concentrates. Transfus Med Hemother. 2013;40:126–32.
32. Samad F, Pandey M, Loskutoff DJ. Tissue factor gene expression in the adipose tissues of obese mice. Proc Natl Acad Sci USA. 1998;95:7591–6.
33. Boden G, Vaidyula VR, Homko C, Cheung P, Rao AK. Circulating tissue factor procoagulant activity and thrombin generation in patients with type 2 diabetes: effects of insulin and glucose. J Clin Endocrinol Metab. 2007;92:4352–8.
34. Wang J, Ciaraldi TP, Samad F. Tissue factor expression in obese type 2 diabetic subjects and its regulation by antidiabetic agents. J Obes. 2015;2015:291209. https://doi.org/10.1155/2015/291209
35. Stegenga ME, van der Crabben SN, Levi M, de Vos AF, Tanck MW, Sauerwein HP, van der Poll T. Hyperglycemia stimulates coagulation, whereas hyperinsulinemia impairs fibrinolysis in healthy humans. Diabetes. 2006;55:1807–12.
36. Singh A, Boden G, Rao AK. Tissue factor and Toll-like receptor (TLR)4 in hyperglycaemia-hyperinsulinaemia. Effects in healthy subjects, and type 1 and type 2 diabetes mellitus. Thromb Haemost. 2015;113(4):750–8. https://doi.org/10.1160/TH14-10-0884
37. Gerrits AJ, Koekman CA, van Haeften TW, Akkerman JW. Platelet tissue factor synthesis in type 2 diabetes patients is resistant to inhibition by insulin. Diabetes. 2010;59:1487–95.
38. Min C, Kang E, Yu S, Shinn SH, Kim YS. Advanced glycation end products induce apoptosis and procoagulant activity in cultured human umbilical vein endothelial cells. Diabetes Res Clin Pract. 1999;46:197–202.
39. Meade TW, Mellows S, Brozovic M, Miller GJ, Chakrabarti RR, North WR, Haines AP, Stirling Y, Imeson JD, Thompson SG. Haemostatic function and ischaemic heart disease: principal results of the Northwick Park Heart Study. Lancet. 1986;2:533–7.

40. Kario K, Miyata T, Sakata T, Matsuo T, Kato H. Fluorogenic assay of activated factor VII. Plasma factor VIIa levels in relation to arterial cardiovascular diseases in Japanese. Arterioscler Thromb. 1994;14:265–74.
41. Folsom AR, Wu KK, Rosamond WD, Sharrett AR, Chambless LE. Prospective study of hemostatic factors and incidence of coronary heart disease: the Atherosclerosis Risk in Communities (ARIC) Study. Circulation. 1997;96:1102–8.
42. Green D, Foiles N, Chan C, Schreiner P, Liu K. Elevated fibrinogen levels and subsequent subclinical atherosclerosis: the CARDIA Study. Atherosclerosis. 2009;202:623–31.
43. Heinrich J, Balleisen L, Schulte H, Assmann G, van de Loo J. Fibrinogen and factor VII in the prediction of coronary risk. Results from the PROCAM study in healthy men. Arterioscler Thromb. 1994;14:54–9.
44. Klein OL, Okwuosa T, Chan C, Schreiner P, Kanaya AM, Liu K, Green D. Changes in procoagulants track longitudinally with insulin resistance: findings from the coronary artery risk development in young adults (CARDIA) study. Diabet Med. 2014;31:462–5.
45. Bruckert E, Carvalho de Sousa J, Giral P, Soria C, Chapman MJ, Caen J, de Gennes JL. Interrelationship of plasma triglyceride and coagulant factor VII levels in normotriglyceridemic hypercholesterolemia. Atherosclerosis. 1989;75:129–34.
46. Mansfield MW, Heywood DM, Grant PJ. Circulating levels of factor VII, fibrinogen, and von Willebrand factor and features of insulin resistance in first-degree relatives of patients with NIDDM. Circulation. 1996;94:2171–6.
47. Vambergue A, Rugeri L, Gaveriaux V, Devos P, Martin A, Fermon C, Fontaine P, Jude B. Factor VII, tissue factor pathway inhibitor, and monocyte tissue factor in diabetes mellitus: influence of type of diabetes, obesity index, and age. Thromb Res. 2001;101:367–75.
48. Patrassi GM, Vettor R, Padovan D, Girolami A. Contact phase of blood coagulation in diabetes mellitus. Eur J Clin Invest. 1982;12(4):307–11. Retrieved from http://www.ncbi.nlm.nih.gov/pubmed/6814921
49. Barillari G, Fabbro E, Pasca S, Bigotto E. Coagulation and oxidative stress plasmatic levels in a type 2 diabetes population. Blood Coagul Fibrinolysis. 2009;20(4):290–6. https://doi.org/10.1097/MBC.0b013e328329e49b
50. Vischer UM. von Willebrand factor, endothelial dysfunction, and cardiovascular disease. J Thromb Haemost. 2006;4:1186–93.
51. Frankel DS, Meigs JB, Massaro JM, Wilson PW, O'Donnell CJ, D'Agostino RB, Tofler GH. Von Willebrand factor, type 2 diabetes mellitus, and risk of cardiovascular disease: the framingham offspring study. Circulation. 2008;118:2533–9.
52. Kistorp C, Chong AY, Gustafsson F, Galatius S, Raymond I, Faber J, Lip GY, Hildebrandt P. Biomarkers of endothelial dysfunction are elevated and related to prognosis in chronic heart failure patients with diabetes but not in those without diabetes. Eur J Heart Fail. 2008;10:380–7.
53. Rumley A, Lowe GD, Sweetnam PM, Yarnell JW, Ford RP. Factor VIII, von Willebrand factor and the risk of major ischaemic heart disease in the Caerphilly Heart Study. Br J Haematol. 1999;105:110–6.
54. Hess K, Grant PJ. Inflammation and thrombosis in diabetes. Thromb Haemost. 2011;105 (Suppl):S43–54.
55. Grant PJ. Diabetes mellitus as a prothrombotic condition. J Intern Med. 2007;262:157–72.
56. Mahendra JV, Kumar SD, Anuradha TS, Talikoti P, Nagaraj RS, Vishali V. Plasma fibrinogen in type 2 diabetic patients with metabolic syndrome and its relation with ischemic heart disease (IHD) and retinopathy. J Clin Diagn Res. 2015;9(1):21. https://doi.org/10.7860/JCDR/2015/10712.5449
57. Neergaard-Petersen S, Hvas AMM, Kristensen SD, Grove EL, Larsen SB, Phoenix F, Kurdee Z, Grant PJ, Ajjan RA. The influence of type 2 diabetes on fibrin clot properties in patients with coronary artery disease. Thromb Haemost. 2014;112(6):1142–50. https://doi.org/10.1160/TH14-05-0468
58. Dunn EJ, Ariëns RA. Fibrinogen and fibrin clot structure in diabetes. Herz. 2004;29:470–9.

59. Jacquemin B, Antoniades C, Nyberg F, Plana E, Müller M, Greven S, Salomaa V, Sunyer J, Bellander T, Chalamandaris AG, Pistelli R, Koenig W, Peters A. Common genetic polymorphisms and haplotypes of fibrinogen alpha, beta, and gamma chains affect fibrinogen levels and the response to proinflammatory stimulation in myocardial infarction survivors: the AIRGENE study. J Am Coll Cardiol. 2008;52:941–52.
60. Dunn EJ, Ariëns RA, Grant PJ. The influence of type 2 diabetes on fibrin structure and function. Diabetologia. 2005;48:1198–206.
61. Lütjens A, te Velde AA, vd Veen EA, vd Meer J. Glycosylation of human fibrinogen in vivo. Diabetologia. 1985;28:87–9.
62. Pieters M, Covic N, van der Westhuizen FH, Nagaswami C, Baras Y, Toit Loots D, Jerling JC, Elgar D, Edmondson KS, van Zyl DG, Rheeder P, Weisel JW. Glycaemic control improves fibrin network characteristics in type 2 diabetes – a purified fibrinogen model. Thromb Haemost. 2008;99:691–700.
63. Kohler HP, Stickland MH, Ossei-Gerning N, Carter A, Mikkola H, Grant PJ. Association of a common polymorphism in the factor XIII gene with myocardial infarction. Thromb Haemost. 1998;79:8–13.
64. Wang G, Zou Z, Ji X, Ni Q, Ma Z. Factor XIII-A Val34Leu polymorphism might be associated with myocardial infarction risk: an updated meta-analysis. Int J Clin Exp Med. 2014;7 (12):5547–52. Retrieved from http://www.ncbi.nlm.nih.gov/pubmed/25664069
65. Hethershaw EL, Cilia La Corte AL, Duval C, Ali M, Grant PJ, Ariëns RA, Philippou H. The effect of blood coagulation factor XIII on fibrin clot structure and fibrinolysis. J Thromb Haemost. 2014;12:197–205.
66. Mansfield MW, Kohler HP, Ariëns RA, McCormack LJ, Grant PJ. Circulating levels of coagulation factor XIII in subjects with type 2 diabetes and in their first-degree relatives. Diabetes Care. 2000;23:703–5.
67. Schneider DJ, Nordt TK, Sobel BE. Attenuated fibrinolysis and accelerated atherogenesis in type II diabetic patients. Diabetes. 1993;42:1–7.
68. Brazionis L, Rowley K, Jenkins A, Itsiopoulos C, O'Dea K. Plasminogen activator inhibitor-1 activity in type 2 diabetes: a different relationship with coronary heart disease and diabetic retinopathy. Arterioscler Thromb Vasc Biol. 2008;28:786–91.
69. Nikolajsen CL, Scavenius C, Enghild JJ. Human complement C3 is a substrate for transglutaminases. A functional link between non-protease-based members of the coagulation and complement cascades. Biochemistry. 2012;51(23):4735–42. https://doi.org/10.1021/bi3004022
70. Richardson VR, Schroeder V, Grant PJ, Standeven KF, Carter AM. Complement C3 is a substrate for activated factor XIII that is cross-linked to fibrin during clot formation. Br J Haematol. 2013;160(1):116–9. https://doi.org/10.1111/bjh.12096
71. Hess K, Alzahrani SH, Mathai M, Schroeder V, Carter AM, Howell G, Koko T, Strachan MW, Price JF, Smith KA, Grant PJ, Ajjan RA. A novel mechanism for hypofibrinolysis in diabetes: the role of complement C3. Diabetologia. 2012;55(4):1103–13. https://doi.org/10.1007/s00125-011-2301-7
72. Hess K, Alzahrani SH, Price JF, Strachan MW, Oxley N, King R, Gamlen T, Schroeder V, Baxter PD, Ajjan RA. Hypofibrinolysis in type 2 diabetes: the role of the inflammatory pathway and complement C3. Diabetologia. 2014;57:1737–41.
73. Howes JM, Richardson VR, Smith KA, Schroeder V, Somani R, Shore A, Hess K, Ajjan R, Pease RJ, Keen JN, Standeven KF, Carter AM. Complement C3 is a novel plasma clot component with anti-fibrinolytic properties. Diab Vasc Dis Res. 2012;9(3):216–25. https://doi.org/10.1177/1479164111432788
74. Brown GE, Ritter LS, McDonagh PF, Cohen Z. Functional enhancement of platelet activation and aggregation by erythrocytes: role of red cells in thrombosis. Peer J PrePrints. 2014;2: e351v351.

75. Wohner N. Role of cellular elements in thrombus formation and dissolution. Cardiovasc Hematol Agents Med Chem. 2008;6(3):224–8. Retrieved from http://www.ncbi.nlm.nih.gov/pmc/articles/PMC2496953/
76. Gersh KC, Nagaswami C, Weisel JW. Fibrin network structure and clot mechanical properties are altered by incorporation of erythrocytes. Thromb Haemost. 2009;102(6):1169–75. https://doi.org/10.1160/TH09-03-0199.
77. Soma P, Pretorius E. Interplay between ultrastructural findings and atherothrombotic complications in type 2 diabetes mellitus. Cardiovasc Diabetol. 2015;14:96. https://doi.org/10.1186/s12933-015-0261-9
78. Schwartz RS, Madsen JW, Rybicki AC, Nagel RL. Oxidation of spectrin and deformability defects in diabetic erythrocytes. Diabetes. 1991;40(6):701–8. Retrieved from http://www.ncbi.nlm.nih.gov/pubmed/2040386
79. Singh M, Shin S. Changes in erythrocyte aggregation and deformability in diabetes mellitus: a brief review. Indian J Exp Biol. 2009;47(1):7–15. Retrieved from http://www.ncbi.nlm.nih.gov/pubmed/19317346
80. Pretorius E. The adaptability of red blood cells. Cardiovasc Diabetol. 2013;12:63. https://doi.org/10.1186/1475-2840-12-63
81. Feng B, Chen Y, Luo Y, Chen M, Li X, Ni Y. Circulating level of microparticles and their correlation with arterial elasticity and endothelium-dependent dilation in patients with type 2 diabetes mellitus. Atherosclerosis. 2010;208:264–9.
82. Puddu P, Puddu GM, Cravero E, Muscari S, Muscari A. The involvement of circulating microparticles in inflammation, coagulation and cardiovascular diseases. Can J Cardiol. 2010;26:140–5.
83. Diamant M, Tushuizen ME, Sturk A, Nieuwland R. Cellular microparticles: new players in the field of vascular disease? Eur J Clin Investig. 2004;34:392–401.
84. Cai X, Hagedorn CH, Cullen BR. Human microRNAs are processed from capped, polyadenylated transcripts that can also function as mRNAs. RNA. 2004;10:1957–66.
85. Ratajczak J, Miekus K, Kucia M, Zhang J, Reca R, Dvorak P, Ratajczak MZ. Embryonic stem cell-derived microvesicles reprogram hematopoietic progenitors: evidence for horizontal transfer of mRNA and protein delivery. Leukemia. 2006;20:847–56.
86. Semeraro F, Ammollo CT, Morrissey JH, Dale GL, Friese P, Esmon NL, Esmon CT. Extracellular histones promote thrombin generation through platelet-dependent mechanisms: involvement of platelet TLR2 and TLR4. Blood. 2011;118:1952–61.
87. Stakos DA, Kambas K, Konstantinidis T, Mitroulis I, Apostolidou E, Arelaki S, Tsironidou V, Giatromanolaki A, Skendros P, Konstantinides S, Ritis K. Expression of functional tissue factor by neutrophil extracellular traps in culprit artery of acute myocardial infarction. Eur Heart J. 2015;7(22):1405–14.
88. Wong SL, Demers M, Martinod K, Gallant M, Wang Y, Goldfine AB, Kahn CR, Wagner DD. Diabetes primes neutrophils to undergo NETosis, which impairs wound healing. Nat Med. 2015;21(7):815–9. https://doi.org/10.1038/nm.3887
89. Joshi MB, Lad A, Bharath Prasad AS, Balakrishnan A, Ramachandra L, Satyamoorthy K. High glucose modulates IL-6 mediated immune homeostasis through impeding neutrophil extracellular trap formation. FEBS Lett. 2013;587(14):2241–6. https://doi.org/10.1016/j.febslet.2013.05.053
90. Menegazzo L, Ciciliot S, Poncina N, Mazzucato M, Persano M, Bonora B, Albiero M, Vigili de Kreutzenberg S, Avogaro A, Fadini GP. NETosis is induced by high glucose and associated with type 2 diabetes. Acta Diabetol. 2015;52(3):497–503. https://doi.org/10.1007/s00592-014-0676-x

Microparticles: Surrogate Markers and Promoters of Cardiovascular Diseases

8

Martin Moser and Philipp Diehl

Abstract

Microparticles are small cell vesicles which are released from several different cells e.g. during cell activation and cellular stress and that can be quantified using flow cytometry. Several studies have found that circulating microparticles can be used as biomarkers indicating the state of activation of the corresponding maternal cells. However, there is strong evidence that besides their diagnostic value mircoparticles furthermore function as circulating vectors transferring biological information from the cells they initially were released to distinct target cells.

This chapter aims to briefly summarize the role of microparticles in cardiovascular diseases such as coronary heart diseases, arterial hypertension, or atherosclerosis.

Contents

M. Moser, MD • P. Diehl, MD, PhD (✉)
Department of Cardiology and Angiology I, Heart Center, Freiburg University, Hugstetterstr. 55, 79106 Freiburg, Germany
e-mail: philipp.diehl@universitaets-herzzentrum.de

A. Zirlik et al. (eds.), *Platelets, Haemostasis and Inflammation*,
Cardiac and Vascular Biology 5, https://doi.org/10.1007/978-3-319-66224-4_8

8.1 Biology of Microparticles

8.1.1 Introduction

Microparticles (MP) are small cell vesicles that can be released by many different cells, such as blood cells and vascular cells but also tumor cells during multiple stress conditions [1]. The main stimulus leading to MP release is cell activation, but it is quite likely that several other stimuli induce MP release, too [2]. Microparticles contain cytoplasm, RNA molecules, as well as surface receptors of their parental cells [1]. Thus, they can be associated by their specific surface receptors with the cell type they initially were released from [3].

MP were first described in plasma by Wolf et al. over 50 years ago [4]. In the initial microparticle studies, it was suggested that they are a kind of cellular debris without any known pathophysiologic function. However, further studies found that they do have strong pro-inflammatory and procoagulatory effects in circulation and are elevated in conditions of strong platelet activation and systemic inflammation [5, 6]. Within the last two decades, the numbers of publications investigating microparticles in different diseases have continuously grown and microparticles have turned into the focus of cardiovascular research [7].

This chapter aims to review current literature of microparticles in cardiovascular disease. Due to the overwhelming number of published microparticle studies, only few of them are discussed here.

8.1.2 MP as Circulating Biological Vectors

The main stimulus leading to microparticle release is cellular stress and cell activation. Thus, microparticles are elevated in multiple pathophysiologic conditions that are associated with platelet, leukocyte, and endothelial cell activation, such as thrombosis, acute systemic inflammation, or endothelial cell dysfunction [6, 8, 9]. Due to the fact that platelets are the most abundant cell type in circulation, platelet microparticles (PMP) have the highest concentration of all microparticle types in the blood.

The stimuli that lead to MP shedding are quite complex and only partially understood. However, there is evidence that MP release starts with a Ca^{2+} influx into the maternal cell causing a deactivation of the enzyme flippase, which is responsible for the asymmetric distribution of phosphatidylserine (PS) in the lipid bilayer of the cell, and an activation of calpain, which is a Ca^{2+}-dependent proteolytic enzyme. As a result, the cell releases cellular blebs with an outer membrane rich of phosphatidylserine (PS), which is a typical characteristic of microparticles [2]. Phosphatidylserine is a phospholipid with a negative charge that on the one hand interacts with the plasmatic coagulation system causing strong procoagulatory effects and on the other hand allows detection of microparticles using annexin V in flow cytometry [10].

As discussed below, microparticles can be used as surrogate markers for blood cell activation in several diseases [7]. However, it has become evident that microparticles are furthermore circulating biological vectors. Once released from their maternal cells into circulation, microparticles can be purified from blood for gene expression analysis, binding assays, or proteomic studies [11, 12]. We and others have found that microparticles have the ability to bind and fuse with distinct to target cells [1, 11]. There is evidence that the adhesion of circulating microparticles to their destination cells is at least to some extent receptor mediated and thus specific. Jy et al. investigated the binding behavior of endothelial microparticles to leukocyte subpopulations in vitro [13]. They found that EMP bind to monocytes and activated those. Blockage of CD54 reduced binding of EMP to leukocytes by approximately 80%. This impressive study shows that the effect of microparticles to their target cells can be therapeutically inhibited by specific antibodies. That microparticles affect the phenotype of specific target cells was also found by Barry and colleagues [14]. They showed that microparticles released from platelets increase adhesion of monocytes to endothelial cells in a dose-dependent manner thus giving new evidence for the pro-inflammatory potency of platelet microparticles. That microparticles affect the phenotype of their target cells was further confirmed by Sabatier et al. [15]. They incubated endothelial microparticles with monocytotic cells and found that they were transformed into a procoagulatory state. This interaction between EMP and monocytotic cells was inhibited by blocking intercellular adhesion molecule 1 (ICAM-19 on EMP and β_2 on target cells).

In conclusion, microparticles are biological, circulating vectors that can bind and fuse with distinct target cells influencing their phenotype far away from the location of their initial release. First studies were able to show that the interaction of microparticles to their target cells can be inhibited by specific receptor blockage. However, future studies will need to address the question whether the effect of microparticles on disease progressions can be inhibited by blocking MP surface receptors.

8.2 Microparticles in Cardiovascular Disease

Vascular inflammation is a strong promoter of several cardiovascular diseases, such as atherosclerosis, myocarditis, valve disease, heart failure, and pulmonary hypertension [16–20]. As microparticles are surrogate markers for vascular inflammation, increased levels of different microparticle types have been described in several cardiovascular diseases [7]. The aim of the following sections is to summarize the role of microparticles as surrogate markers in atherosclerosis, myocardial infarction, heart failure, arterial hypertension, and after cardiopulmonary resuscitation.

8.2.1 Atherosclerosis

Activated platelets and leukocytes as well as inflamed endothelial cells play a major role in the etiology of atherosclerosis [21]. Therefore, several studies have investigated whether atherosclerosis can be predicted by circulating microparticles released from platelets, leukocytes, and endothelial cells. Chironi et al. suggested that circulating microparticles might be increased in patients without clinical signs for atherosclerosis but with early atherosclerotic lesions [22]. They measured different types of microparticles in 216 patients without cardiovascular diseases and found that the amount of circulating $CD11a^+$ leukocyte microparticles correlated well with the extent of subclinical atherosclerosis as detected with ultrasound of the carotid arteries, the abdominal aorta, and femoral arteries. This data suggests that leukocyte microparticles might be potential surrogate markers detecting subclinical stages of atherosclerosis.

Numbers of activated platelets are increased in patients with atherosclerosis [23]. Zeiger et al. investigated if increased platelet activation is associated with enhanced levels of platelet microparticles in patients with peripheral artery disease (PAD) [24]. Measuring platelet microparticles in 50 healthy subjects and 50 PAD patients, they found increased PMP in patients with peripheral artery disease. However, this study was not designed to correlate microparticle numbers with the severity of PAD. This question was addressed by Tan and colleagues who presumed that microparticles released from activated platelets ($CD61^+$, $CD42b^+$) might be increased in PAD patients and furthermore correlate with the extent of peripheral artery disease, too [25]. Measuring circulating platelet microparticles in 30 healthy controls, 36 patients with moderate PAD, and 23 patients with severe peripheral artery disease, they found that PMP are generally increased in patients with PAD and furthermore correlate with the disease severity. However, future studies need to assess whether increased levels of circulating microparticles are the underlying cause or the effect of atherosclerosis.

Ischemic stroke is one of the most feared complications of patients with progressive atherosclerosis and carotid artery stenosis [26]. Many patients with ischemic stroke did not know in advance that they suffer from atherosclerotic stenosis with vulnerable plaque of their carotid arteries. It is therefore of great clinical interest to establish diagnostic tests that allow differentiation of patients with stable carotid plaque from those with unstable plaque and a high risk for ischemic stroke consecutively. Sarlon-Bartoli et al. investigated in 42 patients with >70% carotid artery stenosis before and after thrombendarterectomy if $CD11b^+/CD66b^+$ leukocyte microparticles predict carotid plaque instability [27]. They found that numbers of leukocyte-derived microparticles in blood samples of patients with unstable carotid artery stenosis were significantly higher than those of blood samples of patients with stable carotid artery stenosis. Hence, numbers of circulating $CD11b^+/CD66b^+$ microparticles in patients with high-grade carotid artery stenosis might help in the future to distinguish between those that benefit from thrombendarterectomy versus those that can be treated medically.

Acute occlusion of a cerebral artery results in cerebral ischemia and hypoxia distal to the vessel occlusion leading to endothelial injury and ischemic stroke. Simak and colleagues hypothesized that this endothelial injury after acute cerebral vessel occlusion results in increased levels of circulating endothelial microparticles [28]. To confirm this hypothesis, they measured several EMP phenotypes in 20 patients with mild stroke and compared data with 21 patients suffering from moderate to severe ischemic stroke. It was found that circulating EMP were significantly increased in patients with moderate to severe stroke in comparison to mild ischemic stroke. Furthermore, EMP correlated with the brain lesion volume as assessed with MRI. Thus, circulating endothelial microparticles may be useful predictors for the size of cerebral injury in patients with ischemic stroke.

8.2.2 Coronary Artery Disease and Acute Coronary Syndrome

After it had been found that circulating microparticles are surrogate markers for atherosclerosis, strong effort was undertaken to investigate whether blood MP also can be used to diagnose coronary artery disease [29]. One of the early studies in which microparticles were assessed in CAD patients was performed by Koga et al., in which $CD144^+$ endothelial microparticles were found in patients with diabetes mellitus and coronary artery disease [30]. These data were confirmed by Werner and colleagues having found that $CD31^+$/annexin V^+ microparticles correlate with endothelial cell dysfunction in patients with coronary artery disease [31].

After it had been shown that different phenotypes of microparticles can be detected in the blood of patients with stable coronary artery diseases, Sinning et al. aimed to assess whether microparticles also predict clinical outcomes of CAD patients. They measured circulating $CD31^+$/annexin V^+ microparticles in patients with stable coronary artery disease and performed a follow-up for major adverse cardiovascular and cerebral events (MACCE)-free survival approximately 6 years later [32]. It was found that patients with initially increased levels for $CD31^+$/annexin V^+ microparticles suffered significantly more often from MACCE, indicating that this endothelial microparticle subtype can be used as a diagnostic marker predicting the outcome of patients with stable coronary artery disease.

From a clinical perspective, it is most important not only to detect CAD but to assess the risk for acute coronary syndromes (ACS) in patients with coronary artery diseases. Therefore, several studies have tried to correlate numbers and types of microparticles with the vulnerability of atherosclerotic plaques and with the risk for acute coronary syndrome respectively. Bernal-Mizrachi et al. hypothesized that endothelial activation of coronary artery disease might be reflected by circulating endothelial cells [33]. They therefore measured circulating endothelial microparticles as defined by the surface markers $CD31^+$ and $CD51^+$ in patients with different stages of coronary artery diseases. It was found that both types of endothelial microparticles were higher in patients with CAD than in control patients indicating increased endothelial cell activation in CAD. However, most interestingly $CD31^+$ microparticles discriminated patients with stable angina pectoris from those with an acute coronary syndrome. In the same line of argumentation, Min et al. investigated numbers of microparticles in stable CAD patients and correlated them with the extent of necrotic cores as assessed by virtual histology intravascular ultrasound (VH-IVUS) [34]. They found that circulating microparticles correlated well with a high content of necrotic cores, suggesting that circulating microparticles might be surrogate markers for vulnerability of atherosclerotic plaques.

As the underlying mechanism of a myocardial infarction is most often a prothrombotic condition with platelet clot formation, vessel occlusion, and downstream located ischemia, platelet function is enhanced in patients with acute myocardial infarction [35]. Taking into account that microparticles are surrogate markers for cell activation and increased in patients with coronary artery diseases, it was not particularly surprising that procoagulant microparticles are elevated in patients with myocardial infarction, too [36, 37]. It can be suggested that besides their role as diagnostics markers, microparticles released during acute myocardial infarction have pathophysiological effects on the circulatory system. Boulanger et al. quantified microparticles in AMI patients and found that they impair the endothelial NO pathway in endothelial cells of rat aortic rings and therefore might contribute to the vasomotor dysfunction that often can be observed in patients with acute myocardial infarction [38]. However, future studies will need to investigate whether a therapeutic inhibition of MP release is associated with a better clinical outcome of patients with acute myocardial infarction.

8.2.3 Chronic Heart Failure

Chronic heart failure (CHF) is characterized by a reduced ventricular pump function and is one of the leading causes of cardiovascular death worldwide. Recent data indicate that HF is associated with an impairment of the vascular/endothelial system causing an increased mortality risk [39–41]. Hence, there is a need for circulating biomarker that quantifies this endothelial dysfunction in heart failure patients noninvasively.

Hypothesizing that endothelial microparticles might reflect the endothelial dysfunction in patients with heart failure, Nozaki et al. measured $CD144^+$ EMP in 169 HF patients [42]. They found that HF patients with high EMP levels more often developed cardiovascular events than those with less circulating EMP. Additionally, EMP predicted future cardiovascular complications in this patient group. Berezin and colleagues assessed whether different blood parameters predict the clinical outcome in HF patients [43]. They were able to show that regardless of age, gender, and comorbidities, the measurement of NT-pro-BNP, galectin-3, hs-CRP, osteoprotegerin, $CD31^+$/annexin V^+ EMP, and EMP/$CD14^+CD309^+$ MPC ratio predicts the survival of patients with chronic heart failure. However, it still remains unclear whether these microparticles are a consequence of CHF or if they actively influence the disease progression.

8.2.4 Microparticles in Arterial Hypertension

First described by the Framingham Study, arterial hypertension is one of the risk factors that promotes cardiovascular diseases such as atherosclerosis [44]. Several studies have shown in the past that arterial hypertension causes endothelial inflammation leading to enhanced monocyte recruitment to sites of inflammation with consecutive atherosclerotic plaque formation [45, 46]. In order to allow early detection and thus treatment of endothelial dysfunction in patients who have not developed clinical signs of atherosclerosis yet, easy detectable surrogate markers are needed. As activated, inflamed endothelial cells release microparticles in circulation, Preston and colleagues hypothesized that patients with arterial hypertension might have different patterns of circulating microparticles than normotensive controls [47]. Therefore, they measured levels of $CD31^+$/$CD42^-$ endothelial microparticles and CD41$^+$ platelet microparticles in patients with untreated severe arterial hypertension ($n = 24$), mild arterial hypertension ($n = 19$), and normotensive controls. They found that endothelial and platelet microparticles were significantly increased in patients with severe arterial hypertension. Furthermore, endothelial microparticles correlated with the systolic and diastolic blood pressures. Due to their results, Breston et al. suggest that EMP and PMP might be markers as well as mediators of endothelial and platelet activation in arterial hypertension and might presumably promote hypertensive target organ injury.

One of the most common end-organ damages caused by arterial hypertension is hypertensive nephropathy with impaired kidney function consecutively. Hsu et al. presumed that endothelial microparticles are involved in impaired renal function in patients with arterial hypertension [48]. Therefore, they measured EMP and endothelial progenitor cells (EPC) in 100 patients with arterial hypertension and a glomerular filtration rate of $\geq$30 mL min^{-1}/1.73 m^2. They found that the ratio of EMP to EPC was associated with a subsequent decline of the glomerular filtration rate in hypertensive patients. These data underline the importance of endothelial cell dysfunction as quantified by EMP in patients with arterial hypertension. However, as the investigated study population is comparably small, well-powered studies need to confirm their results.

8.2.5 Resuscitation

Acute myocardial infarction can lead to cardiac arrest with the need for cardiopulmonary resuscitation (CPR) [49]. Patients after successful CPR often develop postcardiac arrest syndromes with symptoms of a severe systemic inflammation and endothelial dysfunction [50, 51]. Hypothesizing that this inflammatory response might be associated with enhanced levels of circulating microparticles, Fink et al. investigated the abundance of microparticles in patients after cardiopulmonary resuscitation and found highly increased numbers of microparticles of various origins in CPR patients versus controls [9]. Furthermore, microparticles of patients after resuscitation induced endothelial dysfunction and apoptosis ex vivo and thus might contribute to systemic vascular dysfunction as often found after CPR [10]. Interestingly, selenium treatment reduced ICAM-1- and VCAM-1-related monocyte adhesion induced by plasma microparticles of patients after cardiopulmonary resuscitation [52]. Hence, selenium might impair the pro-inflammatory effects of circulating microparticles on leukocytes and endothelial cells.

8.2.6 Conclusion

Microparticles are surrogate markers for several cardiovascular diseases that are associated with pro-inflammation and procoagulation. However, due to a lack of standardized assays for microparticle detection, results from different MP studies are hard to compare with each other. Besides their function as surrogate markers, microparticles are circulating biovectors transferring cytoplasm, RNA, lipids, and proteins from their maternal cells to the destination cells. Thereby, microparticles can actively change the phenotype and function of cells far away from the location the microparticles were released. First studies have shown that the interaction between microparticles and their target cells are receptor specific and can be inhibited by antibody treatments. However, larger studies are needed showing that decreased interaction between microparticles and their target cells affects disease progression.

Compliance with Ethical Standards

Conflict of Interest: Martin Moser and Philipp Diehl declares that they have no conflict of interest.

Ethical Approval: This article does not contain any studies with human participants or animals performed by any of the authors.

References

1. Yuana Y, Sturk A, Nieuwland R. Extracellular vesicles in physiological and pathological conditions. Blood Rev. 2012;27:31–9.
2. Hugel B, Martinez MC, Kunzelmann C, Freyssinet JM. Membrane microparticles: two sides of the coin. Physiology (Bethesda). 2005;20:22–7.
3. Orozco AF, Lewis DE. Flow cytometric analysis of circulating microparticles in plasma. Cytometry A. 2010;77:502–14.
4. Wolf P. The nature and significance of platelet products in human plasma. Br J Haematol. 1967;13:269–88.
5. Wakefield TW, Myers DD, Henke PK. Mechanisms of venous thrombosis and resolution. Arterioscler Thromb Vasc Biol. 2008;28:387–91.
6. Ogura H, Tanaka H, Koh T, Fujita K, Fujimi S, Nakamori Y, Hosotsubo H, Kuwagata Y, Shimazu T, Sugimoto H. Enhanced production of endothelial microparticles with increased binding to leukocytes in patients with severe systemic inflammatory response syndrome. J Trauma. 2004;56:823–30. discussion 830-821
7. Bank IE, Timmers L, Gijsberts CM, Zhang YN, Mosterd A, Wang JW, Chan MY, De Hoog V, Lim SK, Sze SK, Lam CS, De Kleijn DP. The diagnostic and prognostic potential of plasma extracellular vesicles for cardiovascular disease. Expert Rev Mol Diagn. 2015;15:1577–88.
8. van Es N, Bleker S, Sturk A, Nieuwland R. Clinical significance of tissue factor-exposing microparticles in arterial and venous thrombosis. Semin Thromb Hemost. 2015;41:718–27.
9. Fink K, Schwarz M, Feldbrugge L, Sunkomat JN, Schwab T, Bourgeois N, Olschewski M, von Zur MC, Bode C, Busch HJ. Severe endothelial injury and subsequent repair in patients after successful cardiopulmonary resuscitation. Crit Care. 2010;14:R104.
10. Fink K, Feldbrugge L, Schwarz M, Bourgeois N, Helbing T, Bode C, Schwab T, Busch HJ. Circulating annexin v positive microparticles in patients after successful cardiopulmonary resuscitation. Crit Care. 2011;15:R251.
11. Diehl P, Fricke A, Sander L, Stamm J, Bassler N, Htun N, Ziemann M, Helbing T, El-Osta A, Jowett JB, Peter K. Microparticles: major transport vehicles for distinct micrornas in circulation. Cardiovasc Res. 2012;93:633–44.
12. Mause SF, Weber C. Microparticles: protagonists of a novel communication network for intercellular information exchange. Circ Res. 2010;107:1047–57.

13. Jy W, Minagar A, Jimenez JJ, Sheremata WA, Mauro LM, Horstman LL, Bidot C, Ahn YS. Endothelial microparticles (emp) bind and activate monocytes: elevated emp-monocyte conjugates in multiple sclerosis. Front Biosci. 2004;9:3137–44.
14. Barry OP, Pratico D, Savani RC, FitzGerald GA. Modulation of monocyte-endothelial cell interactions by platelet microparticles. J Clin Invest. 1998;102:136–44.
15. Sabatier F, Roux V, Anfosso F, Camoin L, Sampol J, Dignat-George F. Interaction of endothelial microparticles with monocytic cells in vitro induces tissue factor-dependent procoagulant activity. Blood. 2002;99:3962–70.
16. Libby P, Ridker PM, Maseri A. Inflammation and atherosclerosis. Circulation. 2002;105:1135–43.
17. Parrillo JE. Inflammatory cardiomyopathy (myocarditis): which patients should be treated with anti-inflammatory therapy? Circulation. 2001;104:4–6.
18. Diehl P, Nagy F, Sossong V, Helbing T, Beyersdorf F, Olschewski M, Bode C, Moser M. Increased levels of circulating microparticles in patients with severe aortic valve stenosis. Thromb Haemost. 2008;99:711–9.
19. Adamopoulos S, Parissis J, Kroupis C, Georgiadis M, Karatzas D, Karavolias G, Koniavitou K, Coats AJ, Kremastinos DT. Physical training reduces peripheral markers of inflammation in patients with chronic heart failure. Eur Heart J. 2001;22:791–7.
20. Diehl P, Aleker M, Helbing T, Sossong V, Germann M, Sorichter S, Bode C, Moser M. Increased platelet, leukocyte and endothelial microparticles predict enhanced coagulation and vascular inflammation in pulmonary hypertension. J Thromb Thrombolysis. 2011;31:173–9.
21. Massberg S, Brand K, Gruner S, Page S, Muller E, Muller I, Bergmeier W, Richter T, Lorenz M, Konrad I, Nieswandt B, Gawaz M. A critical role of platelet adhesion in the initiation of atherosclerotic lesion formation. J Exp Med. 2002;196:887–96.
22. Chironi G, Simon A, Hugel B, Del Pino M, Gariepy J, Freyssinet JM, Tedgui A. Circulating leukocyte-derived microparticles predict subclinical atherosclerosis burden in asymptomatic subjects. Arterioscler Thromb Vasc Biol. 2006;26:2775–80.
23. May AE, Seizer P, Gawaz M. Platelets: inflammatory firebugs of vascular walls. Arterioscler Thromb Vasc Biol. 2008;28:s5–10.
24. Zeiger F, Stephan S, Hoheisel G, Pfeiffer D, Ruehlmann C, Koksch M. P-selectin expression, platelet aggregates, and platelet-derived microparticle formation are increased in peripheral arterial disease. Blood Coagul Fibrinolysis. 2000;11:723–8.
25. Tan KT, Tayebjee MH, Lynd C, Blann AD, Lip GY. Platelet microparticles and soluble p selectin in peripheral artery disease: relationship to extent of disease and platelet activation markers. Ann Med. 2005;37:61–6.
26. Goldstein LB, Adams R, Alberts MJ, Appel LJ, Brass LM, Bushnell CD, Culebras A, Degraba TJ, Gorelick PB, Guyton JR, Hart RG, Howard G, Kelly-Hayes M, Nixon JV, Sacco RL. Primary prevention of ischemic stroke: a guideline from the american heart association/american stroke association stroke council: cosponsored by the atherosclerotic peripheral vascular disease interdisciplinary working group; cardiovascular nursing council; clinical cardiology council; nutrition, physical activity, and metabolism council; and the quality of care and outcomes research interdisciplinary working group: The american academy of neurology affirms the value of this guideline. Stroke. 2006;37:1583–633.
27. Sarlon-Bartoli G, Bennis Y, Lacroix R, Piercecchi-Marti MD, Bartoli MA, Arnaud L, Mancini J, Boudes A, Sarlon E, Thevenin B, Leroyer AS, Squarcioni C, Magnan PE, Dignat-George F, Sabatier F. Plasmatic level of leukocyte-derived microparticles is associated with unstable plaque in asymptomatic patients with high-grade carotid stenosis. J Am Coll Cardiol. 2013;62:1436–41.
28. Simak J, Gelderman MP, Yu H, Wright V, Baird AE. Circulating endothelial microparticles in acute ischemic stroke: a link to severity, lesion volume and outcome. J Thromb Haemost. 2006;4:1296–302.
29. Rautou PE, Vion AC, Amabile N, Chironi G, Simon A, Tedgui A, Boulanger CM. Microparticles, vascular function, and atherothrombosis. Circ Res. 2011;109:593–606.

30. Koga H, Sugiyama S, Kugiyama K, Watanabe K, Fukushima H, Tanaka T, Sakamoto T, Yoshimura M, Jinnouchi H, Ogawa H. Elevated levels of ve-cadherin-positive endothelial microparticles in patients with type 2 diabetes mellitus and coronary artery disease. J Am Coll Cardiol. 2005;45:1622–30.
31. Werner N, Wassmann S, Ahlers P, Kosiol S, Nickenig G. Circulating cd31+/annexin v+ apoptotic microparticles correlate with coronary endothelial function in patients with coronary artery disease. Arterioscler Thromb Vasc Biol. 2006;26:112–6.
32. Sinning JM, Losch J, Walenta K, Bohm M, Nickenig G, Werner N. Circulating cd31+/annexin v+ microparticles correlate with cardiovascular outcomes. Eur Heart J. 2011;32:2034–41.
33. Bernal-Mizrachi L, Jy W, Jimenez JJ, Pastor J, Mauro LM, Horstman LL, de Marchena E, Ahn YS. High levels of circulating endothelial microparticles in patients with acute coronary syndromes. Am Heart J. 2003;145:962–70.
34. Min PK, Cho M, Hong SY, Kim JY, Choi EY, Yoon YW, Lee BK, Hong BK, Rim SJ, Kwon HM. Circulating microparticles and coronary plaque components assessed by virtual histology intravascular ultrasound of the target lesion in patients with stable angina. PLoS One. 2016;11: e0148128.
35. Frossard M, Fuchs I, Leitner JM, Hsieh K, Vlcek M, Losert H, Domanovits H, Schreiber W, Laggner AN, Jilma B. Platelet function predicts myocardial damage in patients with acute myocardial infarction. Circulation. 2004;110:1392–7.
36. Morel O, Hugel B, Jesel L, Mallat Z, Lanza F, Douchet MP, Zupan M, Chauvin M, Cazenave JP, Tedgui A, Freyssinet JM, Toti F. Circulating procoagulant microparticles and soluble gpv in myocardial infarction treated by primary percutaneous transluminal coronary angioplasty. A possible role for gpiib-iiia antagonists. J Thromb Haemost. 2004;2:1118–26.
37. Stepien E, Stankiewicz E, Zalewski J, Godlewski J, Zmudka K, Wybranska I. Number of microparticles generated during acute myocardial infarction and stable angina correlates with platelet activation. Arch Med Res. 2012;43:31–5.
38. Boulanger CM, Scoazec A, Ebrahimian T, Henry P, Mathieu E, Tedgui A, Mallat Z. Circulating microparticles from patients with myocardial infarction cause endothelial dysfunction. Circulation. 2001;104:2649–52.
39. Chong AY, Blann AD, Patel J, Freestone B, Hughes E, Lip GY. Endothelial dysfunction and damage in congestive heart failure: relation of flow-mediated dilation to circulating endothelial cells, plasma indexes of endothelial damage, and brain natriuretic peptide. Circulation. 2004;110:1794–8.
40. Katz SD, Hryniewicz K, Hriljac I, Balidemaj K, Dimayuga C, Hudaihed A, Yasskiy A. Vascular endothelial dysfunction and mortality risk in patients with chronic heart failure. Circulation. 2005;111:310–4.
41. Fujisue K, Sugiyama S, Matsuzawa Y, Akiyama E, Sugamura K, Matsubara J, Kurokawa H, Maeda H, Hirata Y, Kusaka H, Yamamoto E, Iwashita S, Sumida H, Sakamoto K, Tsujita K, Kaikita K, Hokimoto S, Matsui K, Ogawa H. Prognostic significance of peripheral microvascular endothelial dysfunction in heart failure with reduced left ventricular ejection fraction. Circ J. 2015;79:2623–31.
42. Nozaki T, Sugiyama S, Sugamura K, Ohba K, Matsuzawa Y, Konishi M, Matsubara J, Akiyama E, Sumida H, Matsui K, Jinnouchi H, Ogawa H. Prognostic value of endothelial microparticles in patients with heart failure. Eur J Heart Fail. 2010;12:1223–8.
43. Berezin AE, Kremzer AA, Martovitskaya YV, Berezina TA, Samura TA. The utility of biomarker risk prediction score in patients with chronic heart failure. Clin Hypertens. 2015;22:3.
44. Kannel WB, Dawber TR, Kagan A, Revotskie N, Stokes J 3rd. Factors of risk in the development of coronary heart disease—six year follow-up experience. The framingham study. Ann Intern Med. 1961;55:33–50.
45. Dharmashankar K, Widlansky ME. Vascular endothelial function and hypertension: insights and directions. Curr Hypertens Rep. 2010;12:448–55.
46. Libby P. Inflammation in atherosclerosis. Nature. 2002;420:868–74.

47. Preston RA, Jy W, Jimenez JJ, Mauro LM, Horstman LL, Valle M, Aime G, Ahn YS. Effects of severe hypertension on endothelial and platelet microparticles. Hypertension. 2003;41:211–7.
48. Hsu CY, Huang PH, Chiang CH, Leu HB, Huang CC, Chen JW, Lin SJ. Increased circulating endothelial apoptotic microparticle to endothelial progenitor cell ratio is associated with subsequent decline in glomerular filtration rate in hypertensive patients. PLoS One. 2013;8: e68644.
49. Ehlenbach WJ, Barnato AE, Curtis JR, Kreuter W, Koepsell TD, Deyo RA, Stapleton RD. Epidemiologic study of in-hospital cardiopulmonary resuscitation in the elderly. N Engl J Med. 2009;361:22–31.
50. Adams JA. Endothelium and cardiopulmonary resuscitation. Crit Care Med. 2006;34: S458–65.
51. Adrie C, Laurent I, Monchi M, Cariou A, Dhainaou JF, Spaulding C. Postresuscitation disease after cardiac arrest: a sepsis-like syndrome? Curr Opin Crit Care. 2004;10:208–12.
52. Fink K, Moebes M, Vetter C, Bourgeois N, Schmid B, Bode C, Helbing T, Busch HJ. Selenium prevents microparticle-induced endothelial inflammation in patients after cardiopulmonary resuscitation. Crit Care. 2015;19:58.

Mechanisms of Platelet Activation in Diabetes Mellitus

9

Florian Willecke, Prabhakara R. Nagareddy, and Andrew J. Murphy

Abstract

Diabetes mellitus is a multifactorial disease that substanially increases the risk for cardiovascular disease. Increased platelet activation has been indentified as a major factor contributing to increased CVD risk in diabetes by enhancing platelet adhesion and aggregation. The exact contribution of factors such as insulin resistance, hyperglycemia, inflammation, and hyperlipidemia is still under investigation. Here, we review these factors and how they contribute to platelet hyperreactivity in patients with diabetes mellitus and highlight possible pharmacological interventions.

Contents

F. Willecke (✉)
Department of Cardiology and Angiology I, University Heart Center Freiburg, 79106 Freiburg, Germany
e-mail: florian.willecke@uniklinik-freiburg.de

P.R. Nagareddy
Department of Internal Medicine, University of Kentucky, Lexington, KY 40514, USA

A.J. Murphy
Haematopoiesis and Leukocyte Biology, Baker IDI Heart and Diabetes Institute, Melbourne, VIC 3004, Australia

A. Zirlik et al. (eds.), *Platelets, Haemostasis and Inflammation*, Cardiac and Vascular Biology 5, https://doi.org/10.1007/978-3-319-66224-4_9

9.1 Introduction

9.1.1 Diabetes Mellitus and Cardiovascular Risk

A seemingly relentless increase in the incidence of diabetes finds us in the midst of a global diabetes epidemic. More than 382 million people are currently affected worldwide, and this number is expected to rise to 592 million by 2035 [1]. Once a disease of Western countries, diabetes has now become an epidemic of developing countries: 80% of people with diabetes live in low- and middle-income countries [1]. Consequently, diabetes-associated micro- and macrovascular complications are rising: Diabetes is a major risk factor for cardiovascular disease (CVD), and CVD is the most common cause of death in people with diabetes mellitus (DM). Although rates of death attributable to CVD have declined in people without diabetes [2], the burden of CVD in those with diabetes remains high, and implementation of preventive strategies is frequently not adequate [3–5]. The substantially increased risk for CVD of patients with diabetes was first highlighted by the landmark study of Haffner et al. who demonstrated that diabetic patients without prior CVD have the same rate of myocardial infarction as nondiabetic subjects who had a previous event [6]. In addition, the presence of CVD in subjects with diabetes increases the rate of all-cause death nearly threefold and the rate of cardiovascular death nearly fivefold compared to nondiabetic subjects [7].

The majority of diabetic patients have evidence of underlying insulin resistance, which is characterized by a reduction in sensitivity to the action of insulin preceding the development of beta-cell failure and hyperglycemia, the latter being a hallmark of diabetes. Unlike the diabetes-specific microvasculopathy, neuropathy, nephropathy, and retinopathy, the macroangiopathic process in patients with diabetes represents an accelerated but pathophysiological process similar to atherosclerosis in nondiabetic subjects. What are the factors that contribute to this accelerating atherosclerosis in diabetic subjects? Patients with DM not only have a greater atheromatous plaque burden but also a thrombotic diathesis that is in part due to changes in the coagulation system with elevated intravascular thrombin formation, increased levels of plasma fibrinogen, and reduced fibrinolytic potential (see Chap. 7). At the same time, however, platelets from subjects with diabetes display an increased capacity to activate and aggregate after stimulation (platelet hyperreactivity). Besides its acute role in the pathophysiology of myocardial infarction by thrombus formation, platelets contribute to the progression of local vascular lesions by the release of oxidative, constrictive, and mitogenic substances.

Diabetes is a multifactorial disease associated with biochemical factors such as insulin resistance, inflammation, oxidative stress, hyperlipidemia, and hyperglycemia. We aim to review these factors and how they contribute to platelet hyperreactivity in patients with diabetes mellitus (Fig. 9.1).

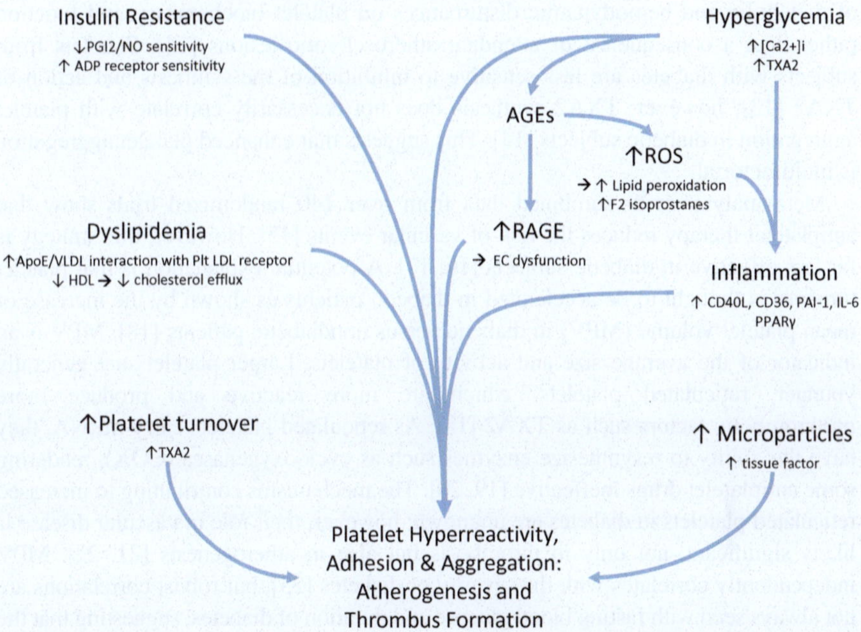

Fig. 9.1 Simplified scheme of pathways involved in platelet hyperreactivity in diabetes mellitus: hyperglycemia and insulin resistance drive platelet activation and thrombus formation through mechanism involving inflammation, dyslipidemia, platelet turnover, ROS production, and increased microparticles. Refer to text for details and abbreviations

9.2 Platelet Activation

9.2.1 Diabetes Mellitus-Induced Thromboxane-Dependent Platelet Activation, Aggregation, and Turnover

Activation and aggregation of platelets is one of the first steps following plaque rupture. In diabetic subjects, platelets display enhanced adhesion, aggregation, and activation [8, 9] as well as decreased platelet sensitivity to anti-aggregating agents (e.g., nitric oxide (NO) and prostacyclin (PGI2)) [10]. After activation, platelets activate prostaglandin synthesis leading to the production of thromboxane A2 (TXA2). TXA2 potentiates platelet aggregation and causes vasoconstriction. In diabetic subjects, platelets produce more TXA2 than platelets from healthy subjects in response to various stimuli [11, 12]. The rate of TXA2 biosynthesis appears to reflect the influence of coexisting disorders such as diabetes mellitus, hypercholesterolemia, and hypertension on platelet biochemistry and function. For example, in an animal model of streptozotocin-induced diabetes, enhanced platelet aggregation and TXA2 synthesis were detected within days of induction of diabetes before the onset of any vascular disease [9]. It has been concluded that platelet activation reflects the influence of metabolic and hemodynamic disturbances on platelet biochemistry and function rather than a consequence of attendant atherosclerotic lesions [14]. Platelets from subjects with diabetes are less sensitive to inhibition of the synthesis and action of TXA2 [13]; however, TXA2 synthesis does not necessarily correlate with platelet aggregation in diabetic subjects [14]. This suggests that enhanced platelet aggregation is multifactorial.

Meta-analyses with combined data from over 140 randomized trials show that antiplatelet therapy reduces the risk of vascular events [15]. However, this strategy is far less effective in diabetic subjects [16, 17]. A potential explanation is that platelet turnover is thought to be accelerated in diabetic patients as shown by the increase of mean platelet volume (MPV) in diabetic versus nondiabetic patients [18]. MPV is an indicator of the average size and activity of platelets. Larger platelets are generally younger, reticulated platelets, which are more reactive and produce more prothrombotic factors such as TXA2 [18]. As reticulated platelets carry mRNA, they have the ability to resynthesize enzymes such as cyclooxygenases (COX), rendering some antiplatelet drugs ineffective [19, 20]. The mechanisms contributing to increased reticulated platelets in diabetes are unknown; however, their role in vascular disease is likely significant, not only in thrombosis but also in atherogenesis [21, 22]. MPV independently correlates with the severity of diabetes [23], but robust correlations are not always seen with fasting blood glucose and duration of diabetes, suggesting that the increase in MPV may be not due to the diabetic state alone. Interestingly, significantly elevated platelet counts in people with diabetes do indeed correlate with CVD [24]. Additionally, it has been shown that the increase of MPV in the late phase of myocardial infarction is an independent predictor for recurrent myocardial infarction [25] and that coronary heart disease is associated with MPV in T2DM patients [26, 27]. Following activation, platelets and endothelial cells—among other cell types—release microparticles (MPs). MPs represent a heterogeneous population of vesicles with a diameter of 100–1000 nm that are released by budding of the plasma membrane and express antigens specific of their parental cells. Elevated circulating MP levels are found in various diseases, including acute coronary syndrome, peripheral artery disease, systemic inflammation, and diabetes. In type 2 diabetes mellitus (T2DM), patients circulating MPs are larger than in healthy subjects and mostly originate from MPs of platelet origin [28]. This increase of MPs is independent of the obesity status [29]. Circulating MPs carry abundant procoagulant tissue factor, thereby likely contributing to thrombus formation at sites of vascular injury [28].

Platelet activation causes changes in the expression of surface glycoproteins (GP), which act as receptors for platelet agonists and for adhesive proteins involved in platelet aggregation. Following platelet activation, P-selectin translocates from the membrane of α-granules to the plasma membrane and the GPIIb/IIIa complex on platelets exposes a fibrinogen binding site, thereby facilitating platelet aggregation through platelet–fibrin and platelet–platelet binding [30]. Platelets of diabetic patients show higher expression of surface receptors (e.g., GPIIb/IIIa) and activation markers (e.g., P-selectin and PCAM-1), thereby facilitating increased binding of von Willebrand factor in diabetic patients [31].

9.2.2 Causes of Platelet Activation

9.2.2.1 Insulin Resistance

About 90% of all DM patients have T2DM, characterized by reduced tissue sensitivity to insulin. Initially, pancreatic β-cells compensate insulin resistance by increasing insulin production. However, in the long term, pancreatic β-cells undergo apoptosis leading to a reduction in β-cells. Hence, the hyperinsulinemia characteristic of early stages of T2DM progressively gives way to relative and absolute insulin deficiency [32]. Insulin regulates platelet function via insulin receptors found on human platelets [33]. Binding of insulin to the insulin receptor leads to activation of the insulin receptor substrate 1 (IRS-1) through tyrosine phosphorylation and its association with the $G_i\alpha$ subunit. This results in the inhibition of $G_i\alpha$ activity and impaired suppression of cyclic adenosine monophosphate (cAMP), thus inhibiting downstream signaling of the adenosine diphosphate receptor $P2Y_{12}$ and reducing platelet activity in vitro [34, 35]. Furthermore, it has been shown in vivo that insulin inhibits platelet interaction with collagen and attenuates the platelet aggregation effect of agonists in healthy individuals [36]. Moreover, insulin increases the surface expression of PGI2 receptors, thereby increasing sensitivity to the anti-aggregating activity of PGI2 [30]. PGI2 and NO are produced by the intact endothelium and retard platelet activation by increasing intraplatelet concentrations of cAMP. In patients with T2DM, platelets lose their responsiveness to insulin leading to increased adhesion, aggregation, and procoagulant activity [37]. Others have reported decreased platelet insulin receptor number and affinity in subjects with T2DM, suggesting that reduced insulin sensitivity may account for platelet hyperactivity in T2DM [38]. With insulin resistance, platelets display reduced sensitivity to NO, further enhancing platelet reactivity [39]. Restoration of insulin sensitivity restores platelet sensitivity to NO and PGI2 [40].

9.2.2.2 Hyperglycemia

Hyperglycemia, resulting from defects in insulin secretion, insulin action, or both, is the diagnostic hallmark finding in diabetes mellitus and plays a significant role in the development of DM-associated CVD and the prothrombotic state [41]. The detrimental effects of glucose already occur with glycemic levels below the threshold for the diagnosis of diabetes. Acute hyperglycemia increases platelet reactivity and markers of platelet activation such as soluble P-selectin. Hyperglycemia leads to increased intracellular Ca^{2+} levels [42] and causes in vivo activation of calcium-sensitive protein kinase C, a mediator for pro-aggregatory platelet agonists [43].

It has been suggested that hyperglycemic spikes may trigger ischemic cardiovascular complications in diabetes mellitus [44]. Improved glycemic control reduces platelet reactivity in T2DM following percutaneous coronary intervention (PCI) [45]. This is of clinical relevance for patients with T2DM and an acute coronary event: intensive glucose-lowering treatment in diabetic patients significantly reduced mortality in acute MI [46]. In addition, in diabetic patients undergoing elective PCI, optimal glycemic control (HbA1c $\leq 7\%$) was associated with a lower rate of restenosis, cardiac rehospitalization, and recurrent angina

[47]. However, more recent clinical studies (ACCORD and NICE-SUGAR) have challenged intensive glucose regimens. Intensive glucose-lowering regimen versus a standard regimen increased mortality in diabetic subjects likely due to an increased rate of hypoglycemia [48, 49].

Increased plasma levels of glucose lead to nonenzymatic protein glycation yielding a heterogeneous class of compounds, collectively termed advanced glycation end products (AGEs) [50]. AGEs may contribute to the development and progression of CVD in diabetes by acting via specific "receptors for AGE" (RAGE) or through other mechanisms. AGEs elicit externalization of phosphatidylserine on platelet membranes leading to activation of surface clotting factor and enhanced thrombogenic state. [51]. The increased glycosylation of platelet membrane proteins in diabetes appears to be related to reduced membrane fluidity [52], which modulates cell function, possibly through alterations in receptor availability and increased sensitivity to agonists [53]. Also, enhanced glycosylation of endothelial proteins quenches endothelial NO production and contributes to reduced platelet inhibition [54].

Hyperglycemia induces mitochondrial dysfunction and endoplasmic reticulum stress, thereby promoting reactive oxygen species (ROS) accumulation. ROS enhance the interaction of sugars with proteins and increase the formation of AGEs. In addition, ROS activates signaling molecules in endothelial cells, including protein kinase C (PKC) and nuclear factor κB (NFκB) leading to enhanced expression of pro-inflammatory and prothrombotic molecules [55]. In DM, production of ROS and potent free radicals enhances platelet activation [56, 57]. Increased ROS production can induce the formation of F2-isoprostanes, such as 8-iso-prostaglandin (PG)F2α, a nonenzymatic oxidation product of circulating LDL and arachidonic acid [58]. In T2DM, enhanced production of 8-iso-PGF2α correlates with the rate of TXA2 biosynthesis and improved metabolic control is associated with a significant reduction in 8-iso-PGF2alpha and 11-dehydro-thromboxane B2 excretion [58].

9.2.2.3 Dyslipidemia

Dyslipidemia is one of the key risk factors for cardiovascular disease (CVD) in DM. The characteristic features of diabetic dyslipidemia are high plasma triglyceride concentration, reduced high-density lipoprotein cholesterol (HDL-C) concentration, and increased concentration of small dense LDL particles. There is a considerable body of evidence supporting an association between dyslipidemia, a hypercoagulable state, and atherothrombosis. Familial hypertriglyceridemia shows increased platelet activation in response to adenosine diphosphate (ADP) and collagen, an effect that might be mediated by the interaction of ApoE on triglyceride-rich VLDL particles with the platelet LDL receptor [59]. VLDL also upregulates expression of the plasminogen activator inhibitor-1 gene and plasminogen activator inhibitor-1 antigen and activity, a process accompanied by platelet aggregation and clot formation [60, 61]. Reconstituted high-density lipoprotein (HDL) attenuates platelet aggregation in individuals with T2DM by promoting cholesterol efflux [62]. In atherosclerotic mice, HDL infusion reduced platelet counts by increasing cholesterol efflux in platelet-generating megakaryocytes

[21]. As discussed above, glycation induces compositional and structural changes in LDL-C, leading to impaired NO production and increased intraplatelet calcium concentration further contributing to platelet hyperactivity [63]. Finally, hypertriglyceridemia and insulin resistance are also features of polycystic ovary syndrome (PCOS). Acute hypertriglyceridemia in patients with PCOS induces platelet hyperactivity but is not attenuated by insulin, implying a more relevant role of triglycerides for platelet activation at least in these patients with PCOS [64].

9.2.2.4 Inflammation

The common soil hypothesis, originally put forward by Stern, suggests that diabetes and CVD are the same condition with common antecedents [65]. Factors like hyperglycemia, insulin resistance, oxidative stress, and dyslipidemia strengthen this association as discussed above. All of the above also trigger low-grade inflammation which is now widely accepted to be one link between insulin resistance, T2DM, and CVD [55, 66]. Inflammation is characterized by increased plasma levels of cytokines, chemokines, and acute-phase proteins, such as C-reactive protein (CRP), and all of these are increased in T2DM patients [67]. Leukocytes induce platelet activation by platelet-activating factor (PAF), a potent phospholipid activator, mediating molecular and cellular interactions between inflammation and platelet activation [68].

CD40L is an inflammatory mediator derived from platelets and expands the functional repertoire of platelets from players of hemostasis and thrombosis to powerful amplifiers of inflammation by promoting the release of cytokines and chemokines, cell activation, and cell–cell interactions [69]. Increased plasma levels of CD40L have been described in both T1DM and T2DM [70] with more than 95% of circulating CD40L deriving from platelets [71]. CD40L signaling increases tissue factor expression [72] and stimulates resting platelets by binding to its constitutively expressed receptor CD40, thereby eliciting prothrombotic and pro-inflammatory responses. Release of CD40L in diabetic patients is likely mediated by AGEs [73]. CD40L, in turn, increases platelet release of ROS through activation of Akt and p38 MAP kinase signaling pathways [74]. Plasma CD40L correlates with urinary thromboxane levels suggesting CD40L release during TXA2-dependent platelet activation [75].

In obesity, which is associated with both diabetes and CVD, visceral adipose tissue is a major source for inflammatory cytokines like monocyte chemotactic protein (MCP-1), tumor necrosis factor (TNF)α, interleukin (IL)-6, plasminogen activator inhibitor (PAI)-1 [67], and damage-associated molecular patterns (DAMPs) such as S100A8/A9 [76]. The expression of DAMPs is elevated in platelets of patients presenting with acute MI and could promote thrombosis [77]. PAI-1 inhibits plasminogen activator and hence is an inhibitor of fibrinolysis, the physiological process that degrades blood clots. The increased secretion of these cytokines substantially drives fibrinogen production and induces a prothrombotic setting [55].

CD36 belongs to the class of type B scavenger receptors and functions as a multifunctional protein involved in the uptake of apoptotic cells, transport of lipids and fatty acids, adhesion, and modulation of inflammation, which are all affected under conditions of CVD and diabetes [78]. There is abundant evidence linking CD36 with diabetes, inflammation, and platelet activation: increased CD36 expression is believed to be a marker of macrophage activation and inflammation [79], and hyperglycemia upregulates CD36 expression on the surface of monocytes in T2DM patients [80]. CD36 is also abundantly expressed on platelets, and interactions of platelet CD36 with oxLDL on monocytes result in increased platelet activation and enhanced thrombus formation [81]. There is also evidence that a specific CD36-dependent signaling pathway is required for platelet activation by oxLDL [82]. Circulating soluble CD36 is associated with glucose metabolism and interleukin-6 in patients with impaired glucose tolerance [83]. Finally, similar to CD40, plasma CD36 levels correlate with the urinary excretion rate of thromboxane and 8-iso-PGF2α, a sensitive marker of in vivo lipid peroxidation due to circulating oxLDL levels [69]. 8-iso-PGF2α, in turn, may amplify the aggregation response to subthreshold concentrations of platelet agonists [5]. In summary, CD36 plays an important role in linking macrophage activation with platelet activation, particularly in setting of increased oxidative stress and lipid peroxidation like DM.

Peroxisome proliferator-activated receptor gamma (PPARγ) is a ligand-activated transcription factor important in lipid metabolism, diabetes, and inflammation. Although PPARγ is considered to be a nuclear receptor, enucleate platelets also highly express this receptor and both synthetic and natural PPARγ ligands inhibit platelet activation and release of bioactive mediators. In particular, release of soluble CD40 ligand (sCD40L) and TXA2 was inhibited by PPARγ ligands in thrombin-activated platelets [84]. Other described effects included reduction of platelet aggregation, suppression of thrombin-induced protein kinase C and beta activation, decrease in plasma P-selectin and platelet P-selectin expression, increase in NO production, and inhibition of tissue factor- and platelet-activating factor-induced morphological changes in macrophages. These findings appeared in parallel with reduction of the plasma concentrations of pro-inflammatory risk markers [85].

9.3 Antiplatelet Treatment in Diabetes

Aspirin irreversibly inhibits COX-1, the key enzyme in the conversion of arachidonic acid into TXA2, thereby limiting the platelet response to agonists such as ADP and collagen. However, some patients' platelets remain reactive to agonist despite aspirin therapy increasing the risk for atherothrombotic events. Although often termed "aspirin resistance," a more appropriate term is high on-treatment platelet reactivity (HTPR) as failure of aspirin to inhibit COX-1 is quite rare (less than 5%) [86]. In contrast, HTPR is particularly prominent in DM patients, with a prevalence of about 20% [87]. As discussed above, patients with DM display platelet hyperactivity and are more sensitive to activation by agonists. Although aspirin may effectively block COX-1, platelets of affected patients with DM continue to manifest high on-treatment platelet reactivity (HTPR) that leads to elevated thrombotic risk [32]. What are possible determinants of HTPR in diabetes? First, low-grade inflammation can induce extra-platelet generation of thromboxane via COX-2 that is hardly sensitive to low-dose aspirin. Secondly, high platelet turnover in diabetic patients may contribute to HTPR through enhanced reactivity of younger platelets, increased COX-2 expression, and partial inhibition of COX-1. Young, reticulated platelets provide RNA to resynthesize COX-1, thereby limiting the effectiveness of aspirin. Third, an excess in F2-isoprostane production can partly activate the thromboxane receptor in a COX-independent way [5]. Finally, hyperglycemia may reduce sensitivity of platelets to aspirin by increased protein glycation [88]. As aspirin has a short half-life, it has been advocated that multiple dosing intervals rather than an increase in the once-daily dose may increase the bioavailability of aspirin and enhance platelet inhibition especially in diabetic patients [5, 89].

Similar to TXA2, ADP also induces platelet aggregation. The thienopyridine clopidogrel limits platelet aggregation by irreversibly inhibiting the ADP receptor $P2Y_{12}$ on platelets. Dual platelet therapy consisting of aspirin and clopidogrel effectively prevents recurrent cardiovascular events [90]. Polymorphisms in genes encoding cytochrome P450 enzymes cause ineffective or even absent metabolic activation of clopidogrel in some patients resulting in HTPR. Again this is particularly eminent in DM patients. In the OPTIMUS trial, two-thirds of diabetic patients were considered to have suboptimal response to aspirin and clopidogrel likely due to HTPR. In patients with DM on clopidogrel therapy, HTPR was associated with a fourfold increase in periprocedural myocardial infarction compared to DM with normal adequately suppressed platelet reactivity [91]. An increase of 75 mg clopidogrel to a daily dose of 150 mg induced platelet reactivity suppression in poor responders [92]. Medical treatment of T2DM might further interfere with HTPR by competition for metabolism by CYP2C9 cytochrome. For instance, concomitant treatment with sulfonylureas might be associated with decreased platelet inhibition by clopidogrel in T2DM patients on dual antiplatelet therapy undergoing elective coronary stent implantation [93]. A proposed mechanism is that platelets from diabetic patients have lower levels of cAMP compared with nondiabetics due to insulin resistance [94]. Lower cAMP levels lead to upregulated $P2Y_{12}$ signaling, thereby decreasing platelet inhibition by $P2Y_{12}$ antagonists. Increasing baseline platelet cAMP by phosphodiesterase-3 inhibitors (e.g., cilostazol) has been shown to improve platelet response to $P2Y_{12}$ inhibition in DM patients, but this has not been implemented into clinical use due to safety concerns [32]. Newer antiplatelet agents like prasugrel and ticagrelor have been approved for patients undergoing percutaneous intervention in the past decade. Both substances are unaffected by cytochrome polymorphisms, leading to more consistent antiplatelet effects compared to clopidogrel. Subgroup analysis of patients with DM from the PLATO (ticagrelor) and the TRITON-TIMI 38 (prasugrel) study showed a significant reduction of the primary end point compared to DM patients treated with clopidogrel [95, 96].

According to the joint position statement of the American Diabetes Association and American Heart Association, aspirin use for primary prevention should be prescribed to DM patients with a more than 10% risk for a fatal CVD within 10 years and can be considered in patients at intermediate (5–10%) risk. Standard aspirin treatment is not recommended in DM patients with a 10-year CVD risk of less than 5% [97]. The same guidelines, however, recommend the general therapy of cardiovascular risk factors in all DM patients. These include the management of high blood pressure, dyslipidemia, and hyperglycemia. In turn, as discussed above, adequate management of these risk factors will also affect platelet reactivity and aggregation. As shown in the UKPDS study, metformin reduced the risk for ischemic heart disease compared to other hypoglycemic agents [98]. Possible mechanisms include reduced levels of PAI-1 and fibrinogen and improved lysis of clots with metformin treatment [99]. The recently published EMPA-REG study has shown that the sodium/glucose cotransporter 2 (SGLT2) inhibitor empagliflozin significantly decreases the rate of cardiovascular events in DM patients [100]. SGLT2 inhibitors decrease the reabsorption of glucose in the kidney and therefore lower blood sugar. The mechanisms for decreased cardiovascular events in DM are currently under investigation, and there has been no study so far investigating the effect of SGLT2 inhibitors in platelet function.

Some, if not all, of these agents also impact underlying low-grade inflammation, thereby influencing platelet function. The pleiotropic effects of statins include anti-inflammatory properties, characterized by a reduction of high sensitive CRP and a decrease of inflammatory cells within the atherosclerotic plaque [101]. PPARγ agonists decrease inflammatory molecules including CRP, TNFα, and IL-6, independent of improving insulin sensitivity [102]. Finally, metformin has favorable effects on some inflammatory markers such as CRP [103].

9.4 Conclusion

Diabetes is a multifactorial disease associated with biochemical factors such as insulin resistance, inflammation, oxidative stress, hyperlipidemia, and hyperglycemia. The detrimental metabolic state that accompanies diabetes is responsible for abnormal platelet (and other cell) function, thereby contributing to accelerating CVD by enhanced adhesion, activation, and aggregation. Consequently, treatment of those risk factors by various agents in DM patients affects platelet function and limits accelerated CVD. Deciphering pathways in platelets that are particularly activated in diabetes might lead to novel treatment strategies.

Compliance with Ethical Standards

Conflict of Interest: Florian Willecke, Prabhakara R. Nagareddy, and Andrew J. Murphy declares that they have no conflict of interest.

Ethical Approval: This article does not contain any studies with human participants or animals performed by any of the authors.

References

1. Group IDFDA. Update of mortality attributable to diabetes for the IDF Diabetes Atlas: estimates for the year 2013. Diabetes Res Clin Pract. 2015;109:461–5.
2. Gregg EW, Cheng YJ, Saydah S, Cowie C, Garfield S, et al. Trends in death rates among U.S. adults with and without diabetes between 1997 and 2006: findings from the National Health Interview Survey. Diabetes Care. 2012;35:1252–7.
3. Farkouh ME, Boden WE, Bittner V, Muratov V, Hartigan P, et al. Risk factor control for coronary artery disease secondary prevention in large randomized trials. J Am Coll Cardiol. 2013;61:1607–15.
4. Saydah SH, Fradkin J, Cowie CC. Poor control of risk factors for vascular disease among adults with previously diagnosed diabetes. JAMA. 2004;291:335–42.
5. Santilli F, Simeone P, Liani R, Davi G. Platelets and diabetes mellitus. Prostaglandins Other Lipid Mediat. 2015;120:28–39.
6. Haffner SM, Lehto S, Ronnemaa T, Pyorala K, Laakso M. Mortality from coronary heart disease in subjects with type 2 diabetes and in nondiabetic subjects with and without prior myocardial infarction. N Engl J Med. 1998;339:229–34.
7. Beckman JA, Paneni F, Cosentino F, Creager MA. Diabetes and vascular disease: pathophysiology, clinical consequences, and medical therapy: part II. Eur Heart J. 2013;34:2444–52.
8. Davi G, Patrono C. Platelet activation and atherothrombosis. N Engl J Med. 2007;357:2482–94.
9. Gerrard JM, Stuart MJ, Rao GH, Steffes MW, Mauer SM, et al. Alteration in the balance of prostaglandin and thromboxane synthesis in diabetic rats. J Lab Clin Med. 1980;95:950–8.
10. Davi G, Rini GB, Averna M, Novo S, Di Fede G, et al. Thromboxane B2 formation and platelet sensitivity to prostacyclin in insulin-dependent and insulin-independent diabetics. Thromb Res. 1982;26:359–70.
11. Mustard JF, Packham MA. Platelets and diabetes mellitus. N Engl J Med. 1984;311:665–7.
12. Halushka PV, Lurie D, Colwell JA. Increased synthesis of prostaglandin-E-like material by platelets from patients with diabetes mellitus. N Engl J Med. 1977;297:1306–10.
13. Halushka PV, Rogers RC, Loadholt CB, Colwell JA. Increased platelet thromboxane synthesis in diabetes mellitus. J Lab Clin Med. 1981;97:87–96.
14. Inui Y, Suehiro T, Kumon Y, Hashimoto K. Platelet volume and urinary prostanoid metabolites in non-insulin-dependent diabetes mellitus. J Atheroscler Thromb. 1994;1:108–12.
15. Collaborative overview of randomised trials of antiplatelet therapy—I: prevention of death, myocardial infarction, and stroke by prolonged antiplatelet therapy in various categories of patients. Antiplatelet Trialists' Collaboration. BMJ. 1994;308:81–106.
16. Nicolucci A, Standl E. Antiplatelet therapy for every diabetic person? Diabetes Care. 2011;34 (Suppl 2):S150–4.
17. Angiolillo DJ. Antiplatelet therapy in diabetes: efficacy and limitations of current treatment strategies and future directions. Diabetes Care. 2009;32:531–40.
18. Hekimsoy Z, Payzin B, Ornek T, Kandogan G. Mean platelet volume in Type 2 diabetic patients. J Diabetes Complicat. 2004;18:173–6.
19. Saur SJ, Sangkhae V, Geddis AE, Kaushansky K, Hitchcock IS. Ubiquitination and degradation of the thrombopoietin receptor c-Mpl. Blood. 2010;115:1254–63.
20. Sungaran R, Markovic B, Chong BH. Localization and regulation of thrombopoietin mRNa expression in human kidney, liver, bone marrow, and spleen using in situ hybridization. Blood. 1997;89:101–7.
21. Murphy AJ, Bijl N, Yvan-Charvet L, Welch CB, Bhagwat N, et al. Cholesterol efflux in megakaryocyte progenitors suppresses platelet production and thrombocytosis. Nat Med. 2013;19:586–94.
22. Huo Y, Schober A, Forlow SB, Smith DF, Hyman MC, et al. Circulating activated platelets exacerbate atherosclerosis in mice deficient in apolipoprotein E. Nat Med. 2003;9:61–7.

23. Shah B, Sha D, Xie D, Mohler ER 3rd, Berger JS. The relationship between diabetes, metabolic syndrome, and platelet activity as measured by mean platelet volume: the National Health And Nutrition Examination Survey, 1999–2004. Diabetes Care. 2012;35:1074–8.
24. Sokunbi DO, Wadhwa NK, Suh H. Vascular disease outcome and thrombocytosis in diabetic and nondiabetic end-stage renal disease patients on peritoneal dialysis. Adv Perit Dial. 1994;10:77–80.
25. Pabon Osuna P, Nieto Ballesteros F, Morinigo Munoz JL, Sanchez Fernandez PL, Arribas Jimenez A, et al. The effect of the mean platelet volume on the short-term prognosis of acute myocardial infarct. Rev Esp Cardiol. 1998;51:816–22.
26. Tavil Y, Sen N, Yazici H, Turfan M, Hizal F, et al. Coronary heart disease is associated with mean platelet volume in type 2 diabetic patients. Platelets. 2010;21:368–72.
27. Hendra TJ, Oswald GA, Yudkin JS. Increased mean platelet volume after acute myocardial infarction relates to diabetes and to cardiac failure. Diabetes Res Clin Pract. 1988;5:63–9.
28. Tsimerman G, Roguin A, Bachar A, Melamed E, Brenner B, et al. Involvement of microparticles in diabetic vascular complications. Thromb Haemost. 2011;106:310–21.
29. Zhang X, McGeoch SC, Johnstone AM, Holtrop G, Sneddon AA, et al. Platelet-derived microparticle count and surface molecule expression differ between subjects with and without type 2 diabetes, independently of obesity status. J Thromb Thrombolysis. 2014;37:455–63.
30. Vinik AI, Erbas T, Park TS, Nolan R, Pittenger GL. Platelet dysfunction in type 2 diabetes. Diabetes Care. 2001;24:1476–85.
31. Eibl N, Krugluger W, Streit G, Schrattbauer K, Hopmeier P, et al. Improved metabolic control decreases platelet activation markers in patients with type-2 diabetes. Eur J Clin Investig. 2004;34:205–9.
32. Kakouros N, Rade JJ, Kourliouros A, Resar JR. Platelet function in patients with diabetes mellitus: from a theoretical to a practical perspective. Int J Endocrinol. 2011;2011:742719.
33. Falcon C, Pfliegler G, Deckmyn H, Vermylen J. The platelet insulin receptor: detection, partial characterization, and search for a function. Biochem Biophys Res Commun. 1988;157:1190–6.
34. Trovati M, Anfossi G, Massucco P, Mattiello L, Costamagna C, et al. Insulin stimulates nitric oxide synthesis in human platelets and, through nitric oxide, increases platelet concentrations of both guanosine-3′, 5′-cyclic monophosphate and adenosine-3′, 5′-cyclic monophosphate. Diabetes. 1997;46:742–9.
35. Ferreira IA, Eybrechts KL, Mocking AI, Kroner C, Akkerman JW. IRS-1 mediates inhibition of Ca2+ mobilization by insulin via the inhibitory G-protein Gi. J Biol Chem. 2004;279:3254–64.
36. Trovati M, Anfossi G, Cavalot F, Massucco P, Mularoni E, et al. Insulin directly reduces platelet sensitivity to aggregating agents. Studies in vitro and in vivo. Diabetes. 1988;37:780–6.
37. Ferreira IA, Mocking AI, Feijge MA, Gorter G, van Haeften TW, et al. Platelet inhibition by insulin is absent in type 2 diabetes mellitus. Arterioscler Thromb Vasc Biol. 2006;26:417–22.
38. Udvardy M, Pfliegler G, Rak K. Platelet insulin receptor determination in non-insulin dependent diabetes mellitus. Experientia. 1985;41:422–3.
39. Betteridge DJ, El Tahir KE, Reckless JP, Williams KI. Platelets from diabetic subjects show diminished sensitivity to prostacyclin. Eur J Clin Investig. 1982;12:395–8.
40. Russo I, Traversa M, Bonomo K, De Salve A, Mattiello L, et al. In central obesity, weight loss restores platelet sensitivity to nitric oxide and prostacyclin. Obesity (Silver Spring). 2010;18:788–97.
41. Paneni F, Beckman JA, Creager MA, Cosentino F. Diabetes and vascular disease: pathophysiology, clinical consequences, and medical therapy: part I. Eur Heart J. 2013;34:2436–43.
42. Vazzana N, Ranalli P, Cuccurullo C, Davi G. Diabetes mellitus and thrombosis. Thromb Res. 2012;129:371–7.
43. Assert R, Scherk G, Bumbure A, Pirags V, Schatz H, et al. Regulation of protein kinase C by short term hyperglycaemia in human platelets in vivo and in vitro. Diabetologia. 2001;44:188–95.

44. Temelkova-Kurktschiev TS, Koehler C, Henkel E, Leonhardt W, Fuecker K, et al. Postchallenge plasma glucose and glycemic spikes are more strongly associated with atherosclerosis than fasting glucose or HbA1c level. Diabetes Care. 2000;23:1830–4.
45. Yngen M, Norhammar A, Hjemdahl P, Wallen NH. Effects of improved metabolic control on platelet reactivity in patients with type 2 diabetes mellitus following coronary angioplasty. Diab Vasc Dis Res. 2006;3:52–6.
46. Malmberg K, Ryden L, Efendic S, Herlitz J, Nicol P, et al. Randomized trial of insulin-glucose infusion followed by subcutaneous insulin treatment in diabetic patients with acute myocardial infarction (DIGAMI study): effects on mortality at 1 year. J Am Coll Cardiol. 1995;26:57–65.
47. Corpus RA, George PB, House JA, Dixon SR, Ajluni SC, et al. Optimal glycemic control is associated with a lower rate of target vessel revascularization in treated type II diabetic patients undergoing elective percutaneous coronary intervention. J Am Coll Cardiol. 2004;43:8–14.
48. Gerstein HC, Miller ME, Byington RP, Goff DC Jr, Bigger JT, et al. Effects of intensive glucose lowering in type 2 diabetes. N Engl J Med. 2008;358:2545–59.
49. Finfer S, Chittock DR, Su SY, Blair D, Foster D, et al. Intensive versus conventional glucose control in critically ill patients. N Engl J Med. 2009;360:1283–97.
50. Wautier JL, Schmidt AM. Protein glycation: a firm link to endothelial cell dysfunction. Circ Res. 2004;95:233–8.
51. Wang Y, Beck W, Deppisch R, Marshall SM, Hoenich NA, et al. Advanced glycation end products elicit externalization of phosphatidylserine in a subpopulation of platelets via 5-HT2A/2C receptors. Am J Phys Cell Physiol. 2007;293:C328–36.
52. Winocour PD, Watala C, Kinglough-Rathbone RL. Membrane fluidity is related to the extent of glycation of proteins, but not to alterations in the cholesterol to phospholipid molar ratio in isolated platelet membranes from diabetic and control subjects. Thromb Haemost. 1992;67:567–71.
53. Watala C, Boncer M, Golanski J, Koziolkiewcz W, Trojanowski Z, et al. Platelet membrane lipid fluidity and intraplatelet calcium mobilization in type 2 diabetes mellitus. Eur J Haematol. 1998;61:319–26.
54. Bucala R, Tracey KJ, Cerami A. Advanced glycosylation products quench nitric oxide and mediate defective endothelium-dependent vasodilatation in experimental diabetes. J Clin Invest. 1991;87:432–8.
55. Hess K, Grant PJ. Inflammation and thrombosis in diabetes. Thromb Haemost. 2011;105 (Suppl 1):S43–54.
56. Jardin I, Redondo PC, Salido GM, Pariente JA, Rosado JA. Endogenously generated reactive oxygen species reduce PMCA activity in platelets from patients with non-insulin-dependent diabetes mellitus. Platelets. 2006;17:283–8.
57. Freedman JE. Oxidative stress and platelets. Arterioscler Thromb Vasc Biol. 2008;28:s11–6.
58. Davi G, Ciabattoni G, Consoli A, Mezzetti A, Falco A, et al. In vivo formation of 8-iso-prostaglandin f2alpha and platelet activation in diabetes mellitus: effects of improved metabolic control and vitamin E supplementation. Circulation. 1999;99:224–9.
59. Pedreno J, Hurt-Camejo E, Wiklund O, Badimon L, Masana L. Platelet function in patients with familial hypertriglyceridemia: evidence that platelet reactivity is modulated by apolipoprotein E content of very-low-density lipoprotein particles. Metabolism. 2000;49:942–9.
60. Sironi L, Mussoni L, Prati L, Baldassarre D, Camera M, et al. Plasminogen activator inhibitor type-1 synthesis and mRNA expression in HepG2 cells are regulated by VLDL. Arterioscler Thromb Vasc Biol. 1996;16:89–96.
61. Nilsson L, Gåfvels M, Musakka L, Ensler K, Strickland DK, et al. VLDL activation of plasminogen activator inhibitor-1 (PAI-1) expression: involvement of the VLDL receptor. J Lipid Res. 1999;40:913–9.

62. Calkin AC, Drew BG, Ono A, Duffy SJ, Gordon MV, et al. Reconstituted high-density lipoprotein attenuates platelet function in individuals with type 2 diabetes mellitus by promoting cholesterol efflux. Circulation. 2009;120:2095–104.
63. Ferretti G, Rabini RA, Bacchetti T, Vignini A, Salvolini E, et al. Glycated low density lipoproteins modify platelet properties: a compositional and functional study. J Clin Endocrinol Metab. 2002;87:2180–4.
64. Aye MM, Kilpatrick ES, Aburima A, Wraith KS, Magwenzi S, et al. Acute hypertriglyceridemia induces platelet hyperactivity that is not attenuated by insulin in polycystic ovary syndrome. J Am Heart Assoc. 2014;3:e000706.
65. Stern MP. Diabetes and cardiovascular disease. The "common soil" hypothesis. Diabetes. 1995;44:369–74.
66. Olefsky JM, Glass CK. Macrophages, inflammation, and insulin resistance. Annu Rev Physiol. 2010;72:219–46.
67. Donath MY, Shoelson SE. Type 2 diabetes as an inflammatory disease. Nat Rev Immunol. 2011;11:98–107.
68. Prescott SM, Zimmerman GA, Stafforini DM, McIntyre TM. Platelet-activating factor and related lipid mediators. Annu Rev Biochem. 2000;69:419–45.
69. Santilli F, Basili S, Ferroni P, Davi G. CD40/CD40L system and vascular disease. Intern Emerg Med. 2007;2:256–68.
70. Lajer M, Tarnow I, Michelson AD, Jorsal A, Frelinger AL, et al. Soluble CD40 ligand is elevated in type 1 diabetic nephropathy but not predictive of mortality, cardiovascular events or kidney function. Platelets. 2010;21:525–32.
71. Andre P, Prasad KS, Denis CV, He M, Papalia JM, et al. CD40L stabilizes arterial thrombi by a beta3 integrin—dependent mechanism. Nat Med. 2002;8:247–52.
72. Zhou L, Stordeur P, de Lavareille A, Thielemans K, Capel P, et al. CD40 engagement on endothelial cells promotes tissue factor-dependent procoagulant activity. Thromb Haemost. 1998;79:1025–8.
73. Varo N, Libby P, Nuzzo R, Italiano J, Doria A, et al. Elevated release of sCD40L from platelets of diabetic patients by thrombin, glucose and advanced glycation end products. Diab Vasc Dis Res. 2005;2:81–7.
74. Chakrabarti S, Varghese S, Vitseva O, Tanriverdi K, Freedman JE. CD40 ligand influences platelet release of reactive oxygen intermediates. Arterioscler Thromb Vasc Biol. 2005;25:2428–34.
75. Inwald DP, McDowall A, Peters MJ, Callard RE, Klein NJ. CD40 is constitutively expressed on platelets and provides a novel mechanism for platelet activation. Circ Res. 2003;92:1041–8.
76. Nagareddy PR, Kraakman M, Masters SL, Stirzaker RA, Gorman DJ, et al. Adipose tissue macrophages promote myelopoiesis and monocytosis in obesity. Cell Metab. 2014;19:821–35.
77. Wang Y, Fang C, Gao H, Bilodeau ML, Zhang Z, et al. Platelet-derived S100 family member myeloid-related protein-14 regulates thrombosis. J Clin Invest. 2014;124:2160–71.
78. Silverstein RL, Febbraio M. CD36, a scavenger receptor involved in immunity, metabolism, angiogenesis, and behavior. Sci Signal. 2009;2:re3.
79. Tuomisto TT, Riekkinen MS, Viita H, Levonen AL, Yla-Herttuala S. Analysis of gene and protein expression during monocyte-macrophage differentiation and cholesterol loading—cDNA and protein array study. Atherosclerosis. 2005;180:283–91.
80. Sampson MJ, Davies IR, Braschi S, Ivory K, Hughes DA. Increased expression of a scavenger receptor (CD36) in monocytes from subjects with Type 2 diabetes. Atherosclerosis. 2003;167:129–34.
81. Podrez EA, Byzova TV, Febbraio M, Salomon RG, Ma Y, et al. Platelet CD36 links hyperlipidemia, oxidant stress and a prothrombotic phenotype. Nat Med. 2007;13:1086–95.

82. Chen K, Febbraio M, Li W, Silverstein RL. A specific CD36-dependent signaling pathway is required for platelet activation by oxidized low-density lipoprotein. Circ Res. 2008;102:1512–9.
83. Handberg A, Lopez-Bermejo A, Bassols J, Vendrell J, Ricart W, et al. Circulating soluble CD36 is associated with glucose metabolism and interleukin-6 in glucose-intolerant men. Diab Vasc Dis Res. 2009;6:15–20.
84. Akbiyik F, Ray DM, Gettings KF, Blumberg N, Francis CW, et al. Human bone marrow megakaryocytes and platelets express PPARgamma, and PPARgamma agonists blunt platelet release of CD40 ligand and thromboxanes. Blood. 2004;104:1361–8.
85. Borchert M, Schondorf T, Lubben G, Forst T, Pfutzner A. Review of the pleiotropic effects of peroxisome proliferator-activated receptor gamma agonists on platelet function. Diabetes Technol Ther. 2007;9:410–20.
86. Gluckman TJ, McLean RC, Schulman SP, Kickler TS, Shapiro EP, et al. Effects of aspirin responsiveness and platelet reactivity on early vein graft thrombosis after coronary artery bypass graft surgery. J Am Coll Cardiol. 2011;57:1069–77.
87. Mehta SS, Silver RJ, Aaronson A, Abrahamson M, Goldfine AB. Comparison of aspirin resistance in type 1 versus type 2 diabetes mellitus. Am J Cardiol. 2006;97:567–70.
88. Watala C, Pluta J, Golanski J, Rozalski M, Czyz M, et al. Increased protein glycation in diabetes mellitus is associated with decreased aspirin-mediated protein acetylation and reduced sensitivity of blood platelets to aspirin. J Mol Med (Berl). 2005;83:148–58.
89. DiChiara J, Bliden KP, Tantry US, Hamed MS, Antonino MJ, et al. The effect of aspirin dosing on platelet function in diabetic and nondiabetic patients: an analysis from the aspirin-induced platelet effect (ASPECT) study. Diabetes. 2007;56:3014–9.
90. Yusuf S, Zhao F, Mehta SR, Chrolavicius S, Tognoni G, et al. Effects of clopidogrel in addition to aspirin in patients with acute coronary syndromes without ST-segment elevation. N Engl J Med. 2001;345:494–502.
91. Mangiacapra F, Heyndrickx GR, Puymirat E, Peace AJ, Wijns W, et al. Comparison of drug-eluting versus bare-metal stents after rotational atherectomy for the treatment of calcified coronary lesions. Int J Cardiol. 2012;154:373–6.
92. Angiolillo DJ, Shoemaker SB, Desai B, Yuan H, Charlton RK, et al. Randomized comparison of a high clopidogrel maintenance dose in patients with diabetes mellitus and coronary artery disease: results of the Optimizing Antiplatelet Therapy in Diabetes Mellitus (OPTIMUS) study. Circulation. 2007;115:708–16.
93. Harmsze AM, Van Werkum JW, Moral F, Ten Berg JN, Hackeng CM, et al. Sulfonylureas and on-clopidogrel platelet reactivity in type 2 diabetes mellitus patients. Platelets. 2011;22:98–102.
94. Angiolillo DJ, Suryadevara S. Aspirin and clopidogrel: efficacy and resistance in diabetes mellitus. Best Pract Res Clin Endocrinol Metab. 2009;23:375–88.
95. James S, Angiolillo DJ, Cornel JH, Erlinge D, Husted S, et al. Ticagrelor vs. clopidogrel in patients with acute coronary syndromes and diabetes: a substudy from the PLATelet inhibition and patient Outcomes (PLATO) trial. Eur Heart J. 2010;31:3006–16.
96. Wiviott SD, Braunwald E, Angiolillo DJ, Meisel S, Dalby AJ, et al. Greater clinical benefit of more intensive oral antiplatelet therapy with prasugrel in patients with diabetes mellitus in the trial to assess improvement in therapeutic outcomes by optimizing platelet inhibition with prasugrel-Thrombolysis in Myocardial Infarction 38. Circulation. 2008;118:1626–36.
97. Pignone M, Alberts MJ, Colwell JA, Cushman M, Inzucchi SE, et al. Aspirin for primary prevention of cardiovascular events in people with diabetes: a position statement of the American Diabetes Association, a scientific statement of the American Heart Association, and an expert consensus document of the American College of Cardiology Foundation. Diabetes Care. 2010;33:1395–402.
98. Effect of intensive blood-glucose control with metformin on complications in overweight patients with type 2 diabetes (UKPDS 34). UK Prospective Diabetes Study (UKPDS) Group. Lancet. 1998;352:854–65.

99. Grant PJ. Beneficial effects of metformin on haemostasis and vascular function in man. Diabetes Metab. 2003;29:6S44–52.
100. Zinman B, Wanner C, Lachin JM, Fitchett D, Bluhmki E, et al. Empagliflozin, cardiovascular outcomes, and mortality in type 2 diabetes. N Engl J Med. 2015;373:2117–28.
101. Takemoto M, Liao JK. Pleiotropic effects of 3-hydroxy-3-methylglutaryl coenzyme a reductase inhibitors. Arterioscler Thromb Vasc Biol. 2001;21:1712–9.
102. Haffner SM, Greenberg AS, Weston WM, Chen H, Williams K, et al. Effect of rosiglitazone treatment on nontraditional markers of cardiovascular disease in patients with type 2 diabetes mellitus. Circulation. 2002;106:679–84.
103. Chu NV, Kong AP, Kim DD, Armstrong D, Baxi S, et al. Differential effects of metformin and troglitazone on cardiovascular risk factors in patients with type 2 diabetes. Diabetes Care. 2002;25:542–9.

Platelet Chemokines in New Modes of Action

10

Madhumita Chatterjee and Meinrad Gawaz

Abstract

Platelets are an enriched source of growth factors, pro- or anti-inflammatory agents, and pro- or anti-angiogenic mediators which are differentially sorted and readily available to be released from their granular repertoire upon receiving adequate stimulus. Platelet-derived mediators either through an autocrine or paracrine mode of action regulate systemic and vascular inflammation, immune defence, also contribute to regenerative mechanisms. Lately, platelet-associated chemokines have been ascribed rather unconventional roles in thrombo-inflammation, lipid uptake, and antimicrobial defence. This chapter highlights the impact of platelet-associated *CXC* chemokine *l*igands and their receptors in modulating haemostasis-thrombosis and platelet life span, and in influencing platelet-induced inflammatory or regenerative processes. We further highlight the contribution of thrombocidins, which are platelet-derived modified chemokines, in executing antimicrobial actions. The recently discovered multifaceted aspects of platelet chemokines as emphasised in this chapter encourages further experimental and clinical investigations in this expansive but still largely uncharted area of research in platelet biology.

Contents

M. Chatterjee
Innere Medizin III, Kardiologie und Kreislauferkrankungen, Eberhard Karls Universität, Tübingen, Germany

M. Gawaz (✉)
Department of Cardiology, University Hospital Tübingen, Tübingen, Germany
e-mail: Meinrad.Gawaz@med.uni-tuebingen.de

A. Zirlik et al. (eds.), *Platelets, Haemostasis and Inflammation*,
Cardiac and Vascular Biology 5, https://doi.org/10.1007/978-3-319-66224-4_10

List of Abbreviations

SDF-1α	Stromal cell-derived factor-α
GRO-α	Growth-regulated oncogene-α
MIP-1	Macrophage inflammatory peptide 1α
MCP-1,3	Monocyte chemotactic protein-1,3
MDC	Macrophage-derived chemokine
IL-8	Interleukin-8
MIF	Macrophage migration inhibitory factor
CRP	C-reactive protein
IL-6	Interleukin-6
vWF	Von Willebrand factor
ADP	Adenosine diphosphate
ATP	Adenosine triphosphate
PGI_2	Prostaglandin I2
PLC	Phospholipase C
CyPA	Cyclophilin A
CyPA-PPIase	Cyclophilin A peptidylprolyl isomerase
TXA2	Thromboxane A2
PRP	Platelet-rich plasma
BAD	Bcl-2 antagonist of cell death
CAD	Coronary artery disease
ASC	Acute coronary syndrome
TC	Thrombocidin
MI	Myocardial infarction
AMI	Acute MI
LVEF%	Left ventricular ejection fraction

10.1 Introduction

Platelets, the critical players in mediating haemostasis and thrombosis [1–3], have emerged as significant cellular participants in immune and inflammatory response in diverse pathophysiologies [4–7]. Upon activation, platelets might secrete a wide arsenal of ready to be released stored mediators from their α-granules that regulate a wide spectrum of responses [8–11]. Thus, platelets, being the initial and prime responders to vascular or tissue injury, are celebrated accomplice in vascular inflammation, atheroprogression [1–3], and atherosclerosis [4–7]. Recent extensive proteomic analysis of platelet α-granules has identified about 300 target molecular species which include growth factors, chemokines, cytokines, adhesion molecules, coagulation factors, fibrinolytic agents, angiogenic regulators, etc., with a wide range of functional attributes [12]. Encompassing a battery of pro-/anti-inflammatory and growth factors in their granular repertoire and being the most abundant blood corpuscle, they are readily positioned to act instantaneously to pathophysiological changes in the systemic or vascular environment and also contribute substantially to circulating levels of pro-/anti-inflammatory mediators to reach target organs and tissues [9, 10]. Moreover, activated platelet-derived microparticles enhance distal coagulation and inflammatory processes. CXCL4 (platelet factor 4 (PF4)) and CXCL7 are the most abundant of platelet-derived chemokines, closely followed by CCL5 (regulated on activation, normal T-cell expressed and secreted (RANTES)), macrophage migration inhibitory factor (MIF), CXCL12 (stromal cell-derived factor 1 (SDF-1)), and CXCL5 (epithelial neutrophil-activating peptide (ENA-78)) [4–7]. Evidential literature reporting the presence and functional significance of platelet-derived factors show much variability in terms of their relative expression detected by various means like proteomic, transcriptomic, and immunologic approaches [4–6]. These factors are differentially sorted in the α-granules, show preferential release upon receiving activating stimuli [14, 23], as presented in Fig. 10.1, also vary considerably in their release kinetics and relative stability. Once released, they mediate chemotaxis, proliferation, and differentiation of inflammatory and progenitor cells, instigate pro- or anti-angiogenic response, promote or retard thrombotic potential of platelets, and modulate their circulatory life span. Such functions are executed through the active engagement of the cognate receptors and a complex network of downstream signalling pathways. Platelets express several chemokine receptors [3–6] such as CCR1, CCR3, CCR4, CXCR2, CXCR4, CXCR6, CXCR7, and CX3CR1 [13–19], which render them susceptible to both autocrine and paracrine modulation by factors like CCL3 (MIP-1), CCL5, CCL7 (MCP-3), CCL17, CXCL1, CXCL5 (ENA-78), CXCL16 [15], CXCL8 (IL-8), CXCL12 [16, 17], and MIF [18, 19]. Therefore, these chemokines are designated as potential platelet agonists with diverse functional influence.

Platelet-derived chemokines are intricately associated with atherosclerotic progression or in directing the course and resolution of vascular inflammation. Platelets adherent to the site of vascular damage not only serve as a substrate for subsequent leukocyte interaction but secrete chemotactic factors, which drive infiltration of a variety of inflammatory and regenerative cells. Activated platelet secreted chemokine CXCL12 provides migratory signals that recruit CXCR4-expressing bone

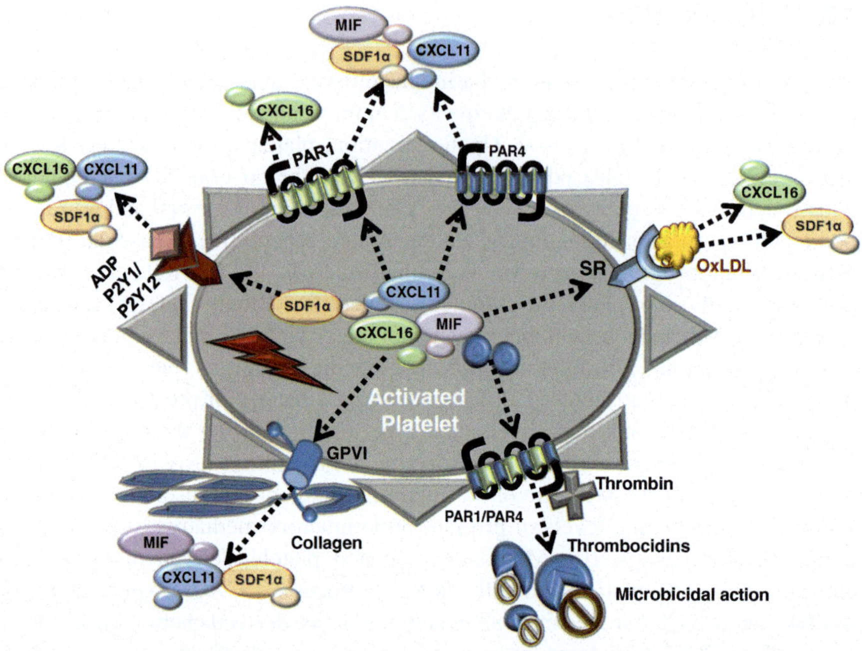

Fig. 10.1 Preferential release of platelet chemokines from the platelet granular repertoire. Platelets store a battery of chemokines or chemokine-like cytokines in their α-granules, which are differentially stored and released preferentially upon receiving distinct activating stimulus. Thus, platelet activation through purinergic receptors P2Y1 and P2Y12 by ADP leads to the release of CXCL11, SDF-1α/CXCL12, and CXCL16 in the external milieu, but not MIF. Thrombin receptor activation or collagen-induced GPVI activation leads to detectable release of MIF in the platelet microenvironment along with SDF-1α/CXCL12 and CXCL11. Platelet activation induced by OxLDL binding to the scavenger receptors on platelet surface leads to the release of soluble CXCL16 and SDF-1α/CXCL12. Microbicidal thrombocidins TC-1 and TC-2 are truncated chemokines released following thrombin stimulation, which might kill microbes in the surrounding milieu

marrow (BM)-derived progenitors of smooth muscle progenitor cells (SPCs) and endothelial progenitor cells (EPCs), monocytes, macrophages, thereby maintain the balance between vascular regeneration as opposed to vascular inflammation and remodelling [4–7]. Thus, platelets may promote neovascularisation or vessel regeneration, yet under pathological conditions also neointima formation [4]. Platelets induce migration of CXCR2-expressing EPC to the sites of vascular injury by releasing epithelial neutrophil-activating peptide-78 (ENA-78, CXCL5) and platelet basic protein (PBP, CXCL7). CXCL7 secreted upon platelet activation is ultimately processed into neutrophil-activating peptide 2 by regulated proteolysis. These chemokines might execute CXCR2-dependent homing to initiate vascular remodelling [4, 6, 7]. Apart from circulating progenitor cells, platelets bear the potential to attract a variety of leukocytes to exaggerate vascular inflammation. During their transient interaction with injured or inflamed endothelial layer, platelets can deposit chemokines onto endothelial cells and thereby promote subsequent monocyte recruitment [2–4]. Similarly, deposition of RANTES (CCL5) onto the vessel wall during rolling in a platelet P-selectin-dependent manner contributes to neointima formation. JAM-A and integrin $\alpha_{IIb}\beta_{III}$ interactions also facilitate the deposition of chemokines by activated platelets or platelet-derived microparticles [2–4], thereby increase monocyte arrest and neointima formation in vivo. Furthermore, infusion of activated platelets in transgenic hyperlipidemic *apoe*$^{-/-}$ mice leads to the delivery of RANTES and PF4 (CXCL4) to the vessel wall and substantiates atherosclerotic progression [4]. These attributes have established platelet and platelet-derived chemokines as celebrate partners in the crime of atherosclerotic disposition which has been extensively reviewed before. This chapter highlights the recent research concerning the CXCL chemokines and CXCRs in modulating haemostatic, thrombotic, and immune attributes of platelets.

10.2 The Privilege of Differential Storage and Preferential Release

Platelet α-granules are enriched with α-chemokines like CXCL4, CXCL12, CXCL16, and CXCL11 and also an extended range of β-chemokines, which include CCL5 (RANTES), CXCL1 (GRO−α) [3–6], or cytokines like MIF with chemokine properties [18–20]. They mediate assorted platelet moderated functions in the pathophysiological processes of wound healing, inflammation [2, 3, 5], atherothrombosis [1], atherosclerosis [2, 4–6], and angiogenesis [9–11]. Platelets contain between 40 and 80 α-granules harbouring most of the platelet-derived factors encapsulated within a membrane-bound space of 200–500 nm in diameter [8]. Heterogeneity of internal contents divides the α-granules into distinct subpopulations which are responsive to preferential agonist-induced activating stimuli [9–11]. α-granules obtain their protein content predominantly during platelet biogenesis as they are synthesised in the precursor megakaryocytes; they might also accumulate through active endocytic process of mature circulating platelets [8, 21, 22, 24]. Granule constituents are not homogenously or indiscriminately packed into α-granules during platelet biogenesis. α-granules develop from budding vesicles in the Golgi complex within the megakaryocytes and subsequently mature into multivesicular bodies, which can also interact with the endocytic vesicles. Multivesicular bodies are abundant in the immature megakaryocytes, which are thought to be common precursors to both α- and dense granules, and gradually decline with cell maturation. Multivesicular bodies containing numerous small vesicles encapsulated in a membranous sac seem to provide a common sorting compartment for both α- and dense granules, and to segregate functionally opposing proteins (e.g., VEGF-endostatin) into distinct classes of α-granules. So far two types of multivesicular bodies have been identified in megakaryocytes. Type I multivesicular bodies contain only internal vesicles, and type II contain both internal vesicles and an electron-dense matrix [13]. Intricate molecular mechanisms which help these multivesicular bodies to develop into distinct units remain vague, but heterogeneity of the internal membranes possibly plays a decisive role. During the process of megakaryopoiesis, the precursor megakaryocytes remodel their cytoplasm into long pro-platelet extensions, which serve as a portal for the transport and delivery of α-granules into nascent platelets that mature at the pro-platelet tips where a microtubule coil is formed [25]. Segregation of proteins into distinct classes of α-granules occurs before pro-platelet production initiates. The microtubule bundles in the shaft serve as the tracks on which distinct subpopulations of α-granules are sent to the maturing platelet buds, in a bidirectional traffic along the pro-platelet extensions. α-granule movement along pro-platelets has been observed in live cells by loading megakaryocytes with labelled fibrinogen, which is actively taken up and packaged into the α-granules. The filling process occurs sequentially as α-granules are translocated in a single file to the maturing platelet buds. The objective and elegance of this transport process lie in the ability to disperse the cargo throughout the pro-platelet extensions as well as to mix the various granules/organelles within the pro-platelet [26]. This process forms the basis of segregation of proteins into distinct subpopulations of α-granules [8, 21, 22, 24–26].

Some of the major characteristic angiogenic regulators have been shown to be segregated into different subsets of α-granules [8–11]. Pro-angiogenic VEGF, bFGF, and anti-angiogenic endostatin, thrombospondin-1 [9], show distinct localisation into different α-granule subpopulations which facilitate their differential secretion in response to pro-angiogenic stimulus acting through PAR1 and anti-angiogenic incentive acting through PAR4, respectively [9]. For chemokines like CXCL12 and PF4, the granular or spatial segregation is not absolute, but shows a considerable degree of sorting into different subpopulations of α-granules. RANTES and PF4, which are capable of heterophilic interaction, share spatial co-localisation in the same subset of α-granules [10]. Intracellular CXCL12 [10], MIF [19], and CXCL11 localisation under resting state exhibit random distribution across the cytoplasm and peripherise into a distinct ring-like pattern upon activation, prior to their milieu release or surface binding. Platelets release CXCL12 in response to activation through ADP receptor (P_2Y_{12}), glycoprotein VI (GPVI), and PAR1 ligation, whereas PAR4 activation drives the release of CXCL12 to a lesser extent [10]. The release of CXCL12 from activated platelets (>100 pg/ml) is comparable to other pro-angiogenic release, e.g. VEGF, also triggered by PAR1 ligation [10]. PAR1- and PAR4-driven selective secretory activity involve different signalling intermediates. PAR1-driven CXCL12 release appears to be dependent on Src and the PKC-PI3K-Akt pathway, while PAR4-driven CXCL12 release (lower than PAR1) seems to be mediated through MEK and p38MAPK [10]. Atherogenic stimulus like OxLDL also prompts CXCL12 release [18]. Thus, both PAR1 and PAR4 initiate the milieu release of CXCL12 and PF4, but the pro-angiogenic and pro-inflammatory release of CXCL12 is preferentially carried out through PAR1, GPVI, and purinergic activation, whereas anti-angiogenic PF4 release is largely instigated through PAR4 stimulation [10]. Such secretory preference not only differs in the trigger of release but also the relative degree and ease or kinetics of the release response from platelets. Interestingly, comparative release kinetics and relative stability influence the accumulation of CXCL12 and CXCL4 in activated platelet supernatant, which differs considerably [18, 20]. CXCL4 concentrations remain optimal through 12 h of stimulation, whereas CXCL12 concentrations decline subsequently over 2-h time intervals and are eventually completely abolished, possibly due to proteolytic degradation by circulating neutrophil elastase, cathepsin G, cathepsin K, matrix metalloproteinase-2/matrix metalloproteinase-9, carboxypeptidase N, and/or re-internalisation along with receptors [10, 18, 20]. Moreover, cell surface-associated peptidase like CD26 on $CD34^+$ cells might also potentially degrade platelet-derived CXCL12 and thereby influence its chemotactic efficacy. A considerably stronger stimulus is required to drive detectable milieu release as compared to surface exposure [10]. About 50–75% of platelets become positive for CXCL12 surface expression following activation with different agonists [10, 16, 23]. The relative ease of release depends on both the type and strength of the stimulus [10].

MIF, on the other hand, shows a granular pattern of distribution both in platelets and K562 megakaryocytes and is therefore expected to be sorted in the α-granules like other chemokines and growth factors. But it does not share co-localisation with other α-granule constituents like CXCL4 and angiogenic VEGF; although MIF is established as a known pro-angiogenic factor [18]. MIF is typically secreted by the nonclassical secretory pathway independent of ER-Golgi network. MIF is released upon TNF-α stimulation of leukocytes; however, such inflammatory stimulus fails to trigger MIF release from activated platelets. CXCL12 and MIF show difference in the trigger and release kinetics. Both thrombogenic stimuli like thrombin and collagen trigger MIF and CXCL12 release from activated platelets, while primarily atherogenic stimuli like ADP (0.5–10 μM) [18, 20] and oxidised-LDL promote CXCL12 but not MIF secretion to a detectable extent in the activated platelet supernatant [18, 20]. Release of MIF following thrombin- and collagen-induced GPVI stimulation follows a dose-dependent pattern. MIF release in the activated platelet supernatant could be detected from thrombin-stimulated platelets (0.5–2 U/ml) with a maximal release at 1 U/ml; whereas in a dose-dependent response to 0.1–100 μg/ml of collagen, the maximal release is observed at 10 μg/ml. Both modes of stimulation cause release of 60% of the total MIF reserve in platelets [18]. The magnitude of release differs significantly between MIF, CXCL12, and CXCL4, achieving maximum concentrations of 100 ng/mL, 1300 pg/mL, and 6 μg/mL, respectively. Detectable MIF release exhibits a delayed mode of execution as compared to that of CXCL12 or CXCL4 and takes about 2 h, whereas CXCL12 release occurs within minutes of thrombin activation. MIF concentration gradually builds up in thrombin-activated platelet supernatant, reaching a peak after 8 h of thrombin and between 4 and 8 h of collagen stimulation, and is plateaued, whereas CXCL12 concentration decreases and is ultimately exhausted, possibly due to degradation [18, 20]. This ensures a distinct spatial distribution of the two chemokines to execute differential functional aspects through the receptors they share, CXCR4 and CXCR7; alternatively, they could evidently complement each other.

Heterogeneity of the α-granules is observed not only for secretory granular components but also adhesion molecules like fibrinogen and vWF, which are sorted into distinct subsets of α-granules. While glass activation of platelets induces fibrinogen release, vWF-containing α-granules are retained [8]. Therefore, by virtue of segregated storage and differential release reaction, platelets may specifically deposit high concentrations of active molecules in a regulated and localised fashion to meet physiological demands. The segregated packaging of the angiogenic or inflammatory regulators into pharmacologically and morphologically distinct populations of α-granules offers the possibility that platelets may promote the differential release of a distinct class of α-granules, while retarding the release of functionally antagonistic factors as and when required [8–11, 21, 22]. This is of significant relevance in pharmaceutical approach for novel drug designing, which might preferentially stimulate or retard the exocytosis of a specific class of α-granule component, thereby manipulate platelets and platelet-derived factors for therapeutic benefit.

10.3 Differential Receptor Trafficking Influences the Targets to Hit

Chemokines and their cognate receptors bridge the execution of haemostasis, thrombosis to vascular inflammation, and repair/regeneration. Presence of functional chemokine receptors- CCR1, CCR3, CCR4, CXCR2, CXCR6, CXCR4, and CXCR7 has been demonstrated at transcript, protein levels and on the surface of human and murine platelets (Table 10.1). Of these, CCR1, CCR3, CCR4, and CXCR6 exhibit strong signals at transcript levels, whereas CXCR1, CXCR4, and CXCR7 are detectable to a lesser extent. CCR1, CCR4, CXCR4, CXCR7, and CXCR6 are detected at protein levels in extents comparable to other cells. Surface expression analysis shows the significant abundance of CXCR4, CXCR6, and CXCR7, but relatively low positivity for CCR1, CCR3, and CXCR2, whereas CXCR1, CXCR3, and CCR5 are apparently absent from the platelet surface. Most of these experimental evidence comes from resting human platelets and therefore have to be re-evaluated following platelet activation and in the presence or absence of their respective ligands in the surrounding microenvironment. For example, the surface expression of chemokine receptors like CXCR4 and CXCR7 exhibits a unique dynamism in the presence or absence of their ligands CXCL11 [19], CXCL12 [17], and MIF [19], which in turn influences their relative availability on platelet surface, thereby their frequency of participation in chemokine-mediated effector functions.

Although CXCR4 and CXCR7 are constitutively expressed in human and murine platelet at transcript and protein levels, the relative surface expression of CXCR4 appears to be much higher than that of CXCR7 at resting state [17]. Presence of ligands like CXCL12 [17], CXCL11, and MIF [19] brings about a dynamic alteration in CXCR4/CXCR7 surface expression as CXCR4 is internalised in the presence of CXCL12 and MIF, while CXCR7 is preferentially translocated to the surface in response to CXCL12 but not CXCL11 and MIF [19]. CXCL11 binds exclusively to CXCR7 in the absence of CXCR3 and internalises the receptor [19]. While CXCR2 is relatively low and CD74 being absent, MIF can ligate to CD44, CXCR4, and CXCR7 [19] on the platelet surface, (Fig. 10.2) but does not influence the availability of CXCR7 [19]. CXCL12-induced CXCR4 internalisation precedes CXCR7 surface exposure [17]. CXCL12-induced bidirectional trafficking of CXCR4 and CXCR7 is a coupled process as it is counteracted by CXCR4 blocking or antagonism offered by AMD3100. CXCL12-/CXCR4-triggered CXCR7 externalisation is executed through the downstream signalling

Table 10.1 Platelet-derived chemokines and their cognate receptors on platelets modulate functional response

Platelet-derived chemokines	Counter receptors	Detection in platelets	Detection of receptors	Functional relevance	References
CXCL8 (IL-8)	CXCR1	Human	Weakly at transcript levels, surface expression undetectable	Effect on platelet unknown, modulates megakaryocyte proliferation, differentiation, ploidy in myeloid metaplasia with myelofibrosis	[13, 14, 23]
CXCL8, CXCL1 (GRO-α), CXCL5 (ENA-78), CXCL7 (NAP-2), MIF	CXCR2	Human	Surface expression by flow cytometry	CXCL8 modulates megakaryocyte proliferation, differentiation, ploidy in myeloid metaplasia with myelofibrosis. MIF effect through CXCR2 ligation in platelets unknown	[13, 14, 23]
CXCL12, MIF	CXCR4	Human, murine	Transcript, protein, surface expression	Receptor internalisation, activation, aggregation, thrombus formation, adhesion to immobilised collagen-fibrinogen, migration for CXCL12, receptor internalisation for MIF	[16–19, 23]
CXCL16	CXCR6	Human, murine	Transcript, protein, surface expression	Degranulation, integrin activation, adhesion to endothelium in vitro and in vivo	[15, 23]
CXCL11, CXCL12, MIF	CXCR7	Human, murine	Transcript, protein, surface expression	Receptor externalisation for CXCL12, receptor internalisation for CXCL11 and survival	[16–19, 23]

(continued)

Table 10.1 (continued)

Platelet-derived chemokines	Counter receptors	Detection in platelets	Detection of receptors	Functional relevance	References
CCL-3 (MIP-1α), CCL5 (RANTES), CCL-7 (MCP-1)	CCR1	Human	Transcript, protein, surface expression	RANTES noncompetitively inhibits stimulator effects of CXCL12, synergistic inhibitory effect with PGE1	[13, 14, 23]
CCL5, CCL7	CCR3	Human, murine	Transcript, protein, surface expression	RANTES noncompetitively inhibits stimulator effects of CXCL12, synergistic inhibitory effect with PGE1	[13, 14, 23]
CCL17 (TARC), CCL22 (MDC)	CCR4	Human	Transcript, protein, surface, expression	Platelet activation, aggregation, synergistic effect with low concentrations of ADP, thrombin, adhesion to immobilised collagen-fibrinogen, shape change-formation of blebs for CCL22	[13, 14, 23]
CXCL4 (PF4)	GAG			Synergistic effect on platelet activation-aggregation with subthreshold concentrations of ADP, arachidonic acid, thrombin	[13, 14, 23]

intermediates like Erk1/2 and the PPIase activity of intracellular molecular chaperone cyclophilin A (CyPA), which is abolished in the presence of CyPA-PPIase inhibitor NIM811 and is absent in *Cypa*$^{-/-}$ murine platelets. CXCR7 ubiquitination is an essential prerequisite for cell surface delivery. In the presence of CXCL12, ubiquitin association of CXCR7 is dynamically upregulated, involving Erk1/2, CyPA-PPIase, and E1-ligase activity. Therefore, pharmacological inhibitors of MEK1/2 like U0126 or CyPA-PPIase inhibitor NIM811 and E1-ligase inhibitor PYR-41 significantly reduce CXCL12-driven CXCR7 ubiquitination and subsequent receptor externalisation [17]. CyPA mediates ubiquitination of viral proteins to control influenza virus replication [27] or uncontrolled viral replication among

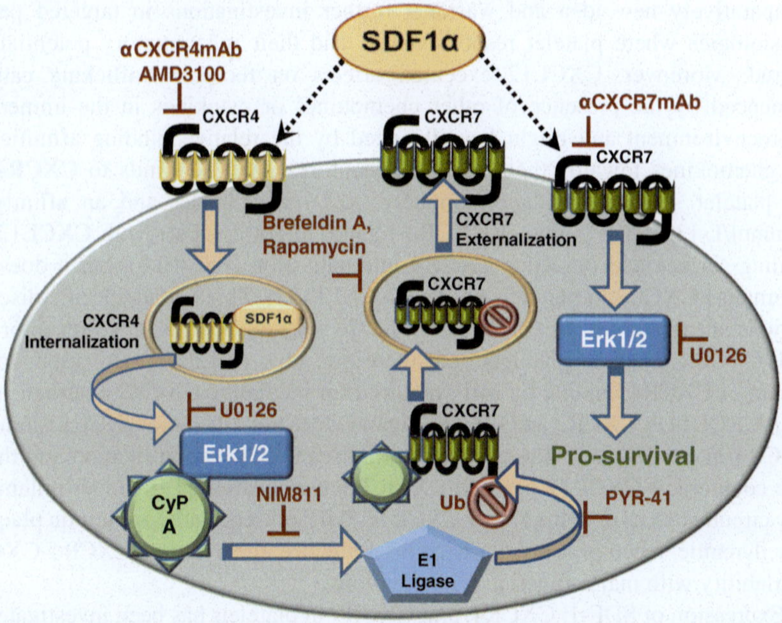

Fig. 10.2 Differential bidirectional receptor trafficking in response to SDF-1α/CXCL12. Schematic representation demonstrating SDF-1α-/CXCL12-induced internalisation of CXCR4 leads to the downstream activation of Erk1/2 and its subsequent interaction with the intracellular molecular chaperone CyPA. CyPA-PPIase activity is involved in enhancing CXCR7 ubiquitination through the enzymatic activity of E1 ligase. Blocking of CXCR4, or CXCR4 antagonism by AMD3100 inhibits the CXCL12-mediated externalisation of CXCR7, which suggests that CXCR4 internalisation is coupled to CXCR7 externalisation. CXCR7 surface translocation is also inhibited in the presence of MEK1/2 inhibitor U0126 and CyPA-PPIase activity inhibitor NIM811. Thereafter, ubiquitinated CXCR7 is translocated to the platelet surface through vesicular transport. This phenomenon is counteracted by the E1-ligase inhibitor PYR-41, and vesicular transport inhibitors brefeldin A and rapamycin. Thus, presence of SDF-1α/CXCL12 in the immediate microenvironment enhances the surface availability of CXCR7 on platelet surface. SDF-1α/CXCL12 subsequently ligates CXCR7 and exerts a prosurvival effect on platelet whereby it rescues platelet from undergoing apoptosis through the participation of the Erk1/2 pathway. SDF-1α/CXCL12 executed prosurvival effects are counteracted by blocking CXCR7 on the platelet surface and inhibitor of the Erk1/2 pathway

myocarditis models in *Cypa*$^{-/-}$ mice [28]. CyPA being involved in CXCR7 ubiquitination and subsequent surface translocation, this phenomenon is absent in *Cypa*$^{-/-}$ mice. Although the exact nature and class of the ubiquitinating enzyme still remains to be clarified, intracellular vesicular transport appears to be involved in the externalisation process since brefeldin A and rapamycin both inhibit CXCL12-induced surface externalisation of CXCR7 [17]. Ubiquitination of CXCR4 [29] and of the β2-adrenergic receptor (β2AR) regulates their lysosomal sorting and degradation [30]. On the contrary, PAR1 which is constitutively ubiquitinated, undergoes de-ubiquitination following activation and is internalised [31]. However, chemokine-induced receptor trafficking in platelets is still a comparatively new idea and warrants further investigations in targeted pathophysiologies where platelet responsiveness and their inflammatory potential are altered. Moreover, CXCL12 executed effects on receptor trafficking can be influenced by the presence of other chemokines or cytokines in the immediate microenvironment and might be influenced by the relative binding affinities of the chemokines towards their cognate receptors. CXCL12 binds to CXCR4 on the platelet surface with approximately 2000 sites/platelet and an affinity of 24 nmol/L [32]. MIF, another CXCR4 ligand, might compete with CXCL12 for binding surface and influence CXCR4 internalisation (Fig. 10.3), but it does not externalise CXCR7 in platelets unlike CXCL12 [32]. This ligand-specific discrepancy is attributed to the absence of CD74 (which acts as a co-receptor for MIF-CXCR4 axis) on platelets and, therefore, lack of Erk1/2 activation downstream of CXCR4 ligation by MIF, required for mediating CXCR7 externalisation [19]. CXCL11, a CXCR3 and CXCR7 ligand, does not affect the CXCL12-induced CXCR4 internalisation, However, it internalises CXCR7 upon ligation and therefore counteracts CXCL12-induced CXCR7 externalisation [19]. Under inflammatory circumstances or at the site of CXCL12-/MIF-enriched atherosclerotic plaques, this dynamic receptor trafficking could influence the relative CXCR4-CXCR7 availability with major functional implications.

Expression of SDF-1, CXCR4, and CXCR7 in platelets has been investigated in patients with acute coronary syndrome (ACS) and stable coronary artery disease (CAD). In a clinical cohort which enrolled 215 patients with symptomatic CAD, platelet CXCR7 surface expression was found to correlate significantly with that of CXCL12 and elevated in ACS patients as compared to stable CAD. Elevated platelet-CXCR7 levels have been found to be associated with functional recovery (improved LVEF%) during an intra-hospital stay of 5 days and a 3-month follow-up period [33]. By contrast, surface expression of CXCR4 was comparable between ACS and CAD patients at baseline evaluation. However, the prognostic significance of platelet CXCR4 surface expression has been revealed in a 12-month follow-up among patients with symptomatic CAD. Baseline CXCR4 levels are significantly lower in patients subsequently falling to all-cause death and/or MI; moreover, both baseline surface expression of CXCR4 and CXCR7 are significantly associated with all-cause mortality in patients with symptomatic CAD [34]. Therefore, it might be speculated that platelet-CXCR7 contributes to short-term myocardial repair mechanism in ACS patients, while platelet-CXCR4 might rather influence long-term outcome in the chronic phase of cardiovascular disease [32, 33]. These clinical observations encourage further thorough mechanistic understanding of the CXCL12/CXCR4/CXCR7 axis. Whether platelet surface expression of CXCR4-CXCR7 shares similar association with other relevant chemokine ligands in the context of CAD remains to be elucidated.

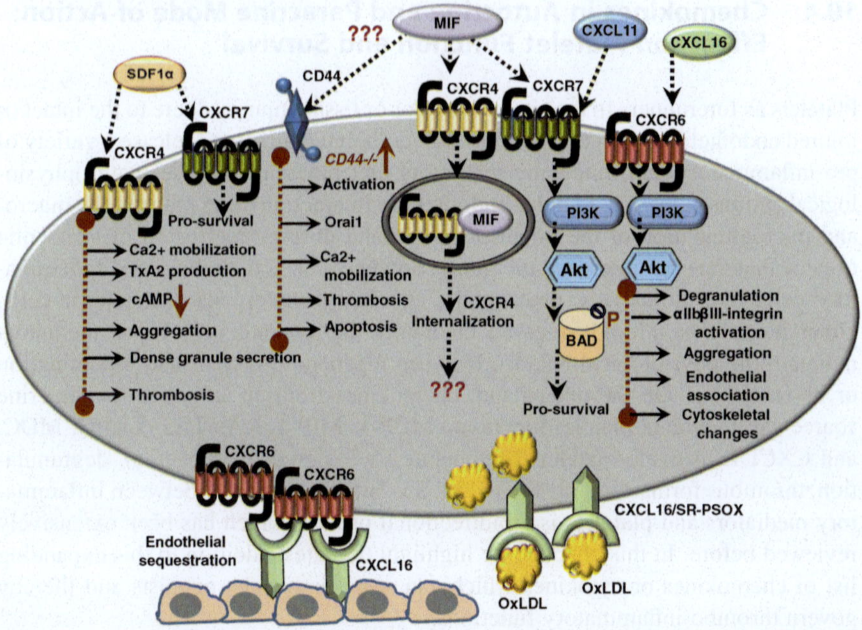

Fig. 10.3 Uncharacteristic functional effects of chemokines on platelets. Chemokines (CXCL12, CXCL11, CXCL16) and chemokine-like cytokines (MIF) exert a profound functional impact on platelets beside their chemotactic potential. CXCL12 through CXCR4 ligation substantiates platelet activation, aggregation, and thrombotic potential. CXCL12 antagonises adenylate cyclase activity and counteracts PGI_2 analogue-induced cAMP levels. CXCL12 released from collagen-GPVI-activated platelets prompts ATP release from the dense granules and TxA2 production, which promotes aggregation and thrombus formation through CXCR4. CXCL12 through CXCR7 ligation rescues platelets from undergoing apoptosis following activation. CXCL11 and MIF also exert a prosurvival effect mediated through CXCR7 ligation and downstream activation of the PI3K-Akt pathway culminating in phosphorylation-mediated inactivation of the pro-apoptotic protein BAD. MIF, like CXCL12, ligates CXCR4 on platelets and induces receptor internalisation, the functional consequences of which are still undefined. Platelets derived from *cd44*$^{-/-}$ mice show exaggerated activation and apoptotic potential when stimulated with thrombin- and collagen-related peptide. MIF can also bind to CD44 on platelets; however, the co-receptor remains to be defined. The membrane-associated form of CXCL16 on platelets functions as a scavenger receptor SR-PSOX for phosphatidylserine exposed on the apoptotic cells and OxLDL, facilitate OxLDL binding to the platelet surface. Platelets express the receptor for CXCL16, i.e. CXCR6 whereby platelets are sequestered from circulation by CXCL16-expressing endothelial cells at atherosclerotic sites. The soluble form of CXCL16 acts through CXCR6 on the platelet surface and the downstream PI3K-Akt pathway to promote degranulation, cytoskeletal reorganisation and shape change, $\alpha_{IIb}\beta_{III}$-integrin activation, and adhesion to the endothelial layer in vitro and to the injured carotid artery of mice in vivo

10.4 Chemokines in Autocrine and Paracrine Mode of Action: Effects on Platelet Function and Survival

Platelets as forerunners to the site of vascular or tissue injury, adhere to the intact or injured endothelium, or the exposed sub-endothelial matrix, and release a variety of pro-inflammatory and mitogenic mediators that modulate diverse (patho)physiological actions [1–5, 35]. Platelet-endothelium interaction takes place in the macro- and microcirculation of the inflamed tissue, and during reperfusion of ischaemic organs. Platelets can intercede the interaction between endothelium and inflammatory cells like monocytes, neutrophils, and lymphocytes; also progenitor cells, either directly or through released mediators like chemokines. These mediators maintain the physiological balance between regeneration repair and inflammation or its resolution. On the other hand, chemokines from an autocrine or paracrine source can modulate platelet functions. MCP-1, MIP-1, RANTES, TARC, MDC, and CXCL12 activate platelets to generate Ca^{2+} signals, aggregation, degranulation, thrombus formation [13, 14, 17, 23, 35, 36]. The interplay between inflammatory mediators and platelets is a bidirectional process which has been extensively reviewed before. In this chapter, we highlight the latest addition to this expanding list of chemokines or cytokines which can serve as platelet agonists and thereby govern thrombo-inflammatory functions.

10.5 Stromal Cell-Derived Factor-1α/CXCL12 Acts as a Pro-thrombotic Platelet Agonist

Platelets inherit CXCL12 transcript and protein from their precursor megakaryocytes [32]. However, experimental evidences accumulated over the years have established the potential of platelets for de novo protein synthesis, and several transcripts have been identified in platelet polysomes. Resting platelets do not exhibit the presence of mature mRNAs for CXCL12; but following activation with thrombin, the maturation process of CXCL12 pre-mRNA is triggered, subsequently leading to translation [37]. A bimodal effect of thrombin stimulation is observed on the presence of CXCL12 protein in activated platelets. Initially, a decrease in CXCL12 immune reactivity is observed in thrombin-activated platelet lysates during the first 30 min of incubation, which is subsequently restored following a prolonged 16-h culture of platelets attributed to de novo protein synthesis. Interestingly, such activation-induced synthesis also shows a selective response as thrombin stimulation triggers de novo synthesis of CXCL12 but not that of angiostatin [37]. Once synthesised, CXCL12 is stored in the α-granules in a ready to be delivered form, upon receiving adequate stimuli. Other vascular cytokines and chemokines, which promote or retard the release of CXCL12 from platelets, are of grave importance in this context. Soluble Kit-ligand, thrombopoietin, erythropoietin, and GM-CSF induce CXCL12 release from platelets, thereby enhance neovascularisation through mobilisation of the $CXCR4^{+}VEGFR^{+}$ haemangiocytes in vivo. Whether they affect maturation of the CXCL12 pre-mRNA and its subsequent translation is not known [32, 35].

Derived from an autocrine or pro-inflammatory paracrine source, CXCL12 can influence platelet functions. The CXCL12/CXCR4 axis induces megakaryocyte progenitor cell migration, and significantly enhances adhesion of mature bone marrow megakaryocyte to the endothelium. CXCL12 shows a bimodal effect on the expression of surface antigens on the early and late megakaryocytes. It upregulates the early megakaryocytic antigen CD41, but later, (days 12–16) induces downregulation of the late megakaryocytic antigen CD42b, which consequentially decreases the number of mature megakaryocytes in cultures supplemented with CXCL12 [16]. Moreover, CXCL12 positively influences the transendothelial migration of mature polyploid megakaryocytes, which might influence pro-platelet formation and finally fragmentation into platelet-like particles [16]. Mature platelets also exhibit transmigration through the endothelial layer towards CXCL12, which triggers platelet activation and phosphorylation of Wiskott-Aldrich syndrome protein. This response is inhibited by the CXCR4 antagonist AMD3100, pertussis toxin, PI3K inhibitor LY294002 or wortmannin, suggesting the involvement of CXCR4-$G_{\alpha i}$-PI3K pathway; also following disruption of actin polymerisation with cytochalasin B, suggesting cytoskeletal re-organisation under the influence of CXCL12 [32, 38]. Platelets preferentially accumulate at areas with high CXCL12 under flow conditions and respond to high shear stress by cellular polarisation, cytoskeletal reorganisation, and flow-directed migration [32, 38, 39]. Following CXCL12-assisted initial adhesion, a certain percentage of adherent platelets demonstrate migratory activity under high shear stress (1500 s^{-1}), associated with the intracellular redistribution of focal adhesion kinase to areas of dynamic focal adhesion contacts. Therefore, mechanotransduction of shear stress in platelets might facilitate platelet extravasation into inflamed tissue like atherosclerotic plaques where CXCL12 is predominant. CXCL12 does induce morphological changes leading to the formation of blebs [35] and enhance adhesion to collagen type IV and fibrinogen under arterial flow conditions [35].

Although considered a weak platelet agonist [40, 41], CXCL12 enhances platelet activation through $G_{\alpha i}$-coupled CXCR4 but not CXCR7. Moreover, CXCL12 acts as a strong antagonist of adenylate cyclase activity in platelets and counteracts PGI_2 analogue-induced cAMP levels [36]. CXCL12-induced platelet aggregation is affected by purinergic ADP receptor antagonists, which suggests that CXCL12 primarily activates platelets through lowering of cAMP levels and further substantiates it by granular release and PLC activation, subsequently leading to full aggregation [36]. CXCL12 at lower concentrations instigates the primary phase of aggregation response, but at increased concentrations triggers both primary and secondary response. Complete inhibition of CXCL12-CXCR4-induced aggregation by wortmannin and LY29004 strongly suggests PI3 kinase involvement in the initial primary phase of aggregation; whereas inhibition of the secondary wave by genistein and aspirin suggests engagement of downstream tyrosine kinases, prostanoids, to achieve maximal irreversible aggregation and granule secretion. CXCL12 as a co-stimulatory molecule further enhances aggregation induced by a subthreshold concentration of ADP and thrombin under arterial and lower shear conditions [16, 31, 35]. Moreover, CXCL12 synergistically and selectively enhances the aggregatory response of serotonin (5HT) but not epinephrine [16, 23, 35, 36]. CXCL12 fails to activate washed platelets; presumably, it requires the presence of plasma components like epinephrine and serotonin or a synergistic effect from TxA_2 or ADP released in the PRP preparations to instigate a biphasic aggregation response. Similarly, CXCL12 does not mobilise intracellular calcium in washed platelet preparations, but triggers TXB_2 and CXCL4 release in PRP preparations under stirring conditions. Although CXCL12 does not alter P-selectin exposure, $\alpha_{IIb}\beta_{III}$ activation under non-stirring conditions, exposure of platelets to CXCL12 in the presence of subthreshold ADP concentrations and under low shear drives P-selectin exposure. CXCL12 does not induce serotonin secretion from the dense granules either alone or in combination with low ADP concentrations [16, 23], but instigates ATP release from dense granules [16, 23, 35, 36, 40, 41]. CXCL12 released from activated platelets following collagen-GPVI stimulation exerts its autocrine effect through CXCR4 to prompt ATP release from the dense granules and TxA_2 production, which promotes aggregation and thrombus formation under dynamic flow conditions [36]. Moreover, platelet-derived chemokines are capable of heterophilic interactions exerting a synergistic or antagonising effect on each other. Heterodimerisation of CXCL12 with CXCL4 and CXCL7 have been reported [16, 23, 42, 43]. For instance, although RANTES neither induces platelet adhesion nor aggregation by itself, it noncompetitively reduces the stimulating effect of CXCL12 on platelet aggregation in PRP, and adhesion to endothelial monolayers under venous flow conditions. On the other hand, pretreatment of PRP with subthreshold concentrations of CXCL12 synergistically potentiates the aggregatory response from subthreshold concentrations of MDC and TARC [42, 43].

Platelet-derived CXCL12 mediates both pro-inflammatory and regenerative functions. Secreted from activated or adherent platelets, it triggers the migration and differentiation of $CD34^+$ progenitor cells into endothelial progenitor cells, thereby promote vascular regeneration [44, 45]; but it can also mediate monocyte migration and their differentiation into macrophages and foam cells, perpetuating vascular inflammation [46]. Activated platelets in circulation, or adherent platelets provide chemotactic cues like CXCL12 in mobilising PCs from the bone marrow or circulation to vascular lesions to form neointimal ECs and SMCs. Adherent platelets substantially induce chemotactic migration of murine embryonic endothelial progenitor cells, i.e. eEPCs (T17b cells) in vitro [47]. Furthermore, fibrin-activated platelets, support the chemotaxis of human $CD34^+$ PCs in vitro [44]. Platelet thrombi as a rich source of CXCL12 recruit $CD34^+$c-Kit^+Sca-1^+Lin-1^- (KSL-BMPCs) in vivo, which subsequently differentiate into neointimal cells, and contribute to vascular remodelling [16, 44–47]. Among inflammatory cells, CXCR4 and CXCR7 on monocytes preferentially respond to platelet-derived CXCL12. Whereas CXCR4 appears to the primary receptor responsive to the chemotactic cues of platelet-derived CXCL12, both CXCR4 and specifically CXCR7 appear to be significant for firm adhesion to collagen-adherent platelet layer, support monocyte-mediated platelet phagocytosis, and prompt subsequent thrombus resolution. Prosurvival effects of platelet-derived CXCL12 on monocytes is primarily mediated through CXCR7, whereas both CXCR4 and CXCR7 are involved in monocyte to macrophage and foam cell differentiation in a co-culture set-up with platelets [48]. Platelet surface expression of CXCL12 is enhanced in patients with CAD and associates with disease severity, as it is particularly elevated in patients with ST-elevation myocardial infarction [49, 50]. On the other hand, enhanced platelet CXCL12 surface expression correlates with the circulating number of $CD34^{+}$ progenitor cells, reduced infarct size, and improved functional recovery after acute myocardial infarction in ACS patients [49–54]. Moreover, circulating CXCL12 levels might also influence the dynamic surface availability of CXCR4-CXCR7 on platelets [17]. CXCL12 enhances CXCR7 receptor availability and upon subsequent CXCR7 ligation executes survival response, rescuing platelets from activation-induced apoptosis (Fig. 10.2). The physiological consequence of the anti-apoptotic recovery of platelets under the influence of CXCL12 remains to be investigated.

10.6 Macrophage Migration Inhibitory Factor (MIF) Functions as an Inflammatory but Antithrombotic Prosurvival Agent

The pleiotropic inflammatory chemokine like cytokine MIF contributes to atherosclerotic plaque development [55, 56], the expression of which is elevated and is associated with lesion formation and course of disease progression [18, 20]. Therapeutic potential of MIF is exemplified by the fact that peripheral MIF depletion in *Apoe*$^{-/-}$ mice reduces the pro-inflammatory response associated with atheroprogression [56]. In the context of CAD, it has been demonstrated that plasma levels of MIF are significantly elevated in patients with ACS as compared to symptomatic CAD, it is further associated with systemic inflammatory markers CRP, IL-6, and correlates with troponin release [57]. Both human and murine platelets are an appreciable source of MIF, which is detected at both transcript and protein levels [18]. Human platelets harbour 0.3fg or 15, 220 copies of MIF per platelet, whereas murine platelets have 0.0006fg or 30 copies of MIF per cell [18]. MIF transcripts are more abundant in human platelets than those of CXCL12, but considerably less than CCL5. Moreover, differentiated K562 cells, which exhibit megakaryocyte-like characteristics, also show expression of MIF at protein and transcript levels [18], suggesting the possibility that MIF in circulating mature platelet could well be derived from the precursor megakaryocytes like other chemokines/cytokines. However, an extracellular signal initiated post-transcriptional or translation process in the mature platelets cannot be ruled out. Once released, MIF can engage in autocrine or paracrine modes of action. MIF does not alter platelet activation, induce degranulation, P-selectin surface exposure, or prompt release of prototype chemokines like CXCL4 or CXCL12, either alone or in combination with other agonists [18–20]. MIF does not modulate aggregation either alone or that elicited by low and high concentrations of ADP and TxA2 analogue, or affect spreading of platelets over fibrinogen coated surface [20]. MIF, unlike CXCL12, is unable to mobilise intracellular calcium pools in TxA2 receptor agonist U46619 pre-sensitised platelets; however, both CXCL12 and MIF significantly block/desensitise increase in calcium in response to ADP [20]. Furthermore, thromboelastography with MIF-treated whole blood when monitored in the presence of tissue plasminogen activator, which destabilises clots by promoting fibrin degradation, reveals that MIF after 24 h exhibits a delayed dose-dependent anti-fibrinolytic effect [20]. Thus, CXCL12 and MIF although sharing receptors, show distinct modes of actions or effects on platelet functionality, possibly executed through distinct intracellular signalling cascades. However, both platelet-derived CXCL12 and MIF share pro-inflammatory attributes in stimulating monocyte chemotaxis [20, 48]. Presence of a neutralising anti-MIF antibody leads to a significant reduction in the chemotactic response induced by activated platelet supernatant obtained from thrombin-activated platelets following 8 h of stimulation, ensuring maximal and predominant MIF release [20]. Supernatants from activated platelets following 15 min of thrombin stimulation predominantly contains CXCL12 and induce monocyte migration. The significance of platelet-derived MIF is emphasised in experimentation with $mif^{-/-}$ mice. The chemotactic potential of the activated platelet supernatant derived from $mif^{-/-}$ mice is drastically reduced as compared to their wild-type counterparts [20]. Moreover, endothelial cell monolayers incubated with conditioned supernatants from platelets stimulated with thrombin for 8 h show significantly enhanced monocyte adhesion, further substantiating the potential of MIF as a pro-inflammatory mediator from platelets [20].

Autocrine or paracrine platelet functions are influenced by their activation status, survival vs apoptotic potential in circulation, since apoptotic platelets have reduced functional capacities. CXCL11, CXCL12, and MIF as survival factors, rescue platelets from activation, and BH3-mimetic-induced apoptosis, also prolong survival of circulating platelets through CXCR7 engagement. Enhanced availability of CXCR7 following CXCL12 exposure further perpetuates these survival benefits. Pharmacological inhibition of the Erk1/2 pathway (U0126) and CyPA-PPIase activity (NIM811) uncouples the dynamic trafficking of CXCR4-CXCR7 from the resultant anti-apoptotic effect of CXCL12, as mediated through CXCR7 [17] (Fig. 10.2). The CXCL11-CXCR7-/MIF-CXCR7-triggered anti-apoptotic effect is mediated through the PI3K-Akt pathway which culminates in the phosphorylation and thereby inactivation of the pro-apoptotic effector BAD [19] (Fig. 10.3). Therefore, the anti-apoptotic benefits of MIF and CXCL11 are lacking in $Akt^{-/-}$ murine platelets [19]. Administration of MIF to mice in vivo also enhances the relative longevity of circulatory platelets, an effect lacking in $Akt^{-/-}$ mice [19]. The relative functional potential of these aged platelets in circulation remains to be verified as compared to newly liberated platelets in circulation. But enhanced platelet survival might substantiate their regenerative mechanisms required for sustained repair processes.

As a consequence of CXCR7 ligation by MIF, there is an attenuation of pro-thrombotic phosphatidylserine exposure on the platelet surface, whereby MIF-CXCR7 also exerts an antithrombotic effect in vitro and checks thrombus build-up following arterial injury among mice in vivo, as deciphered by prolonged time to occlusion of the FeCl3-injured vessel [19]. Since *mif*$^{-/-}$ mice also respond to the antithrombotic effects of MIF during thrombus formation assay ex vivo, it appears that MIF from an external paracrine source could supplement for the lack of intracellular platelet reserve [19]. Recently, the expression of another potential receptor for MIF, CD44, has been demonstrated in platelets. Although the co-receptor for CD44 on platelets remains to ascertained, and it is still not clear whether MIF executed effects on platelet function and survival involve CD44, subtle functional differences have been demonstrated in platelets derived from *cd44*$^{-/-}$ and *cd44*$^{+/+}$ mice. *cd44*$^{-/-}$ and *cd44*$^{+/+}$ platelets are comparable in terms of relative P-selectin surface expression, $\alpha_{IIb}\beta_{III}$ integrin activation, Orai1 protein abundance, $[Ca^{2+}]_i$, apoptotic potential, denoted by caspase-3 activity, and phosphatidylserine exposure under resting condition [58]. However, thrombin- and collagen-related peptide-induced degranulation, $\alpha_{IIb}\beta_{III}$ integrin activation, intracellular calcium mobilisation, caspase-3 activity, phosphatidylserine exposure, and Orai1 surface abundance appear to be more pronounced in *cd44*$^{-/-}$ than in *cd44*$^{+/+}$ platelets. Moreover, platelet adhesion to fibronectin and ex vivo thrombus formation under high arterial shear rates is significantly augmented in *cd44*$^{-/-}$ mice [58]. These observations point towards a possibility whereby CD44 might act as a negative regulator to keep platelet activation and thrombotic events in check. These effects might be mediated through MIF in co-operation with as of undefined binding partner of CD44 on the platelet surface (Fig. 10.3). Thus, current experimental evidence points towards a functional preference in executing the haemostatic and thrombotic attributes of platelets through CXCR4 (by CXCL12) whereas support platelet survival through CXCR7 (by CXCL12, MIF, CXCL11) (Fig. 10.3). This could provide promising therapeutic alternatives for patients undergoing antiplatelet therapy against cardiovascular syndromes. Agents that check thrombotic potential without compromising the haemostatic and regenerative capacity of platelets are wanted, and factors acting through CXCR7 could emerge as potential candidates.

10.7 Chemokines Influencing Inflammatory Potential: The Significance of CXCL16 as a Scavenger Receptor, Chemokine, and Adhesion Molecule

The multifaceted chemokine CXCL16 is a significant pathogenic mediator in inflammatory conditions associated with rheumatoid arthritis, glomerulonephritis, or cancer [15, 59, 60]. The transmembrane form of CXCL16 serves as a scavenger receptor which binds to phosphatidylserine exposed on apoptotic cells and thereby clears cellular debris; additionally, it also binds to oxidised lipoprotein (OxLDL) and acts as a scavenger receptor [15, 59, 60]. Therefore, the membrane-associated form of CXCL16 is also designated as SR-PSOX. CXCL16 belongs to a specific class G of scavenger receptor due to its structural dissimilarities with the other scavenger receptors. Platelets express CXCL16 as both transcript and protein. Resting platelets show constitutive surface expression of CXCL16, which is further enhanced upon platelet activation with conventional agonists like ADP, TRAP, and also in the presence of OxLDL [59] (Fig. 10.1). Moreover, membrane-associated CXCL16 further perpetuates binding of labelled OxLDL to the surface of TRAP-activated platelets, which is significantly abrogated in the presence of an antibody targeting CXCL16 [59]. CXCL16 surface expression on platelets is significantly elevated among ASC patients as compared to stable angina pectoris (SAP). Moreover, platelet-CXCL16 surface expression shows a positive correlation with plasma C-reactive protein and baseline creatinine kinase, which denotes myocardial infarction [59]. CXCL16 is enhanced in inflammatory cardiomyopathy and turned out as an independent predictor of death in patients with HF undergoing endomyocardial biopsy [61]. Whether the elevated surface expression of CXCL16 contributes to lipid load in circulatory platelets and predisposes to atherothrombotic or atherosclerotic progression remains to be ascertained. Nevertheless, since platelet membrane-associated form of CXCL16 influences lipid binding to activated platelet surface (Fig. 10.3), it appears to be a strong possibility. CXCL16/SR-PSOX also acts as a scavenger receptor for phosphatidylserine on dying cells. Glucose-depleted erythrocytes, while undergoing suicidal death or eryptosis, externalise phosphatidylserine on their surface. Eryptotic erythrocytes adhere to endothelial cells and platelets involving phosphatidylserine at the erythrocyte surface and CXCL16 as well as CD36 at the endothelial cell membrane or on platelets immobilised to a collagen-coated surface under dynamic arterial flow conditions [62]. Adherence of erythrocytes to platelets is interfered with by coating of erythrocytic phosphatidylserine with annexin V or by blocking platelet phosphatidylserine receptors CXCL16/SR-PSOX and CD36 with respective antibodies. Such an association between platelets and erythrocyte might facilitate thrombo-occlusive events [62].

Platelets also release detectable amounts of CXCL16 following PAR1 activation by TRAP (Fig. 10.1) and therefore could be considered as an enriched source of CXCL16 in plasma among ACS patients [59]. Plasma level of soluble CXCL16 serves as a peripheral biomarker in ACS and is associated with long-term motility [59]. However, circulatory levels of CXCL16 and plasma OxLDL do not show a correlation in rheumatoid arthritis patients [59]. Further, in vitro experimentations in liquid phase demonstrate that the soluble form of the chemokine fails to bind to or scavenge OxLDL [60]. Soluble CXCL16, which chiefly functions as a chemokine for inflammatory cells and is generated following proteolytic cleavage of the chemokine domain by disintegrin-like metalloproteinase ADAM10, executes its chemotactic potential on CXCR6-/BONZO-expressing cells [60]. Platelets express CXCR6 as transcript, protein, also on their surface, which mediates CXCL16-driven effects [15]. CXCL16 can induce degranulation, cytoskeletal reorganisation and shape change, $\alpha_{IIb}\beta_{III}$-integrin activation, and adhesion to endothelial layer under arterial flow conditions in vitro and to the injured carotid artery of mice in vivo. CXCL16 enhances the aggregatory response to subthreshold concentration of ADP and further potentiates aggregation induced by fibrinogen. CXCL16 effects mediated through CXCR6 engagement lead to downstream activation of the PI3K-Akt pathway (Fig. 10.3) and are therefore significantly abrogated in *CXCR6*$^{-/-}$ and

$Akt^{-/-}$ mice, and following pharmacological inhibition of the PI3K-Akt pathway [15]. Moreover, CXCL16-triggered degranulation, integrin-$\alpha_{IIb}\beta_{III}$ activation, and adhesion to the vascular wall are diminished in the presence of ADP-degrading apyrase or purinergic P_2Y_1 and P_2Y_{12} receptor antagonists MRS2179 and cangrelor (AR-C69931MX), suggesting a feedback loop mediated through ADP. CXCL16 immobilised or microsphere bound, resembling a membrane-tethered version of the chemokine, enhances intracellular calcium mobilisation and subsequent integrin activation, degranulation, and aggregation responses of platelets to ADP. Therefore, CXCL16 both as a soluble mediator [15] and membrane-bound form can modulate the activation status and haemostatic attributes of platelets through CXCR6 ligation. Platelet-derived CXCL16 might also act a chemotactic substance for inflammatory cells like monocytes and granulocytes.

CXCL16 in its membrane-associated form can mediate adhesion to CXCR6-expressing cells [60]. Thus, CXCL16 through interaction with CXCR6 mediates platelet adhesion to the vessel wall (Fig. 10.3) along with vWF. CXCL16 is widely expressed at vascular sites predisposed to atherosclerosis and therefore detected in the human carotid atherosclerotic lesions from complex carotid endarterectomy specimens. Moreover, high level of CXCL16 expression is observed in the endothelium and in close proximity to mural thrombus enriched in vWF and platelet $GPIb_\alpha$ [60]. On the other hand, the presence of platelet-derived $GPIb_\alpha$ is limited to CXCL16-enriched regions, suggesting a platelet-endothelial association and subsequent platelet activation, driving thrombotic events [60]. CXCL16 sequesters circulating platelets through CXCR6 engagement (Fig. 10.3) in stimulated human radial arteries and supports vWF-mediated platelet associations to CXCL16-vWF-immobilised surfaces, which might bear pro-inflammatory consequences and lead to atheroprogression. Taken together, CXCL16 with its multifaceted character is poised to serve as a scavenger receptor and enhance lipid turnover in platelets; in its solubilised form exert a stimulatory effect on their haemostatic and thrombotic aspects, and also mediate inflammatory associations with endothelium and erythrocytes to contribute substantially to atheroprogressive complications.

10.8 Platelet Chemokines in Antimicrobial Action: The Thrombocidins

Continually expansive research in platelet biology has stretched the horizons beyond thrombosis and haemostasis [63], to the active participation of platelets as regulatory or effector immune cells [64–68]. Platelet-associated immune competencies range from sensing pathogenic intrusion or (non-)pathogenic inflammation [64–68] to direct microbicidal action [69] and also to interaction with innate and adaptive immune cells [68]. Platelets can migrate towards bacterial chemoattractant like *N-f*-MetLeuPhe, express Fc and complement receptors on their surface [68], facilitate complement fixation on microbes [70, 71], generate microbicidal free radicals or peptides [63], release cytokines and chemokines that trigger the adaptive immune wing, and potentiate the antimicrobial mechanisms of leukocytes.

Platelets can interact directly with microorganisms and thereby contribute to their clearance from the bloodstream and also participate in antibody-dependent cell cytotoxicity against microbial pathogens [72]. A number of antibacterial peptides have been characterised from the α-granules of platelets which exert direct antimicrobial activity; these are variously known as platelet microbicidal proteins (PMPs) and kinocidins [69, 73]. Kinocidins are essentially C-terminal truncated variants of the platelet-derived chemokines CXCL7, CXCL4, and CCL5 showing a broad spectrum of antimicrobial action against both Gram-positive and Gram-negative species like *Staphylococcus aureus* and *Escherichia coli* [69, 73]. Generally, these antimicrobial effector proteins arise from five genetically distinct lineages: PF4 and variants, PBP and its proteolytic derivatives CTAP-3 and NAP-2, RANTES, Tb-4, and fibrinopeptides. Such kinocidins can also be designated according to their chemokine nomenclature. Therefore, PF4, platelet basic protein (PBP), CTAP-3, and NAP-2 with CXC-chemokine motifs are referred to as α-kinocidins, whereas CC-chemokine motif containing RANTES is a β-kinocidin [69, 73]. Particularly those microbicidal peptides or bactericidins derived from thrombin-activated platelets are termed thrombocidins, i.e. *thrombo*cyte microbi*ci-d*al protei*ns* (Fig. 10.1). Antibacterial proteins released from thrombin-activated platelets have been found to be effective against viridans streptococci from cardiac vegetations in the experimental infective endocarditis (IE) [73, 74] model in rabbits. Viridans streptococci with low susceptibility to these proteins remain persistent in vegetations, whereas susceptible bacteria are rapidly eliminated [73, 74]. Similarly, strains of *Staphylococcus aureus* and *Candida albicans* which are insusceptible to rabbit platelet microbicidal proteins (PMPs) cause more severe experimental IE than PMP-susceptible strains [73, 74]. Furthermore, thrombocytopenic rabbits or rabbits treated with neutralising antibodies against platelet bactericidal proteins are more susceptible to streptococcal IE than control rabbits. Moreover essentially being chemokines, these peptides might further strengthen host defence by facilitating infiltration of immune effector cells (e.g. T cells) to the site of infection [73, 74].

Thrombocidin (TC)-1 and TC-2, the cationic antibacterial peptides purified from platelet granules when characterised using mass spectrometry and N-terminal sequencing, were found to be variants of the CXC chemokines neutrophil-activating peptide-2 and connective tissue-activating peptide-3, respectively, differing from these chemokines by a C-terminal truncation of two amino acids. In fact, platelet-derived microbicidal proteins can be further processed extracellularly following their release from the activated platelets [73]. Thrombin (a serine protease) which is abundant at the site of infection, besides platelet-derived proteases, and proteases that are activated in response to tissue injury, inflammation (e.g. plasmin), or phagocyte derived (e.g. cathepsin G) proteases, may process native PMPs and kinocidins, generating multiple antimicrobial peptide subspecies. Truncation and charge redistribution are crucial for their antimicrobial potential as the C-terminal part of TCs is involved in the 'cidal' mechanism. The C-termini of all CXC chemokines extend as α-helix, which like other antibacterial α-helical cecropins might insert into the bacterial membrane, thereby killing them [73]. The

N-termini of TC-1 and native NAP-2 are identical. Since TC-1 is directly isolated from platelet granules along with the native untruncated form, some cathepsin G-like protease activity is predicted to be present inside the granules [69, 73]. Moreover, as TC-1 and TC-2 are C-terminal truncated versions of NAP-2 and CTAP-3, respectively, the platelet granules are likely to contain carboxypeptidase activity [69]. Both TCs are effective against Gram-positive *Bacillus subtilis*, *Staphylococcus aureus*, Gram-negative *Escherichia coli*, and *Lactococcus lactis* and fungicidal for *Cryptococcus neoformans*. Furthermore, the native and unfolded TC-1 exert antimicrobial actions in different ways. Reduction in the charge of the TC-1-positive patch reduces antibacterial activity and almost abolishes antifungal activity against *Candida albicans*. Conversely, increasing the positive patch results in a two- to threefold increased activity against *Staphylococcus aureus*, *Escherichia coli*, and *Bacillus subtilis*, but does not affect activity against *C. albicans* [73]. The N-terminal part of TC-1 retains similar antimicrobial activity as the intact TC-1. The positive patch is essential for the activity of folded TC-1. Unfolded TC-1 retains its antimicrobial activity despite the absence of a positive patch [75]. The antibacterial activity of unfolded TC-1 is predictably exerted by a linear peptide stretch in the N-terminal part. Although TCs interact with bacterial membrane, they do not dissipate the bacterial membrane potential $\Delta\psi$ as seen in the case of *L. lactis* or liposomes composed of *E. coli* lipids. Therefore, their target for microbicidal action might be located intracellularly [75, 76]. Moreover, an interspecies difference in the mode of action also needs to be taken into account. Human TC-1 is highly and TC-2 is moderately effective against the fungus *C. neoformans*, whereas neither TC is effective against *Candida* species. However, preparations from rabbit platelets containing PMPs are more active against *Candida* species than against *C. neoformans*, indicating that the antimicrobial spectra of the human TCs and rabbit PMPs are different [69, 73–76].

Circulating platelets can be viewed as quiescent vigils, which sense danger or pathogen-associated molecular patterns indicative of vascular, systemic, or tissue infection and duly respond by delivering anti-infective molecules and wound-healing mediators [77, 78]. Thus, timely delivery of the PMPs/thrombocidins and their efficacy rely on the detection of warning signals by platelets and the prompt processing and release of these molecules to serve in immune defence. PMPs and kinocidins are not only released from platelets upon exposure to infection-relevant stimuli, such as thrombin, but also stimuli from microbial origin such as *S. aureus* α-toxin or those from viridans group streptococci, *S. aureus*, and *C. albicans*. *S. aureus* at a ratio of 10:1 or greater with platelets induces platelet activation, degranulation of ADP/ATP, and an aggregatory response which results in the release of staphylocidal levels of PMPs and kinocidins, sensitive to the presence of inhibitors like apyrase (ADP degradation), suramin (a general P2 receptor antagonist), pyridoxal 5-phosphonucleotide derivative (P2X1 antagonist), and cangrelor (P2Y12 antagonist). This indicates that following initial activation, an ADP-mediated activation of P2X1/P2Y12 receptors on adjacent platelets perpetuates the response and amplifies the release of PMPs and kinocidins from successively activated platelets [78]. Future research in this direction would uncover novel modes of trigger, new platelet-derived microbicidal chemokines, and their mechanistic strategy of microbicidal effects in infections which frequently feature platelets at the site of action.

10.9 Future Perspectives

Platelets as the most abundant circulating blood corpuscle and by virtue of their functional versatility and a plethora of inflammatory mediators, can potentially influence several pathophysiological attributes of the vasculature, beyond their conventional role in haemostasis and thrombosis. Consequentially, anti-platelet therapeutics directed against platelet aggregation (e.g. aspirin, prasugrel) and in the prevention of arterial or venous thrombosis might also affect the inflammatory, regenerative, and immune attributes of these cells as mediated through platelet-derived chemokines. Therefore, selective targeting of platelet-derived molecules to control or resolve inflammation, retard angiogenesis, and promote regenerative mechanisms would be the ideal therapeutic strategy to treat cardiovascular, inflammatory, and infectious diseases. As most platelet-associated chemokines exhibit subtle synergistic effects along with conventional agonist on the haemostatic functions of platelets, they appear to be promising candidates for such targeted therapeutic approaches without increasing the risk of bleeding.

Compliance with Ethical Standards

Conflict of Interest: Madhumita Chatterjee and Meinrad Gawaz declare that they have no conflict of interest.

Ethical Approval: This article does not contain any studies with human participants or animals performed by any of the authors.

References

1. Ruggeri ZM. Platelets in atherothrombosis. Nat Med. 2002;8:1227–34.
2. Gleissner CA, von Hundelshausen P, Ley K. Platelet chemokines in vascular disease. Arterioscler Thromb Vasc Biol. 2008;28:1920–7.
3. Gawaz M, Langer H, May AE. Platelets in inflammation and atherogenesis. J Clin Invest. 2005;115:3378–84.
4. Koenen RR, Weber C. Platelet-derived chemokines in vascular remodeling and atherosclerosis. Semin Thromb Hemost. 2010;36:163–9.
5. von Hundelshausen P, Weber C. Platelets as immune cells: bridging inflammation and cardiovascular disease. Circ Res. 2007;100:27–40.
6. Karshovska E, Weber C, von Hundelshausen P. Platelet chemokines in health and disease. Thromb Haemost. 2013;110:894–902.
7. Nurden AT. Platelets, inflammation and tissue regeneration. Thromb Haemost. 2011;105 (Suppl 1):S13–33.

8. Italiano JE Jr, Battinelli EM. Selective sorting of alpha-granule proteins. J Thromb Haemost. 2009;7(Suppl):173–6.
9. Italiano JE Jr, Richardson JL, Patel-Hett S, Battinelli E, Zaslavsky A, Short S, Ryeom S, Folkman J, Klement GL. Angiogenesis is regulated by a novel mechanism: pro- and anti-angiogenic proteins are organized into separate platelet alpha granules and differentially released. Blood. 2008;111:1227–33.
10. Chatterjee M, Huang Z, Zhang W, et al. Distinct platelet packaging, release, and surface expression of proangiogenic and antiangiogenic factors on different platelet stimuli. Blood. 2011;117:3907–11.
11. Battinelli EM, Markens BA, Italiano JE Jr. Release of angiogenesis regulatory proteins from platelet alpha granules: modulation of physiologic and pathologic angiogenesis. Blood. 2011; 118:1359–69.
12. Maynard DM, Heijnen HF, Horne MK, White JG, Gahl WA. Proteomic analysis of platelet alpha-granules using mass spectrometry. J Thromb Haemost. 2007;5:1945–55.
13. Clemetson KJ, Clemetson JM, Proudfoot AE, Power CA, Baggiolini M, Wells TN. Functional expression of CCR1, CCR3, CCR4, and CXCR4 chemokine receptors on human platelets. Blood. 2000;13:4046–54.
14. Gear AR, Camerini D. Platelet chemokines and chemokine receptors: linking hemostasis, inflammation, and host defense. Microcirculation. 2003;10:335–50.
15. Borst O, Münzer P, Gatidis S, et al. The inflammatory chemokine CXC motif ligand 16 triggers platelet activation and adhesion via CXC motif receptor 6-dependent phosphatidylinositide 3-kinase/Akt signaling. Circ Res. 2012;111:1297–307.
16. Chatterjee M, Gawaz M. Platelet-derived CXCL12 (SDF-1α): basic mechanisms and clinical implications. J Thromb Haemost. 2013;11:1954–67.
17. Chatterjee M, Seizer P, Borst O, et al. SDF-1α induces differential trafficking of CXCR4-CXCR7 involving cyclophilin A, CXCR7 ubiquitination and promotes platelet survival. FASEB J. 2014;28:2864–78.
18. Strüßmann T, Tillmann S, Wirtz T, Bucala R, von Hundelshausen P, Bernhagen J. Platelets are a previously unrecognised source of MIF. Thromb Haemost. 2013;110:1004–13.
19. Chatterjee M, Borst O, Walker B, et al. Macrophage migration inhibitory factor limits activation-induced apoptosis of platelets via CXCR7-dependent Akt signaling. Circ Res. 2014;115:939–49.
20. Wirtz TH, Tillmann S, Strüßmann T, Kraemer S, Heemskerk JW, Grottke O, Gawaz M, von Hundelshausen P, Bernhagen J. Platelet-derived MIF: a novel platelet chemokine with distinct recruitment properties. Atherosclerosis. 2015;239:1–10.
21. Reed GL, Fitzgerald ML, Polgar J. Molecular mechanisms of platelet exocytosis: insights into the "secrete" life of thrombocytes. Blood. 2000;96:3334–42.
22. Harrison P, Cramer EM. Platelet alpha-granules. Blood Rev. 1993;7:52–62.
23. Chatterjee M, Rath D, Gawaz M. Role of chemokine receptors CXCR4 and CXCR7 for platelet function. Biochem Soc Trans. 2015;43:720–6.
24. Heijnen HF, Debili N, Vainchencker W, Breton-Gorius J, Geuze HJ, Sixma JJ. Multivesicular bodies are an intermediate stage in the formation of platelet alpha-granules. Blood. 1998;91: 2313–25.
25. Italiano JE Jr, Lecine P, Shivdasani RA, Hartwig JH. Blood platelets are assembled principally at the ends of proplatelet processes produced by differentiated megakaryocytes. J Cell Biol. 1999;147:1299–312.
26. Richardson JL, Shivdasani RA, Boers C, Hartwig JH, Italiano JE Jr. Mechanisms of organelle transport and capture along proplatelets during platelet production. Blood. 2005;106:4066–75.
27. Liu X, Zhao Z, Xu C, et al. Cyclophilin A restricts influenza A virus replication through degradation of the M1 protein. PLoS One. 2012;7:e31063.
28. Seizer P, Klingel K, Sauter M, et al. Cyclophilin A affects inflammation, virus elimination and myocardial fibrosis in coxsackievirus B3-induced myocarditis. J Mol Cell Cardiol. 2012;53: 6–14.

29. Marchese A, Benovic J. Agonist-promoted ubiquitination of the G protein-coupled receptor CXCR4 mediates lysosomal sorting. J Biol Chem. 2001;276:45509–12.
30. Shenoy SK, McDonald PH, Kohout TA, Lefkowitz RJ. Regulation of receptor fate by ubiquitination of activated β_2-adrenergic receptor and β arrestin. Science. 2001;294:1307–13.
31. Wolfe BL, Marchese A, Trejo J. Ubiquitination differentially regulates clathrin-dependent internalization of protease-activated receptor-1. J Cell Biol. 2007;177:905–16.
32. Rath D, Chatterjee M, Stellos K, et al. Expression of SDF-1 receptors CXCR4 and CXCR7 on circulating platelets from patients with acute coronary syndrome and its association with left ventricular functional recovery. Eur Heart J. 2014;35:386–94.
33. Rath D, Chatterjee M, Borst O, et al. Platelet surface expression of stromal cell-derived factor-1 receptors CXCR4 and CXCR7 is associated with clinical outcomes in patients with coronary artery disease. J Thromb Haemost. 2015;13:719–28.
34. Stellos K, Gawaz M. Platelets and stromal cell-derived factor-1 in progenitor cell recruitment. Semin Thromb Hemost. 2007;33:159–64.
35. Gear AR, Suttitanamongkol S, Viisoreanu D, Polanowska-Grabowska RK, Raha S, Camerini D. Adenosine diphosphate strongly potentiates the ability of the chemokines MDC, TARC, and SDF-1 to stimulate platelet function. Blood. 2001;97:937–45.
36. Walsh TG, Harper MT, Poole AW. SDF-1α is a novel autocrine activator of platelets operating through its receptor CXCR4. Cell Signal. 2014;S0898-6568:325–8.
37. Huang Z, Rahman MF, Jiang L, et al. Thrombin induces de novo protein synthesis of stromal cell-derived factor-1α but not angiostatin in human platelets. J Thromb Haemost. 2012;10:2202–5.
38. Kraemer BF, Borst O, Gehring EM, et al. PI3 kinase-dependent stimulation of platelet migration by stromal cell-derived factor 1 (SDF-1). J Mol Med (Berl). 2010;88:1277–88.
39. Kraemer BF, Schmidt C, Urban B, et al. High shear flow induces migration of adherent human platelets. Platelets. 2011;22:415–21.
40. Kowalska MA, Ratajczak MZ, Majka M, et al. Stromal cell-derived factor-1 and macrophage-derived chemokine: 2 chemokines that activate platelets. Blood. 2000;96:50–7.
41. Abi-Younes S, Sauty A, Mach F, Sukhova GK, Libby P, Luster AD. The stromal cell-derived factor-1 chemokine is a potent platelet agonist highly expressed in atherosclerotic plaques. Circ Res. 2000;86:131–8.
42. Shenkman B, Brill A, Brill G, Lider O, Savion N, Varon D. Differential response of platelets to chemokines: RANTES non-competitively inhibits stimulatory effect of SDF-1 alpha. Thromb Haemost. 2004;2:154–60.
43. von Hundelshausen P, Koenen RR, Sack M, et al. Heterophilic interactions of platelet factor 4 and RANTES promote monocyte arrest on endothelium. Blood. 2005;105:924–30.
44. Stellos K, Langer H, Daub K, et al. Platelet-derived stromal cell-derived factor-1 regulates adhesion and promotes differentiation of human $CD34^+$ cells to endothelial progenitor cells. Circulation. 2008;117:206–15.
45. Stellos K, Langer H, Daub K, et al. Platelets secrete stromal cell-derived factor 1alpha and recruit bone marrow-derived progenitor cells to arterial thrombi in vivo. J Exp Med. 2006;203: 1221–33.
46. Daub K, Langer H, Seizer P, et al. Platelets induce differentiation of human $CD34^+$ progenitor cells into foam cells and endothelial cells. FASEB J. 2006;20:2559–61.
47. Langer H, May AE, Daub K, et al. Adherent platelets recruit and induce differentiation of murine embryonic endothelial progenitor cells to mature endothelial cells in vitro. Circ Res. 2006;98:e2–e10.
48. Chatterjee M, von Ungern-Sternberg SN, Seizer P, et al. Platelet-derived CXCL12 regulates monocyte function, survival, differentiation into macrophages and foam cells through differential involvement of CXCR4-CXCR7. Cell Death Dis. 2015;6:e1989.
49. Stellos K, Bigalke B, Langer H, et al. Expression of stromal-cell-derived factor-1 on circulating platelets is increased in patients with acute coronary syndrome and correlates with the number of $CD34^+$ progenitor cells. Eur Heart J. 2009;30:584–93.

50. Stellos K, Rahmann A, Kilias A, et al. Expression of platelet-bound stromal cell-derived factor-1 in patients with non-valvular atrial fibrillation and ischemic heart disease. J Thromb Haemost. 2012;10:49–55.
51. Stellos K, Ruf M, Sopova K, et al. Plasma levels of stromal cell-derived factor-1 in patients with coronary artery disease: effect of clinical presentation and cardiovascular risk factors. Atherosclerosis. 2011;219:913–6.
52. Geisler T, Fekecs L, Wurster T, et al. Association of platelet-SDF-1 with hemodynamic function and infarct size using cardiac MR in patients with AMI. Eur J Radiol. 2012;81:e486–90.
53. Wurster T, Stellos K, Geisler T, et al. Expression of stromal-cell-derived factor-1 (SDF-1): a predictor of ischaemic stroke? Eur J Neurol. 2012;19:395–401.
54. Wurster T, Stellos K, Haap M, et al. Platelet expression of stromal-cell-derived factor-1 (SDF-1): an indicator for ACS? Int J Cardiol. 2013;164:111–5.
55. Morand EF, Leech M, Bernhagen J. MIF: a new cytokine link between rheumatoid arthritis and atherosclerosis. Nat Rev Drug Discov. 2006;5:399–410.
56. Bernhagen J, Krohn R, Lue H, et al. MIF is a noncognate ligand of CXC chemokine receptors in inflammatory and atherogenic cell recruitment. Nat Med. 2007;13:587–96.
57. Müller II, Müller KA, Karathanos A, et al. Impact of counterbalance between macrophage migration inhibitory factor and its inhibitor Gremlin-1 in patients with coronary artery disease. Atherosclerosis. 2014;237:426–32.
58. Liu G, Liu G, Alzoubi K, Chatterjee M, et al. CD44 sensitivity of platelet activation, membrane scrambling and adhesion under high arterial shear rates. Thromb Haemost. 2015;115
59. Seizer P, Stellos K, Selhorst G, et al. CXCL16 is a novel scavenger receptor on platelets and is associated with acute coronary syndrome. Thromb Haemost. 2011;105:1112–4.
60. Meyer Dos Santos S, Blankenbach K, Scholich K, et al. Platelets from flowing blood attach to the inflammatory chemokine CXCL16 expressed in the endothelium of the human vessel wall. Thromb Haemost. 2015;114:297–312.
61. Borst O, Schaub M, Walker B, et al. CXCL16 is a novel diagnostic marker and predictor of mortality in inflammatory cardiomyopathy and heart failure. Int J Cardiol. 2014;176:896–903.
62. Walker B, Towhid ST, Schmid E, et al. Dynamic adhesion of eryptotic erythrocytes to immobilized platelets via platelet phosphatidylserine receptors. Am J Physiol Cell Physiol. 2014;306:C291–7.
63. Müller K, Chatterjee M, Rath D, Geisler T. Platelets, inflammation and anti-inflammatory effects of antiplatelet drugs in ACS and CAD. Thromb Haemost. 2015;114:498–518.
64. Zhang G, Han J, Welch EJ, et al. Lipopolysaccharide stimulates platelet secretion and potentiates platelet aggregation via TLR4/MyD88 and the cGMP-dependent protein kinase pathway. J Immunol. 2009;182:7997–8004.
65. Tsai JC, Lin YW, Huang CY, Lin FY, Tsai CS. Calpain activity and Toll-like receptor 4 expression in platelet regulate haemostatic situation in patients undergoing cardiac surgery and coagulation in mice. Mediat Inflamm. 2014;2014:484510.
66. Panigrahi S, Ma Y, Hong L, et al. Engagement of platelet toll-like receptor 9 by novel endogenous ligands promotes platelet hyperreactivity and thrombosis. Circ Res. 2013;112:103–12.
67. Koupenova M, Vitseva O, MacKay CR, et al. Platelet-TLR7 mediates host survival and platelet count during viral infection in the absence of platelet-dependent thrombosis. Blood. 2014;124:791–802.
68. Verschoor A, Langer HF. Crosstalk between platelets and the complement system in immune protection and disease. Thromb Haemost. 2013;110:910–9.
69. Tang YQ, Yeaman MR, Selsted ME. Antimicrobial peptides from human platelets. Infect Immun. 2002;70:6524–33.
70. Hamad OA, Bäck J, Nilsson PH, Nilsson B, Ekdahl KN. Platelets, complement, and contact activation: partners in inflammation and thrombosis. Adv Exp Med Biol. 2012;946: 185–205.

71. Del CI, Cruz MA, Zhang H, et al. Platelet activation leads to activation and propagation of the complement system. J Exp Med. 2005;201:871–9.
72. Verschoor A, Neuenhahn M, Navarini AA, et al. A platelet-mediated system for shuttling blood-borne bacteria to CD8α+ dendritic cells depends on glycoprotein GPIb and complement C3. Nat Immunol. 2011;12:1194–201.
73. Krijgsveld J, Zaat SA, Meeldijk J, et al. Thrombocidins, microbicidal proteins from human blood platelets, are C-terminal deletion products of CXC chemokines. J Biol Chem. 2000;275:20374–81.
74. Dankert J, van der Werff J, Zaat SA, Joldersma W, Klein D, Hess J. Involvement of bactericidal factors from thrombin-stimulated platelets in clearance of adherent viridans streptococci in experimental infective endocarditis. Infect Immun. 1995;63:663–71.
75. Kwakman PH, Krijgsveld J, de Boer L, et al. Native thrombocidin-1 and unfolded thrombocidin-1 exert antimicrobial activity via distinct structural elements. J Biol Chem. 2011;286:43506–14.
76. Koo SP, Bayer AS, Kagan BL, Yeaman MR. Membrane permeabilization by thrombin-induced platelet microbicidal protein 1 is modulated by transmembrane voltage polarity and magnitude. Infect Immun. 1999;67:2475–81.
77. Yeaman MR. Platelets: at the nexus of antimicrobial defence. Nat Rev Microbiol. 2014;12:426–37.
78. Yeaman MR. Platelets in defense against bacterial pathogens. Cell Mol Life Sci. 2010;67:525–44.

PI3K-Dependent Platelet Signaling in Vascular Inflammation and Atherothrombosis

11

Oliver Borst, Florian Lang, and Patrick Münzer

Abstract

Platelets are anucleated blood cells responsible for hemostasis and thrombosis after vascular injury. Upon activation, platelets adhere to subendothelial structures like collagen or von Willebrand factor (vWF). Activated platelets execute a dramatic shape change through cytoskeletal reorganization, and the integrin $\alpha_{IIb}\beta_3$ is converted into a high-affinity state so that there is platelet aggregation and thrombus formation. Besides their critical role in arterial thrombosis, platelets are of central importance in inflammatory processes and immunity. By exhibiting a wide variety of immunodulatory receptors and signaling molecules, platelets seem to be a physiological break point between innate/adaptive immunity and hemostasis. Platelet actions in hemostasis and inflammation are mediated by receptors on the platelet surface and intracellular signaling pathways and molecules. A crucial element in the activation-dependent platelet signaling is the phosphoinositide 3-kinase (PI3K) and its downstream targets. This chapter aims to summarize the most important PI3K-dependent signaling pathways and molecules in platelets.

Contents

O. Borst (✉) • P. Münzer
Department of Cardiology and Cardiovascular Medicine, University of Tübingen, Otfried-Müller-Straße 10, 72076 Tübingen, Germany
e-mail: oliver.borst@med.uni-tuebingen.de

F. Lang
Department of Physiology, University of Tübingen, Tübingen, Germany

A. Zirlik et al. (eds.), *Platelets, Haemostasis and Inflammation*, Cardiac and Vascular Biology 5, https://doi.org/10.1007/978-3-319-66224-4_11

11.1 Introduction

Platelets are small anucleated blood cells which develop through fragmentation of the cytoplasm of precursor megakaryocytes [1]. By executing adhesion, aggregation, and subsequent thrombus formation at sites of vascular injury, platelets are a prerequisite for the primary hemostasis. In detail, after vascular injury platelets adhere to the corrupted endothelium via interaction with subendothelial collagen and vWF. Integrin $\alpha_{IIb}\beta_3$ gets activated and the content of platelet α- and dense granules is secreted with subsequent platelet aggregation and cell recruitment. Platelet activation at sites of injured vasculature is followed by thrombus formation and blood clotting so that the vascular integrity is kept intact and the blood loss is minimized. Simultaneously, platelet activation can also lead to thrombo-occlusive disorders like myocardial infarction and ischemic stroke [2]. Moreover, platelets participate in a wide variety of inflammatory processes like hepatitis [3], encephalomyelitis [4], myocarditis [5], and especially in vascular inflammation underlying atherogenesis [6]. Pathophysiological atherothrombotic events commonly occur at sites of atherosclerotic lesion where platelets form a serious thrombotic occlusion after a plaque rupture [7].

Besides their involvement in atherothrombosis, platelets represent a physiological break point between immunity and hemostasis [8]. Platelets and leukocytes probably evolutionarily developed from one common precursor cell which simultaneously served hemostatic and immune function in invertebrates, birds, and fish [9, 10]. Thus, platelets retained several immune cell-specific properties and signaling mechanisms as well as immunmodulatory molecules. Along those lines, platelets store, present, and release a broad spectrum of pro- and anti-inflammatory molecules, thereby modulating inflammatory processes. Most of them belong to the chemokine family like chemokine C-X-C motif ligand 16 (CXCL16) [11–13], chemokine C-C motif ligand 5 (CCL5), and regulated on activation, normal T cell expressed and secreted (RANTES), respectively [14]. Furthermore, platelets store and release the chemokine C-X-C motif ligand 12 (CXCL12) or rather the stromal cell-derived factor 1 (SDF-1) [15–18] which mainly act as chemotactic factors. Some of them belong to the cytokine family like interleukin 1β (Il-1β) [19], transforming growth factor β (TGF-β) [10], and the macrophage migration inhibitory factor (MIF) [20]. Other platelet proteins mediate cytokine-related actions like high-mobility group protein B1 (HMGB1) [21]. Cyclophilins depict a further class of immunodulatory enzymes in platelets. So, for example, platelets contain cyclophilin A [22], are activated by extracellular cyclophilin A in vitro and in vivo

[23], and thus contribute to a wide variety of cardiovascular diseases like atherosclerosis, myocardial infarction and myocarditis [24].

Following contact with several inflammatory stimuli, platelets degranulate and efficiently adhere to extracellular matrix at sites of inflamed endothelium [4]. The adhesion is accomplished by special adhesion molecules like cluster of differentiation 40 ligand (CD40L) [25] or the membrane-bound scavenger receptor that binds phosphatidylserines and oxidized lipoprotein (SR-PSOX). Interestingly, this SR-PSOX is the membrane-bound form of the CXCL16 as this chemokine exists in a soluble and surface-bound form and can support the adhesion of cells to the endothelium [13, 26]. SR-PSOX is expressed in atherosclerotic lesions of apolipoprotein E-deficient mice pointing to a link between lipid metabolism, immune activity, and platelet activation in the atherosclerotic lesion [27]. By releasing pro- and anti-inflammatory mediators after activation stimuli, platelets accelerate inflammatory processes and atherothrombotic events [11]. This tremendous platelet activation machinery and granule release is thereby mediated via several distinct platelet surface receptors and intracellular signaling pathways.

In platelets and leukocytes, chemokines and cytokines largely signal via G-protein coupled receptors (GPCR). But due to the vast diversity of immunmodulatory molecules, there are several other surface receptors and signaling pathways which are involved in platelet activation [8]. In general, platelet signaling and activation is triggered by orchestration of several agonists using different specific surface receptors. On the one hand, collagen, collagen-related peptide (CRP), and the snake venom convulxin bind to glycoprotein VI/FcRγ (GPVI/FcRγ) complex [28], while on the other hand, thrombin, adenosine diphosphate (ADP), or thromboxane A_2 (TxA_2) in particular binds to G-protein coupled receptors. Thereby, thrombin recognizes protease-activated receptors (PAR) whereas ADP signals via purinergic receptors. After receptor activation, several distinct signaling molecules or pathways and mechanisms in platelets are activated. Although there are several completely different specific surface receptors and intracellular molecules as well as kinases, all signaling pathways in platelets converge more or less in the intracellular increase of the cytoplasmic calcium which represents the major hallmark of platelet activation [29]. In doing so, the intracellular calcium increase is responsible for the successive platelet activation, degranulation, integrin $\alpha_{IIb}\beta_3$ outside-in signaling, and procoagulant activity. Less is known about the exact platelet activation by immunodulatory chemokines. But as shown in Figs. 11.1 and 11.2, one key molecule in platelet activation and procoagulant activity after consecutive receptor initiation as well as in inflammation is the phosphoinositide 3-kinase (PI3K). Several PI3K downstream targets and subsequent PI3K-signaling actions play a pivotal role in platelet-dependent hemostasis and thrombosis. This chapter focuses on the PI3K-dependent platelet signaling and its role in chemokine-mediated vascular inflammation and atherothrombosis.

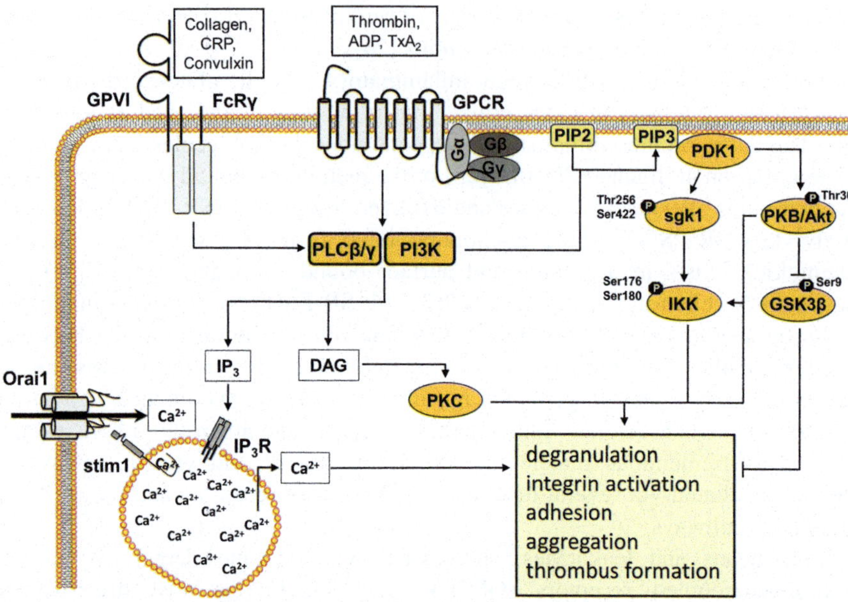

Fig. 11.1 PI3K-dependent signaling in platelets. Activation of ITAM (GPVI/FcRγ) and G-protein coupled receptors (GPCR) lead to phospholipase C (PLC)β/γ and PI3K activation. Subsequently, PLC increases the cytosolic calcium concentration by triggering of intracellular calcium release and subsequent stim1 and Orai1 activation. PI3K stimulates PDK1 leading to activation (sgk1 and PKB/Akt) or inhibition (GSK3β) of downstream targets. Finally, both signaling pathways cross talk and support platelet activation by regulating degranulation, integrin activation, adhesion, aggregation, and thrombus formation

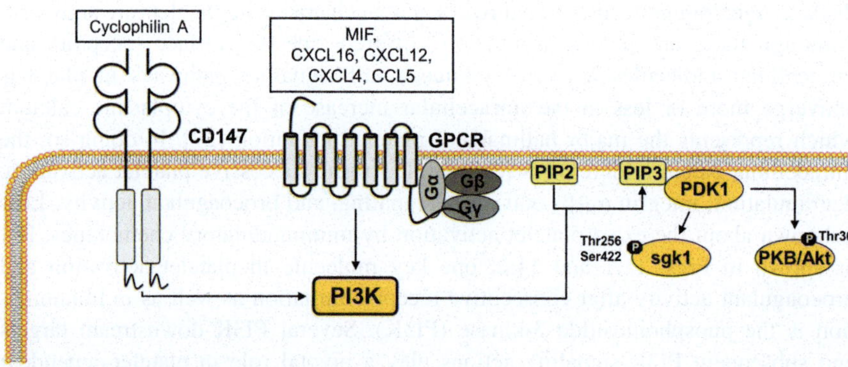

Fig. 11.2 Chemokine-mediated platelet signaling. Immunmodulatory molecules bind to specific platelet surface receptors with subsequent activation of the phosphoinositide 3-kinase (PI3K) and downstream effectors PKB/Akt and sgk1

11.2 Phosphoinositide 3-Kinase (PI3K)

The lipid kinase family of the phosphoinositide 3-kinase (PI3K) consists of numerous isoforms which are subdivided into class I, II, and III. Class I PI3K generates phosphoinositide-(3,4,5)P_3 (PIP3) by phosphorylation of the membrane compound phosphoinositide-(4,5)P_2 (PIP2) [30]. Both phosphoinositides (PI) serve in several cell models as second messengers and modulators of signaling events. Jackson and colleagues showed in 2005 that especially class I PI3K plays a role in platelet integrin $\alpha_{IIb}\beta_3$ activation and thrombosis [30]. The class I PI3Ks are further divided into α, β, δ, and γ subtypes. Platelets express all class I subtypes whereby the δ subtype shows the lowest expression level and the β subtype plays a pivotal role in the regulation of platelets [31]. Less is known about the role of class II and III PI3K in platelet function [32].

Different class I subtypes in platelets discriminate between diverse extracellular signals and receptors. For instance, class I PI3K subtypes in platelets can act in varying compositions downstream of G-protein coupled receptors (GPCR) and receptor-tyrosine kinase (RTK) or immunoreceptor tyrosine-based activation motif (ITAM) signaling receptors, respectively. The β subtype for example is mainly involved in platelet activation after collagen, thrombin, ADP, and TxA_2 stimulation, while PI3Kγ only mediates purinergic ADP signaling. PI3Kβ and PI3Kγ signaling is required for platelet ADP receptor function in dynamic thrombus stabilization [33]. Nevertheless, there is also evidence that particularly the β and γ isoforms have nonredundant roles in glycoprotein VI-induced platelet signaling and thrombus formation [34]. Moreover, PI3Kβ is the mediator in integrin $\alpha_{IIb}\beta_3$ activation and outside-in signaling [31]. Additionally, the β and γ subtypes of class I PI3K are also able to mediate and influence a broad range of downstream effectors. In platelets, the most important PI3K-dependent downstream targets are so-called AGC kinases as these kinases execute various important platelet actions as summarized in Fig. 11.1.

11.2.1 Downstream Effectors of PI3K

11.2.1.1 3-Phosphoinositide-Dependent Protein Kinase-1 (PDK1)

The main downstream target of PI3K-generated phosphoinositide-(3,4,5)P_3 is the ubiquitously expressed 3-phosphoinositide-dependent protein kinase-1 (PDK1). This 63 kDa kinase was first described in 1997 when Alessi and coworkers showed a phosphoinositide-(3,4,5)P_3-dependent kinase in a cell culture approach [35]. As all AGC kinases, PDK1 presents a Ser/Thr kinase activity.

Although PDK1 is a master kinase in a wide variety of cellular processes, to date less is known about the exact role of 3-phosphoinositide-dependent protein kinase-1 in platelet activation. In 2013, it was shown that platelet-specific PDK1 deficiency impairs activation and thrombus formation in mice after platelet-dependent GPCR stimulation with thrombin, ADP, and thromboxane A_2 [36]. Moreover, PDK1 was shown to play a pivotal role in physiological platelet function downstream of glycoprotein VI (GPVI)- and GPCR-mediated activation as in human platelets treated with the specific PDK1 inhibitor BX795, Par4 and collagen-dependent aggregation and clot retraction was significantly blunted [37].

PDK1 is a crucial player in cell migration and chemotaxis as its kinase activity regulates several cytoskeletal dynamics. In endothelial cells [38], T-lymphocytes [39], neutrophils [40], and cancer cells [41], PDK1 is described as mediator of cell migration events in the context of an inflamed microenvironment. Furthermore, it was shown that the chemokine SDF-1α/CXCL12 could influence PDK1 gene expression directly [42]. Interestingly, a few years ago two single nucleotide polymorphisms (SNPs) within the PDK1 gene region were associated with an increased cancer risk [43] showing the relevance for PDK1 in inflammatory diseases.

In platelets, there are two distinct targets of the 3-phosphoinositide-dependent protein kinase-1, namely, the protein kinase B/Akt [11] and the serum- and glucocorticoid-inducible kinase 1 (sgk1) [44].

11.2.1.2 Protein Kinase B (PKB)/Akt

One of the most important downstream targets of the 3-phosphoinositide-dependent protein kinase-1 in platelets is the protein kinase B (PKB) or Akt, respectively. In platelets and other cell types, PDK1 phosphorylates the PKB/Akt at Thr^{308} in vitro and in vivo [35]. There are the three isoforms Akt1, Akt2, and Akt3 which have to some extent overlapping functions [45]. In platelets, mainly the isoforms Akt1 and Akt2 are expressed [46]. In murine platelets, Akt1 signals downstream of GPVI receptor and GPCR as in Akt1-deficient mice, the platelet response to both collagen and thrombin is impaired [47]. In contrast, in Akt2-deficient mice only the GPCR stimulation by thrombin and thromboxane A_2 is impaired, whereas the platelet response to ADP and collagen seems to be almost unaffected [48]. A current study also indicates a role of the Akt3 isoform in platelets as Akt3-deficient mice showed significantly reduced activation-dependent integrin $\alpha_{IIb}\beta_3$ outside-in signaling and aggregation as well as impaired in vivo thrombus formation in a $FeCl_3$-induced approach [49].

In platelets there is accumulating evidence that the PI3K/Akt signaling pathway plays a central role in chemokine-mediated platelet activation and secretion. The CXCL16 chemokine, a newly discovered marker of coronary atherosclerosis, induces platelet activation and adhesion via an Akt1/Akt2-dependent signaling. In platelets from appropriate knockout mice there is a significantly decreased platelet activation and adhesion after CXCL16 stimulation [11]. In addition, the macrophage migration inhibitory factor (MIF) limits activation-dependent apoptosis in platelets via an Akt-dependent mechanism [20], whereas the Par1-driven SDF-1α secretion in platelets is similarly Akt/PKB sensitive [15]. Especially the platelet-derived MIF chemokine was recently described as a principal supporter of inflammatory cardiovascular diseases like artherosclerosis by acting in a paracrine/autocrine manner as clotting modulator, platelet activator, and chemotactic recruiter of monocytes [50]. The peptidyl-prolyl-*cis/trans* isomerase Cyclophilin A (CypA) occurs in platelets as cytoplasmic and extracellular immunmodulatory protein.

Both fractions are described in various cardiovascular inflammatory processes such as atherosclerosis [51], inflammatory cardiomyopathies [52] and myocardial ischemia [53]. Notably especially the extracellular CypA cytokine modulates platelet activation, adhesion and thrombus formation via a CD147 (cluster of differentiation 147)/PI3K/Akt signaling axis [23], while the intracellular CypA fraction is mainly effective through calcium signaling independently of PI3K/Akt [22].

Beyond the mediation of chemokine and cytokine stimulation, Akt also influences several downstream effectors in platelets, thus contributing to atherothrombosis and inflammation. For instance, the NO synthase isoforms in platelets and endothelial cells are regulated in an Akt-dependent manner [54] resulting in an altered platelet granule exocytosis [55]. Above all, in platelets Akt mediates cyclic adenosine monophosphate (cAMP) levels and actions by affecting cAMP-dependent phosphodiesterase (PDE3A) after thrombin and ADP stimulation [56]. One of the further well-known downstream targets of Akt/PKB in platelets is the glycogen synthase kinase 3 [57].

11.2.1.3 Glycogen Synthase Kinase 3 β (GSK3β)

The glycogen synthase kinase 3 β (GSK3β) is a ubiquitously expressed Serine/Threonine kinase regulating several cellular events. In platelets, the β isoform is the predominant isoform, whereas the α isoform seems to play only an ancillary role in platelet function [57]. GSK3β is a negative regulator of platelet activation incorporating the GPCR as well as the ITAM-mediated stimulation as inhibition of GSK3β leads to enhanced platelet activation [57, 58]. Usually, the constitutive active GSK3β is regulated by phosphorylation at Ser9 which leads to a decreased GSK3β activity and therefore to hyperreactive platelets. The Ser9 phosphorylation in platelets is mainly Akt/PKB dependent [57]. Impaired GSK3β also leads to increased integrin $\alpha_{IIb}\beta_3$ activation and platelet granule secretion after thrombin stimulation [59]. Particularly, thrombus formation and stability at high shear rates is mediated by specific GSK3β platelet events [60], thus promoting the progression of atherothrombotic events. Recently, it was shown that the glycogen synthase kinase 3 β is connected with chemokine and cytokine signaling in inflammation as well as in Toll-like receptor (TLR) [61] and CCL5/RANTES-dependent signaling [62]. Both events are known mechanisms in platelet activation. In cardiac stem/progenitor cells, GSK3β generates the SDF-1/CXCL12-mediated cell migration and quiescence [63, 64]. These findings are of interest as platelets could also be stimulated by all of the mentioned chemokines and signaling pathways [65].

11.2.1.4 Serum- and Glucocorticoid-Inducible Kinase 1 (SGK1)

The serum- and glucocorticoid-inducible kinase 1 (SGK1) is a further member of the AGC kinase family downstream of the 3-phosphoinositide-dependent protein kinase-1 (PDK1) signaling. Phosphorylation of Thr256 and Ser422 in the activation loop of SGK1 is executed by PDK1 directly [66] or via an mTOR complex [67]. Recently, it has been shown that the serum- and glucocorticoid-inducible kinase 1 is strongly expressed in platelets and megakaryocytes and plays a pivotal role in platelet activation upon stimulation with thrombin and collagen-related peptide [44, 68]. SGK1 has originally been cloned as a gene sensitive to glucocorticoids but was later shown to be regulated by a variety of hormones and growth factors such as insulin-like growth factor 1 (IGF-1) and TGF-β, oxidative stress, and ischemia, all factors elevating platelet activity [69]. In platelets, SGK1 is mainly effective by upregulating the expression of the pore-forming Ca^{2+} channel subunit Orai1 [44] facilitating activation-dependent increase of platelet cytosolic Ca^{2+} activity accomplished by store-operated calcium entry (SOCE) [70]. An increase of SGK1 expression and activity in platelets may thus lead to increased platelet SOCE and aggregability linked to pathological thrombus formation. Hyperglycemia or diabetes mellitus as well as metabolic syndrome are strong activators of SGK1 and associated with platelet hyperresponsiveness and atherothrombotic complications, such as myocardial infarction or ischemic stroke [71, 72]. In line with these findings, patients with acute myocardial infarction showed a significantly enhanced platelet Orai1 surface expression [73]. According to platelet proteome analysis in SGK1-deficient platelets, SGK1 regulates further proteins involved in platelet activation and arterial thrombus formation by modulating platelet dense granule secretion, such as the small GTPase Rab27b [68]. A recent study identified a common gain-of-function SGK1 gene variant to be associated with ischemic stroke [74]. Beside its important role in platelet activation, SGK1 signaling also plays a central role in vascular inflammation and atherogenesis in an ApoE-deficient mouse model by participating in the regulation of monocyte/macrophage migration and matrix metalloproteinase 9 (MMP-9) transcription via the modulation of transcription factor nuclear factor-κB [75].

11.2.2 PI3K-Dependent Ca^{2+} Signaling

The activation-dependent increase of the intracellular calcium concentration is a prerequisite for physiological and pathophysiological platelet function as well as thrombus formation and hemostasis. For this reason, the intracellular calcium concentration and influx is regulated precisely in platelets [29]. A major mechanism in platelet calcium signaling is the so-called store-operated calcium entry (SOCE) mediated by the STIM1 and Orai1 coupling machinery [70].

After release of calcium from intracellular stores, the stromal interacting molecule 1 (STIM1) undergoes a conformational change, clusters, and builds, together with the Orai1 protein in the plasma membrane, the so-called calcium release-activated Ca^{2+} channel (CRAC) with subsequent extracellular calcium influx [44]. A R93W point mutation in the Orai1 gene leads to a significant impaired calcium influx in murine platelets with ultimately decreased platelet function [76]. Orai1 and STIM1 expression is regulated in megakaryocytes in a PI3K/SGK1-dependent manner by regulating NF-κB gene expression, major determinants of platelet calcium signaling and platelet function [44, 77]. Along those lines, PI3K inhibitors completely abolish the thrombin-dependent upregulation of Orai1 surface expression and the following calcium influx/release [73, 78] pointing to a significant contribution of PI3K to platelet calcium signaling. Beside its role in atherothrombosis and hemostasis, store-operated Ca^{2+} entry accomplished by STIM1 and Orai1 participates in diverse further functions including immunity and cancer [79, 80].

A further calcium signaling mechanism in platelets is mediated by the activation and regulation of phospholipases (Fig. 11.1). After platelet GPVI and GPCR stimulation by collagen and thrombin, phospholipase C (PLC) is activated. Platelets mainly express the isoforms PLCβ and PLCγ [29]. The GPVI signaling acts via activation and recruitment of the phospholipase Cγ [81], thereby controlling thrombus formation induced by collagen and tissue factor in vitro and in vivo [82]. The PLCγ isoform constitutes a link between PI3K-dependent and independent calcium signaling in platelets as the PI3K-generated phosphoinositide-(3,4,5)P_3 recruits the PLCγ to the plasma membrane and fosters PLCγ activation. This becomes important as collagen-related peptide regulates the PLCγ2 in human platelets via a PI3K-dependent mechanism [83]. After PI3K inhibition by LY294002 and wortmannin, the PLCγ activation is markedly decreased in platelets [32]. The PLCγ2 also acts in platelets downstream of the integrin $\alpha_{IIb}\beta_3$ outside-in signaling by influencing cytoskeletal rearrangement and thus platelet spreading [84]. After fibrinogen stimulation of the integrin $\alpha_{IIb}\beta_3$, the PLCγ2 also contributes to physiological clot retraction as in PLCγ2-deficient mouse platelets there are significantly impaired clot retractions after thrombus formation [85]. The negative regulatory role of the PLCγ2 isoform in platelet function is illustrated by the platelet hyperreactivity and prothrombotic phenotype of mice carrying a gain-of-function mutation in PLCγ2 [86].

The G_q-protein coupled receptors for thrombin, ADP, and TxA_2 signal by activating the PLCβ isoform [87]. Nevertheless, activation of both isoforms leads to the generation of cytosolic second messengers. After PLC activation, inositol 1,4,5 trisphosphate (IP_3) and diacylglycerol (DAG) are formed in platelets. Once generated, IP_3 binds to its appropriate inositol 1,4,5 trisphosphate receptors (IP_3R) in the membrane of intracellular calcium stores. By binding to its appropriate receptors, IP_3 causes the release of calcium with subsequent store depletion and stim1 clustering. The second PLC-generated second messenger DAG binds to protein kinase C, thus strengthening the kinase activity. Furthermore, DAG locates the protein kinase C to plasma membrane where the signaling machinery takes place.

11.3 Protein kinase C (PKC)

The protein kinase C (PKC) is a further popular Serine/Threonine kinase in platelet function. Platelets primarily express highly the classical isoforms α and β as well as the novel isoforms δ and θ [88]. By using different knockout mouse models and PKC inhibitors, Gilio et al. showed that the different PKC isoforms play distinct roles in platelet function and thrombus formation after collagen exposition. According to this study, the collagen-dependent α-granule secretion and thrombus formation is mediated by the conventional PKCα and β isoforms. In addition, the novel PKCθ isoform negatively regulates thrombus formation at shear stress [88]. This is consistent with previous findings where the PKC isoforms balance the pro-aggregatory and procoagulant functions of thrombi as well as thrombin generation and platelet degranulation in a calcium-dependent manner [89] suggesting a role of platelet PKC isoforms in inflammatory atherothrombotic events. These findings are consistent with the fact that the α and β isoforms need DAG and calcium for activation, whereas the δ and θ isoforms only need the DAG as activating factor of kinase activity. The protein kinase C in platelets also operates after stimulation of the G protein-coupled receptors as in a pharmacological approach it was shown that Ca^{2+}/SFKs/PI3K and PKC represent two alternative signaling pathways mediating G(q)-dependent platelet activation [90]. On the contrary, activation of the protease-activated receptor 1 (PAR-1) seems to mediate negatively calcium signaling in platelets as inhibition of PKC after G_q activation leads to increased calcium influx indicating a role of calcium concentration in determination of platelet activation and atherothrombosis [91]. A dual role of the PKC in platelet activation was also shown in 2011 when it was indicated that PKC-dependent mechanisms regulate platelet dense granule secretion after collagen-related peptide (CRP) and thrombin stimulation [92]. GSK3β activation is also controlled by a PKCα-dependent phosphorylation and PKB/Akt and PLC/PKCα play a dual role in inhibition of GSK3β after thrombin-dependent platelet stimulation and thrombus formation [59]. The strictly regulated cross talk between the PI3K and PLC signaling is thus pivotal for the regulation of platelet function and thrombus formation under shear stress, atherothrombosis, hemostasis and inflammation.

11.4 Inflammatory Ligands Triggering Platelet PI3K Signaling

Platelet-derived chemokines and immunmodulatory molecules assume a role in inflammatory cardiovascular diseases, whereby chemokine-dependent platelet activation is strictly PI3K dependent (Fig. 11.2). There are strong clinical studies that link platelet-derived inflammatory molecules with cardiovascular diseases, thus opening novel therapeutic opportunities [93].

Platelet-derived chemokine CXCL16 acts as a pro-inflammatory chemokine which is highly expressed in atherosclerotic lesions [94] and acute coronary syndrome (ACS) [13]. Particularly in ACS, CXCL16 expression is associated with long-term mortality [95]. CXCL16/CXCR6 signaling mainly acts through a PI3K/Akt-dependent pathway in prostate cancer cells [96]. Furthermore, CXCL16 fosters platelet activation by influencing degranulation, thus enhancing autocrine platelet activation and thrombus formation similarly in a PI3K/Akt-dependent manner [11]. Along those lines, the CXCL16 chemokine could be linked to inflammatory cardiomyopathies and heart failure. Therefore, CXCL16 could serve as novel biomarker and predictor of mortality in these diseases [12].

Chemokine C-X3-C motif ligand 1 (CX3CL1/fractalkine) is a further chemokine which could represent a link between platelet activation and inflammatory signaling in atherothrombosis. The CX3CL1 chemokine shows close structural similarities with CXCL16 [96]. CX3CL1 thus influences platelet activation and adhesion [97]. More recent studies show a significant contribution of CX3CL1 to cardiovascular diseases or atherothrombotic events by influencing platelet and monocyte function [98]. CX3CL1 aggravates platelet activation at the vulnerable plaque thus fostering the progression of atherosclerosis [99] and promotes platelet activation and vascular dysfunction in congestive heart failure [100].

A further well-understood chemokine in platelet physiology is the so-called stromal cell-derived factor 1 (SDF-1) or CXCL12 [65]. Abi-Younes and colleagues were able to identify the SDF-1 as a new highly potent platelet agonist that is expressed in atherosclerotic plaques. They also showed that SDF-1 stimulated platelets signals via a PI3K-dependent pathway as after wortmannin and LY294002 treatment there was a significantly decreased SDF-1-dependent platelet aggregation [101]. C-X-C chemokines are described as novel autocrine platelet activators [102]. Platelet-derived SDF-1 regulates monocyte function, survival, and differentiation into macrophages and foam cell, thus contributing to the progression of atherosclerosis and other atherothrombotic disorders [103]. For instance, the platelet-derived SDF-1 level is significantly increased in patients with acute coronary syndrome (ACS) [17]. Furthermore, SDF-1 is connected to the inflammation process of myocarditis as the endomyocardial expression of SDF-1 could predict mortality and outcome of patients with suspected myocarditis [18].

The macrophage migration inhibitory factor (MIF) is released by platelets, thus influencing the monocyte migration in vascular inflammation [104]. MIF evokes an antiapoptotic effect in platelets, thus sustaining platelet life span in the circulation of mice. The effect is dependent on PI3K/Akt signaling [20]. In coronary artery disease, the MIF and the appropriate inhibitor Gremlin-1 influences the progression of these diseases [105], while CypA affects myocardial fibrosis after coxsackievirus B3-induced myocarditis [52].

11.5 Conclusions

Platelet degranulation, integrin $\alpha_{IIb}\beta_3$ activation, and procoagulant events after stimulation of platelet surface receptors lead to primary hemostasis minimizing blood loss or pathologic thrombus formation resulting in acute vascular occlusion. Basic mechanisms in platelet activation include phosphoinositide 3-kinase (PI3K)-dependent signaling pathways. As platelets store and release (in an activation-dependent manner) a vast array of immunmodulatory molecules, the platelet-specific PI3K-dependent pathways have the ability to affect inflammatory atherothrombotic events like atherosclerosis, inflammatory cardiomyopathies, and heart failure. As described above, a broad range of chemokines activate and modulate platelet function and activation, thus mediating inflammatory actions in cardiovascular diseases [106]. Thereby, chemokines and cytokines modulate platelet function in combination with weak platelet agonists like ADP [107] or independently of co-stimulation and preactivation, respectively [97]. Thus, platelet-dependent PI3K signaling events after physiological (Fig. 11.1) or immunodulatory (Fig. 11.2) stimulation can influence atherothrombosis and inflammation. PI3 kinase-dependent signaling may thus become an attractive target for pharmacological treatment of inflammatory cardiovascular diseases and atherothrombotic events.

Compliance with Ethical Standards

Conflict of Interest: Oliver Borst, Florian Lang, and Patrick Münzer declares that they have no conflict of interest.

Ethical Approval: This article does not contain any studies with human participants or animals performed by any of the authors.

References

1. Nieswandt B, Stritt S. Megakaryocyte rupture for acute platelet needs. J Cell Biol. 2015;209:327–8.
2. Jackson SP. Arterial thrombosis—insidious, unpredictable and deadly. Nat Med. 2011;17:1423–36.
3. Lang PA, Contaldo C, Georgiev P, El-Badry AM, Recher M, Kurrer M, Cervantes-Barragan-L, Ludewig B, Calzascia T, Bolinger B, Merkler D, Odermatt B, Bader M, Graf R, Clavien PA, Hegazy AN, Lohning M, Harris NL, Ohashi PS, Hengartner H, Zinkernagel RM, Lang KS. Aggravation of viral hepatitis by platelet-derived serotonin. Nat Med. 2008;14:756–61.
4. Langer HF, Choi EY, Zhou H, Schleicher R, Chung KJ, Tang Z, Gobel K, Bdeir K, Chatzigeorgiou A, Wong C, Bhatia S, Kruhlak MJ, Rose JW, Burns JB, Hill KE, Qu H, Zhang Y, Lehrmann E, Becker KG, Wang Y, Simon DI, Nieswandt B, Lambris JD, Li X, Meuth SG, Kubes P, Chavakis T. Platelets contribute to the pathogenesis of experimental autoimmune encephalomyelitis. Circ Res. 2012;110:1202–10.
5. Negrotto S, Jaquenod de Giusti C, Rivadeneyra L, Ure AE, Mena HA, Schattner M, Gomez RM. Platelets interact with Coxsackieviruses B and have a critical role in the pathogenesis of virus-induced myocarditis. J Thromb Haemost. 2015;13:271–82.
6. Gawaz M, Langer H, May AE. Platelets in inflammation and atherogenesis. J Clin Invest. 2005;115:3378–84.
7. Ruggeri ZM. Platelets in atherothrombosis. Nat Med. 2002;8:1227–34.
8. Herter JM, Rossaint J, Zarbock A. Platelets in inflammation and immunity. J Thromb Haemost. 2014;12:1764–75.
9. Levin J. The evolution of mammalian platelets. Platelets. AD Michelson. New York: Elsevier; 2013.
10. Semple JW, Italiano JE Jr, Freedman J. Platelets and the immune continuum. Nat Rev Immunol. 2011;11:264–74.
11. Borst O, Munzer P, Gatidis S, Schmidt EM, Schonberger T, Schmid E, Towhid ST, Stellos K, Seizer P, May AE, Lang F, Gawaz M. The inflammatory chemokine CXC motif ligand 16 triggers platelet activation and adhesion via CXC motif receptor 6-dependent phosphatidylinositide 3-kinase/Akt signaling. Circ. Res. 2012;111:1297–307.
12. Borst O, Schaub M, Walker B, Sauter M, Muenzer P, Gramlich M, Mueller K, Geisler T, Lang F, Klingel K, Kandolf R, Bigalke B, Gawaz M, Zuern CS. CXCL16 is a novel diagnostic marker and predictor of mortality in inflammatory cardiomyopathy and heart failure. Int J Cardiol. 2014;176:896–903.

13. Seizer P, Stellos K, Selhorst G, Kramer BF, Lang MR, Gawaz M, May AE. CXCL16 is a novel scavenger receptor on platelets and is associated with acute coronary syndrome. Thromb Haemost. 2011;105:1112–4.
14. Sibbing D, Schulz C. Platelet CCL5 links acute coronary syndrome and vascular inflammation. Thromb Haemost. 2014;112:1079.
15. Chatterjee M, Seizer P, Borst O, Schonberger T, Mack A, Geisler T, Langer HF, May AE, Vogel S, Lang F, Gawaz M. SDF-1alpha induces differential trafficking of CXCR4-CXCR7 involving cyclophilin A, CXCR7 ubiquitination and promotes platelet survival. FASEB J. 2014;28:2864–78.
16. Kraemer BF, Borst O, Gehring EM, Schoenberger T, Urban B, Ninci E, Seizer P, Schmidt C, Bigalke B, Koch M, Martinovic I, Daub K, Merz T, Schwanitz L, Stellos K, Fiesel F, Schaller M, Lang F, Gawaz M, Lindemann S. PI3 kinase-dependent stimulation of platelet migration by stromal cell-derived factor 1 (SDF-1). J Mol Med (Berl). 2010;88:1277–88.
17. Rath D, Chatterjee M, Borst O, Muller K, Stellos K, Mack AF, Bongartz A, Bigalke B, Langer H, Schwab M, Gawaz M, Geisler T. Expression of stromal cell-derived factor-1 receptors CXCR4 and CXCR7 on circulating platelets of patients with acute coronary syndrome and association with left ventricular functional recovery. Eur Heart J. 2014;35:386–94.
18. Zuern CS, Walker B, Sauter M, Schaub M, Chatterjee M, Mueller K, Rath D, Vogel S, Tegtmeyer R, Seizer P, Geisler T, Kandolf R, Lang F, Klingel K, Gawaz M, Borst O. Endomyocardial expression of SDF-1 predicts mortality in patients with suspected myocarditis. Clin Res Cardiol. 2015;104:1033–43.
19. Hottz ED, Lopes JF, Freitas C, Valls-de-Souza R, Oliveira MF, Bozza MT, Da Poian AT, Weyrich AS, Zimmerman GA, Bozza FA, Bozza PT. Platelets mediate increased endothelium permeability in dengue through NLRP3-inflammasome activation. Blood. 2013;122:3405–14.
20. Chatterjee M, Borst O, Walker B, Fotinos A, Vogel S, Seizer P, Mack A, Alampour-Rajabi S, Rath D, Geisler T, Lang F, Langer HF, Bernhagen J, Gawaz M. Macrophage migration inhibitory factor limits activation-induced apoptosis of platelets via CXCR7-dependent Akt signaling. Circ Res. 2014;115:939–49.
21. Vogel S, Bodenstein R, Chen Q, Feil S, Feil R, Rheinlaender J, Schaffer TE, Bohn E, Frick JS, Borst O, Munzer P, Walker B, Markel J, Csanyi G, Pagano PJ, Loughran P, Jessup ME, Watkins SC, Bullock GC, Sperry JL, Zuckerbraun BS, Billiar TR, Lotze MT, Gawaz M, Neal MD. Platelet-derived HMGB1 is a critical mediator of thrombosis. J Clin Invest. 2015;125:4638–54.
22. Elvers M, Herrmann A, Seizer P, Munzer P, Beck S, Schonberger T, Borst O, Martin-Romero FJ, Lang F, May AE, Gawaz M. Intracellular cyclophilin A is an important Ca(2+) regulator in platelets and critically involved in arterial thrombus formation. Blood. 2012;120:1317–26.
23. Seizer P, Ungern-Sternberg SN, Schonberger T, Borst O, Munzer P, Schmidt EM, Mack AF, Heinzmann D, Chatterjee M, Langer H, Malesevic M, Lang F, Gawaz M, Fischer G, May AE. Extracellular cyclophilin A activates platelets via EMMPRIN (CD147) and PI3K/Akt signaling, which promotes platelet adhesion and thrombus formation in vitro and in vivo. Arterioscler Thromb Vasc Biol. 2015;35:655–63.
24. Seizer P, Gawaz M, May AE. Cyclophilin A and EMMPRIN (CD147) in cardiovascular diseases. Cardiovasc Res. 2014;102:17–23.
25. Franco AT, Corken A, Ware J. Platelets at the interface of thrombosis, inflammation, and cancer. Blood. 2015;126:582–8.
26. Borst O, Abed M, Alesutan I, Towhid ST, Qadri SM, Foller M, Gawaz M, Lang F. Dynamic adhesion of eryptotic erythrocytes to endothelial cells via CXCL16/SR-PSOX. Am J Physiol Cell Physiol. 2012;302:C644–51.
27. Wuttge DM, Zhou X, Sheikine Y, Wagsater D, Stemme V, Hedin U, Stemme S, Hansson GK, Sirsjo A. CXCL16/SR-PSOX is an interferon-gamma-regulated chemokine and scavenger receptor expressed in atherosclerotic lesions. Arterioscler Thromb Vasc Biol. 2004;24:750–5.

28. Bergmeier W, Stefanini L. Novel molecules in calcium signaling in platelets. J Thromb Haemost. 2009;7(Suppl 1):187–90.
29. Varga-Szabo D, Braun A, Nieswandt B. Calcium signaling in platelets. J Thromb Haemost. 2009;7:1057–66.
30. Jackson SP, Schoenwaelder SM, Goncalves I, Nesbitt WS, Yap CL, Wright CE, Kenche V, Anderson KE, Dopheide SM, Yuan Y, Sturgeon SA, Prabaharan H, Thompson PE, Smith GD, Shepherd PR, Daniele N, Kulkarni S, Abbott B, Saylik D, Jones C, Lu L, Giuliano S, Hughan SC, Angus JA, Robertson AD, Salem HH. PI 3-kinase p110beta: a new target for antithrombotic therapy. Nat Med. 2005;11:507–14.
31. Gratacap MP, Guillermet-Guibert J, Martin V, Chicanne G, Tronchere H, Gaits-Iacovoni F, Payrastre B. Regulation and roles of PI3Kbeta, a major actor in platelet signaling and functions. Adv Enzyme Regul. 2011;51:106–16.
32. Moroi AJ, Watson SP. Impact of the PI3-kinase/Akt pathway on ITAM and hemITAM receptors: haemostasis, platelet activation and antithrombotic therapy. Biochem Pharmacol. 2015;94:186–94.
33. Cosemans JM, Munnix IC, Wetzker R, Heller R, Jackson SP, Heemskerk JW. Continuous signaling via PI3K isoforms beta and gamma is required for platelet ADP receptor function in dynamic thrombus stabilization. Blood. 2006;108:3045–52.
34. Gilio K, Munnix IC, Mangin P, Cosemans JM, Feijge MA, van der Meijden PE, Olieslagers S, Chrzanowska-Wodnicka MB, Lillian R, Schoenwaelder S, Koyasu S, Sage SO, Jackson SP, Heemskerk JW. Non-redundant roles of phosphoinositide 3-kinase isoforms alpha and beta in glycoprotein VI-induced platelet signaling and thrombus formation. J Biol Chem. 2009;284:33750–62.
35. Alessi DR, James SR, Downes CP, Holmes AB, Gaffney PR, Reese CB, Cohen P. Characterization of a 3-phosphoinositide-dependent protein kinase which phosphorylates and activates protein kinase Balpha. Curr Biol. 1997;7:261–9.
36. Chen X, Zhang Y, Wang Y, Li D, Zhang L, Wang K, Luo X, Yang Z, Wu Y, Liu J. PDK1 regulates platelet activation and arterial thrombosis. Blood. 2013;121:3718–26.
37. Dangelmaier C, Manne BK, Liverani E, Jin J, Bray P, Kunapuli SP. PDK1 selectively phosphorylates Thr(308) on Akt and contributes to human platelet functional responses. Thromb Haemost. 2014;111:508–17.
38. Primo L, di Blasio L, Roca C, Droetto S, Piva R, Schaffhausen B, Bussolino F. Essential role of PDK1 in regulating endothelial cell migration. J Cell Biol. 2007;176:1035–47.
39. Finlay DK, Rosenzweig E, Sinclair LV, Feijoo-Carnero C, Hukelmann JL, Rolf J, Panteleyev AA, Okkenhaug K, Cantrell DA. PDK1 regulation of mTOR and hypoxia-inducible factor 1 integrate metabolism and migration of CD8+ T cells. J Exp Med. 2012;209:2441–53.
40. Yagi M, Kantarci A, Iwata T, Omori K, Ayilavarapu S, Ito K, Hasturk H, Van Dyke TE. PDK1 regulates chemotaxis in human neutrophils. J Dent Res. 2009;88:1119–24.
41. Pinner S, Sahai E. PDK1 regulates cancer cell motility by antagonising inhibition of ROCK1 by RhoE. Nat Cell Biol. 2008;10:127–37.
42. Luo Y, Cai J, Xue H, Miura T, Rao MS. Functional SDF1 alpha/CXCR4 signaling in the developing spinal cord. J Neurochem. 2005;93:452–62.
43. Slattery ML, Lundgreen A, Herrick JS, Wolff RK. Genetic variation in RPS6KA1, RPS6KA2, RPS6KB1, RPS6KB2, and PDK1 and risk of colon or rectal cancer. Mutat Res. 2011;706:13–20.
44. Borst O, Schmidt EM, Munzer P, Schonberger T, Towhid ST, Elvers M, Leibrock C, Schmid E, Eylenstein A, Kuhl D, May AE, Gawaz M, Lang F. The serum- and glucocorticoid-inducible kinase 1 (SGK1) influences platelet calcium signaling and function by regulation of Orai1 expression in megakaryocytes. Blood. 2012;119:251–61.
45. Risso G, Blaustein M, Pozzi B, Mammi P, Srebrow A. Akt/PKB: one kinase, many modifications. Biochem J. 2015;468:203–14.
46. Woulfe DS. Akt signaling in platelets and thrombosis. Expert Rev Hematol. 2010;3:81–91.

47. Chen J, De S, Damron DS, Chen WS, Hay N, Byzova TV. Impaired platelet responses to thrombin and collagen in AKT-1-deficient mice. Blood. 2004;104:1703–10.
48. Woulfe D, Jiang H, Morgans A, Monks R, Birnbaum M, Brass LF. Defects in secretion, aggregation, and thrombus formation in platelets from mice lacking Akt2. J Clin Invest. 2004;113:441–50.
49. O'Brien KA, Stojanovic-Terpo A, Hay N, Du X. An important role for Akt3 in platelet activation and thrombosis. Blood. 2011;118:4215–23.
50. Wirtz TH, Tillmann S, Strussmann T, Kraemer S, Heemskerk JW, Grottke O, Gawaz M, von Hundelshausen P, Bernhagen J. Platelet-derived MIF: a novel platelet chemokine with distinct recruitment properties. Atherosclerosis. 2015;239:1–10.
51. Seizer P, Schonberger T, Schott M, Lang MR, Langer HF, Bigalke B, Kramer BF, Borst O, Daub K, Heidenreich O, Schmidt R, Lindemann S, Herouy Y, Gawaz M, May AE. EMMPRIN and its ligand cyclophilin A regulate MT1-MMP, MMP-9 and M-CSF during foam cell formation. Atherosclerosis. 2010;209:51–7.
52. Seizer P, Klingel K, Sauter M, Westermann D, Ochmann C, Schonberger T, Schleicher R, Stellos K, Schmidt EM, Borst O, Bigalke B, Kandolf R, Langer H, Gawaz M, May AE. Cyclophilin A affects inflammation, virus elimination and myocardial fibrosis in coxsackievirus B3-induced myocarditis. J Mol Cell Cardiol. 2012;53:6–14.
53. Seizer P, Ochmann C, Schonberger T, Zach S, Rose M, Borst O, Klingel K, Kandolf R, MacDonald HR, Nowak RA, Engelhardt S, Lang F, Gawaz M, May AE. Disrupting the EMMPRIN (CD147)-cyclophilin A interaction reduces infarct size and preserves systolic function after myocardial ischemia and reperfusion. Arterioscler Thromb Vasc Biol. 2011;31:1377–86.
54. Marjanovic JA, Li Z, Stojanovic A, Du X. Stimulatory roles of nitric-oxide synthase 3 and guanylyl cyclase in platelet activation. J Biol Chem. 2005;280:37430–8.
55. Morrell CN, Matsushita K, Chiles K, Scharpf RB, Yamakuchi M, Mason RJ, Bergmeier W, Mankowski JL, Baldwin WM 3rd, Faraday N, Lowenstein CJ. Regulation of platelet granule exocytosis by S-nitrosylation. Proc Natl Acad Sci USA. 2005;102:3782–7.
56. Zhang W, Colman RW. Thrombin regulates intracellular cyclic AMP concentration in human platelets through phosphorylation/activation of phosphodiesterase 3A. Blood. 2007;110:1475–82.
57. Li D, August S, Woulfe DS. GSK3beta is a negative regulator of platelet function and thrombosis. Blood. 2008;111:3522–30.
58. Moroi AJ, Watson SP. Akt and mitogen-activated protein kinase enhance C-type lectin-like receptor 2-mediated platelet activation by inhibition of glycogen synthase kinase 3alpha/beta. J Thromb Haemost. 2015;13:1139–50.
59. Moore SF, van den Bosch MT, Hunter RW, Sakamoto K, Poole AW, Hers I. Dual regulation of Glycogen Synthase Kinase 3 (GSK3)alpha/beta by Protein Kinase C (PKC)alpha and akt promotes thrombin-mediated Integrin alphaIIbbeta3 activation and granule secretion in platelets. J Biol Chem. 2013;288:3918–28.
60. Laurent PA, Severin S, Hechler B, Vanhaesebroeck B, Payrastre B, Gratacap MP. Platelet PI3Kbeta and GSK3 regulate thrombus stability at a high shear rate. Blood. 2015;125:881–8.
61. Martin M, Rehani K, Jope RS, Michalek SM. Toll-like receptor-mediated cytokine production is differentially regulated by glycogen synthase kinase 3. Nat Immunol. 2005;6:777–84.
62. Chan O, Burke JD, Gao DF, Fish EN. The chemokine CCL5 regulates glucose uptake and AMP kinase signaling in activated T cells to facilitate chemotaxis. J Biol Chem. 2012;287:29406–16.
63. Dimova N, Wysoczynski M, Rokosh G. Stromal cell derived factor-1alpha promotes C-Kit+ cardiac stem/progenitor cell quiescence through casein kinase 1alpha and GSK3beta. Stem Cells. 2014;32:487–99.
64. Lapid K, Itkin T, D'Uva G, Ovadya Y, Ludin A, Caglio G, Kalinkovich A, Golan K, Porat Z, Zollo M, Lapidot T. GSK3beta regulates physiological migration of stem/progenitor cells via cytoskeletal rearrangement. J Clin Invest. 2013;123:1705–17.

65. Chatterjee M, Gawaz M. Platelet-derived CXCL12 (SDF-1alpha): basic mechanisms and clinical implications. J Thromb Haemost. 2013;11:1954–67.
66. Kobayashi T, Cohen P. Activation of serum- and glucocorticoid-regulated protein kinase by agonists that activate phosphatidylinositide 3-kinase is mediated by 3-phosphoinositide-dependent protein kinase-1 (PDK1) and PDK2. Biochem J. 1999;339(Pt 2):319–28.
67. Garcia-Martinez JM, Alessi DR. mTOR complex 2 (mTORC2) controls hydrophobic motif phosphorylation and activation of serum- and glucocorticoid-induced protein kinase 1 (SGK1). Biochem J. 2008;416:375–85.
68. Walker B, Schmid E, Russo A, Schmidt EM, Burk O, Munzer P, Velic A, Macek B, Schaller M, Schwab M, Seabra MC, Gawaz M, Lang F, Borst O. Impact of the serum- and glucocorticoid-inducible kinase 1 on platelet dense granule biogenesis and secretion. J Thromb Haemost. 2015;13:1325–34.
69. Lang F, Bohmer C, Palmada M, Seebohm G, Strutz-Seebohm N, Vallon V. (Patho)physiological significance of the serum- and glucocorticoid-inducible kinase isoforms. Physiol Rev. 2006;86:1151–78.
70. Braun A, Varga-Szabo D, Kleinschnitz C, Pleines I, Bender M, Austinat M, Bosl M, Stoll G, Nieswandt B. Orai1 (CRACM1) is the platelet SOC channel and essential for pathological thrombus formation. Blood. 2009;113:2056–63.
71. Ferreira IA, Mocking AI, Feijge MA, Gorter G, van Haeften TW, Heemskerk JW, Akkerman JW. Platelet inhibition by insulin is absent in type 2 diabetes mellitus. Arterioscler Thromb Vasc Biol. 2006;26:417–22.
72. Lang F, Munzer P, Gawaz M, Borst O. Regulation of STIM1/Orai1-dependent Ca2+ signalling in platelets. Thromb Haemost. 2013;110:925–30.
73. Tolios A, Gatidis S, Munzer P, Liu GX, Towhid ST, Karathanos A, Tavlaki E, Geisler T, Seizer P, May AE, Bigalke B, Borst O, Gawaz M, Lang F. Increased platelet Ca^{2+} channel Orai1 expression upon platelet activation and in patients with acute myocardial infarction. Thromb Haemost. 2013;110:386–9.
74. Dahlberg J, Smith G, Norrving B, Nilsson P, Hedblad B, Engstrom G, Lovkvist H, Carlson J, Lindgren A, Melander O. Genetic variants in serum and glucocortocoid regulated kinase 1, a regulator of the epithelial sodium channel, are associated with ischaemic stroke. J Hypertens. 2011;29:884–9.
75. Borst O, Schaub M, Walker B, Schmid E, Munzer P, Voelkl J, Alesutan I, Rodriguez JM, Vogel S, Schoenberger T, Metzger K, Rath D, Umbach A, Kuhl D, Muller II, Seizer P, Geisler T, Gawaz M, Lang F. Pivotal role of serum- and glucocorticoid-inducible kinase 1 in vascular inflammation and atherogenesis. Arterioscler Thromb Vasc Biol. 2015;35:547–57.
76. Bergmeier W, Oh-Hora M, McCarl CA, Roden RC, Bray PF, Feske S. R93W mutation in Orai1 causes impaired calcium influx in platelets. Blood. 2009;113:675–8.
77. Borst O, Munzer P, Schmid E, Schmidt EM, Russo A, Walker B, Yang W, Leibrock C, Szteyn K, Schmidt S, Elvers M, Faggio C, Shumilina E, Kuro-o M, Gawaz M, Lang F. 1,25 (OH)2 vitamin D3-dependent inhibition of platelet Ca2+ signaling and thrombus formation in klotho-deficient mice. FASEB J. 2014;28:2108–19.
78. Munzer P, Tolios A, Pelzl L, Schmid E, Schmidt EM, Walker B, Frohlich H, Borst O, Gawaz M, Lang F. Thrombin-sensitive expression of the store operated Ca(2+) channel Orai1 in platelets. Biochem Biophys Res Commun. 2013;436:25–30.
79. Bergmeier W, Weidinger C, Zee I, Feske S. Emerging roles of store-operated Ca(2)(+) entry through STIM and ORAI proteins in immunity, hemostasis and cancer. Channels (Austin). 2013;7:379–91.
80. Lang F, Stournaras C. Ion channels in cancer: future perspectives and clinical potential. Philos Trans R Soc Lond B Biol Sci. 2014;369:20130108.
81. Mazharian A, Thomas SG, Dhanjal TS, Buckley CD, Watson SP. Critical role of Src-Syk-PLC{gamma}2 signaling in megakaryocyte migration and thrombopoiesis. Blood. 2010;116:793–800.

82. Munnix IC, Strehl A, Kuijpers MJ, Auger JM, van der Meijden PE, van Zandvoort MA, oude Egbrink MG, Nieswandt B, Heemskerk JW. The glycoprotein VI-phospholipase Cgamma2 signaling pathway controls thrombus formation induced by collagen and tissue factor in vitro and in vivo. Arterioscler Thromb Vasc Biol. 2005;25:2673–8.
83. Pasquet JM, Bobe R, Gross B, Gratacap MP, Tomlinson MG, Payrastre B, Watson SP. A collagen-related peptide regulates phospholipase Cgamma2 via phosphatidylinositol 3-kinase in human platelets. Biochem J. 1999;342(Pt 1):171–7.
84. Wonerow P, Pearce AC, Vaux DJ, Watson SP. A critical role for phospholipase Cgamma2 in alphaIIbbeta3-mediated platelet spreading. J Biol Chem. 2003;278:37520–9.
85. Suzuki-Inoue K, Hughes CE, Inoue O, Kaneko M, Cuyun-Lira O, Takafuta T, Watson SP, Ozaki Y. Involvement of Src kinases and PLCgamma2 in clot retraction. Thromb Res. 2007;120:251–8.
86. Elvers M, Pozgaj R, Pleines I, May F, Kuijpers MJ, Heemskerk JM, Yu P, Nieswandt B. Platelet hyperreactivity and a prothrombotic phenotype in mice with a gain-of-function mutation in phospholipase Cgamma2. J Thromb Haemost. 2010;8:1353–63.
87. Offermanns S. Activation of platelet function through G protein-coupled receptors. Circ Res. 2006;99:1293–304.
88. Gilio K, Harper MT, Cosemans JM, Konopatskaya O, Munnix IC, Prinzen L, Leitges M, Liu Q, Molkentin JD, Heemskerk JW, Poole AW. Functional divergence of platelet protein kinase C (PKC) isoforms in thrombus formation on collagen. J Biol Chem. 2010;285:23410–9.
89. Strehl A, Munnix IC, Kuijpers MJ, van der Meijden PE, Cosemans JM, Feijge MA, Nieswandt B, Heemskerk JW. Dual role of platelet protein kinase C in thrombus formation: stimulation of pro-aggregatory and suppression of procoagulant activity in platelets. J Biol Chem. 2007;282:7046–55.
90. Xiang B, Zhang G, Stefanini L, Bergmeier W, Gartner TK, Whiteheart SW, Li Z. The Src family kinases and protein kinase C synergize to mediate Gq-dependent platelet activation. J Biol Chem. 2012;287:41277–87.
91. Harper MT, Poole AW. PKC inhibition markedly enhances Ca2+ signaling and phosphatidylserine exposure downstream of protease-activated receptor-1 but not protease-activated receptor-4 in human platelets. J Thromb Haemost. 2011;9:1599–607.
92. Konopatskaya O, Matthews SA, Harper MT, Gilio K, Cosemans JM, Williams CM, Navarro MN, Carter DA, Heemskerk JW, Leitges M, Cantrell D, Poole AW. Protein kinase C mediates platelet secretion and thrombus formation through protein kinase D2. Blood. 2011;118:416–24.
93. Muller KA, Chatterjee M, Rath D, Geisler T. Platelets, inflammation and anti-inflammatory effects of antiplatelet drugs in ACS and CAD. Thromb Haemost. 2015;114:498–518.
94. Zernecke A, Shagdarsuren E, Weber C. Chemokines in atherosclerosis: an update. Arterioscler Thromb Vasc Biol. 2008;28:1897–908.
95. Jansson AM, Aukrust P, Ueland T, Smith C, Omland T, Hartford M, Caidahl K. Soluble CXCL16 predicts long-term mortality in acute coronary syndromes. Circulation. 2009;119:3181–8.
96. Wang J, Lu Y, Wang J, Koch AE, Zhang J, Taichman RS. CXCR6 induces prostate cancer progression by the AKT/mammalian target of rapamycin signaling pathway. Cancer Res. 2008;68:10367–76.
97. Schafer A, Schulz C, Eigenthaler M, Fraccarollo D, Kobsar A, Gawaz M, Ertl G, Walter U, Bauersachs J. Novel role of the membrane-bound chemokine fractalkine in platelet activation and adhesion. Blood. 2004;103:407–12.
98. Flierl U, Bauersachs J, Schafer A. Modulation of platelet and monocyte function by the chemokine fractalkine (CX3 CL1) in cardiovascular disease. Eur J Clin Invest. 2015;45:624–33.
99. Flierl U, Schafer A. Fractalkine—a local inflammatory marker aggravating platelet activation at the vulnerable plaque. Thromb Haemost. 2012;108:457–63.

100. Hildemann SK, Schulz C, Fraccarollo D, Schopp C, Flierl U, Wissel K, Pelisek J, Massberg S, Bauersachs J, Schafer A. Fractalkine promotes platelet activation and vascular dysfunction in congestive heart failure. Thromb Haemost. 2014;111:725–35.
101. Abi-Younes S, Sauty A, Mach F, Sukhova GK, Libby P, Luster AD. The stromal cell-derived factor-1 chemokine is a potent platelet agonist highly expressed in atherosclerotic plaques. Circ Res. 2000;86:131–8.
102. Walsh TG, Harper MT, Poole AW. SDF-1alpha is a novel autocrine activator of platelets operating through its receptor CXCR4. Cell Signal. 2015;27:37–46.
103. Chatterjee M, von Ungern-Sternberg SN, Seizer P, Schlegel F, Buttcher M, Sindhu NA, Muller S, Mack A, Gawaz M. Platelet-derived CXCL12 regulates monocyte function, survival, differentiation into macrophages and foam cells through differential involvement of CXCR4-CXCR7. Cell Death Dis. 2015;6:e1989.
104. Strussmann T, Tillmann S, Wirtz T, Bucala R, von Hundelshausen P, Bernhagen J. Platelets are a previously unrecognised source of MIF. Thromb Haemost. 2013;110:1004–13.
105. Muller II, Muller KA, Karathanos A, Schonleber H, Rath D, Vogel S, Chatterjee M, Schmid M, Haas M, Seizer P, Langer H, Schaeffeler E, Schwab M, Gawaz M, Geisler T. Impact of counterbalance between macrophage migration inhibitory factor and its inhibitor Gremlin-1 in patients with coronary artery disease. Atherosclerosis. 2014;237:426–32.
106. Gleissner CA, von Hundelshausen P, Ley K. Platelet chemokines in vascular disease. Arterioscler Thromb Vasc Biol. 2008;28:1920–7.
107. Gear AR, Suttitanamongkol S, Viisoreanu D, Polanowska-Grabowska RK, Raha S, Camerini D. Adenosine diphosphate strongly potentiates the ability of the chemokines MDC, TARC, and SDF-1 to stimulate platelet function. Blood. 2001;97:937–45.

Linking Pathologies: Cyclophilins in Inflammation and Thrombosis

12

David Heinzmann, Andreas E. May, and Peter Seizer

Abstract

Apart from their intracellular function as chaperones in protein folding, cyclophilins have been found to play important roles in the pathogenesis of thrombosis as well as inflammation. With liberation of cyclophilin A (CyPA) into the extracellular space (eCyPA), it acts as a danger-associated molecular pattern. Following interaction with its primary extracellular receptor CD147, eCyPA facilitates platelet activation with subsequent adhesion to the endothelium, degranulation as well as shape change. Furthermore, the eCyPA–CD147 interaction induces leucocyte adhesion and has strong chemotactic effects on leucocytes. In this chapter, we review the effects of cyclophilins in the context of thrombo-inflammation and give insight into current pharmacological strategies targeting cyclophilins.

Contents

D. Heinzmann • A.E. May • P. Seizer (✉)
University Hospital Tübingen, Otfried-Müller-Straße 10, 72076 Tübingen, Germany
e-mail: peter.seizer@med.uni-tuebingen.de

A. Zirlik et al. (eds.), *Platelets, Haemostasis and Inflammation*,
Cardiac and Vascular Biology 5, https://doi.org/10.1007/978-3-319-66224-4_12

12.1 Introduction

Amongst other intracellular chaperones, cyclophilins (CyPs) are of considerable interest. Due to their peptidyl-prolyl cis/trans isomerase (PPIase) activity, which catalyses isomerization of peptidyl-prolyl bonds at proline residues, cyclophilins are regarded as important intracellular players for the proper folding of newly synthesized peptides as well as restoring the three-dimensional shape of damaged proteins in the intracellular matrix [1].

While being strongly conserved amongst most species, there are many subtypes of CyPs, which are not well characterized. In recent years, cyclophilin A (CyPA) has emerged as a pathophysiologically important factor, which contributes to many, especially clinically important mechanisms of various kinds.

As the intracellular target of cyclosporin A (CsA), it has paved the road for modern organ transplantation by inhibiting the rejection of the donor organ. This is mainly achieved by the CsA-CyPA complex inhibiting the NFAT-dependent activation of T-cells [2]. Lately, CyPA has been found to be of great significance as a pro-inflammatory signal, once it reaches the extracellular space (Figs. 12.1 and 12.2).

Cyclophilin B (CyPB), the second cyclophilin to be discovered, shares a great part of its sequence with CyPA. Histological studies have shown that CyPB is mostly located in the nucleus [3]. Due to its differences in the C-terminal domain, it is also found in the ER, from which it can be secreted easily. Here, it is part of the complex protein folding of secreted proteins facilitated by the calnexin cycle [4].

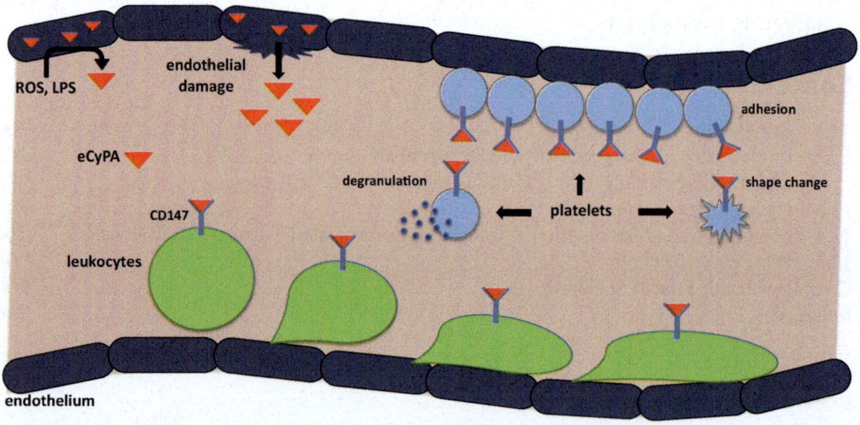

Fig. 12.1 Cyclophilin A as a mediator in thrombo-inflammation. Release of cyclophilin A (CyPA) into the extracellular space (eCyPA) by various stimuli, including reactive oxygen species (ROS), lipopolysaccharides (LPS) and disruption of cellular integrity, facilitates various pro-inflammatory and prothrombotic events. Via the primary receptor CD147 (extracellular matrix metalloproteinase inducer, EMMPRIN), eCyPA induces platelet adhesion to the endothelium, shape change as well as degranulation of platelets. The eCyPA–CD147 interaction furthermore induces leucocyte recruitment to the vessel wall as well as chemotaxis

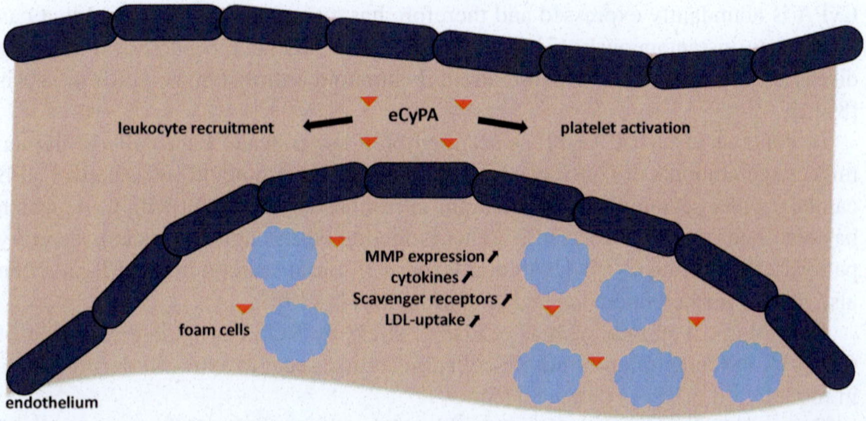

Fig. 12.2 Extracellular cyclophilin A plays a pivotal role in the pathogenesis of atherosclerosis. The role of extracellular cyclophilin A (eCyPA) in atherosclerosis is closely associated with the interaction with CD147. eCyPA has been found to be of great significance for the regulation of MT-1-MMP, MMP-9 and M-CSF during the development of foam cells from CD34+ progenitor cells. Induction of matrix metalloproteinases (MMPs) through eCyPA is considered to be a driver of increased vulnerability in the plaque, leading to rupture and subsequent thrombosis by activation of platelets and coagulation factors. In earlier stages of atherosclerosis, eCyPA furthermore facilitates uptake of low-density lipoprotein (LDL) into the vascular wall. Experimental data suggest that CyPA deficiency leads to a reduced recruitment of leucocytes into the lesions

In the context of inflammation and thrombosis, so far CyPA and CyPB have been the most interesting topics of research amongst the known CyPs. Especially CyPA has been found to be an important pro-inflammatory signal in various pathophysiological conditions.

12.2 Cyclophilins as Danger-Associated Molecular Patterns

The basic idea of danger-associated molecular patterns, alerting the immune system of tissue damage by externalized intracellular molecules, was established by Matzinger in 1994 [5]. Through the presence in the extracellular space, these molecules can interact as a first response with different effector cells of the immune system in order to concert an appropriate response to the occurring danger and initiate restoration processes of the affected tissue once the inflammation is slowing down.

While intracellular CyPA is an important housekeeping gene, its extracellular functions are of great importance for many stress-related responses.

There have been several reports about the release mechanisms of CyPA from the cytoplasm to the extracellular space. Amongst others, release by disruption of the membrane, necrosis and directed secretion in response to various pro-inflammatory stimuli, such as reactive oxygen species (ROS), hypoxia, lipopolysaccharides and others, have been found [6]. With roughly 0.3% of the protein mass of lymphocytes,

CyPA is abundantly expressed and therefore has a great intracellular pool that can be set free instantaneously [7]. Elevated CyPA levels have been shown in many diseases, such as vascular inflammation, rheumatoid arthritis, myocarditis or sepsis [8–11].

In contrast to CyPA, CyPB is secreted on a larger scale. Especially in human milk, large amounts of CyPB can be found [12]. Furthermore, in HeLa cells, CyPB can be rapidly secreted into the medium in response to treatment with CsA, which has also been reported for CyPA. This is facilitated through a classical secretory pathway by forming CyPB-CsA complexes [13]. Basal secretion of CyPB has been also shown for keratinocytes and chondrocytes [14, 15].

Physiological stimuli such as oxidative stress by ROS can initiate secretion of CyPB by vascular smooth muscle cells, indicating specific extracellular functions in modulation of stress defence [16].

Several mechanisms have been elucidated that give insights on how externalized CyPs connect inflammatory responses in various settings.

12.2.1 Extracellular Cyclophilins as Inflammatory Mediators

When released into the extracellular space, CyPA interacts with a broad range of inflammatory cells and orchestrates a pro-inflammatory response with migration and activation of leucocytes as a key feature (Fig. 12.1). Many of these effects seem to be facilitated through the interaction of CyPA with the extracellular domain of *e*xtracellular *m*atrix *m*etallo*pr*oteinase *in*ducer (EMMPRIN, CD147, basigin). CD147 is considered to be the primary receptor molecule for extracellular CyPA signalling. Belonging to the immunoglobulin superfamily, it is expressed on many cell types, including leucocytes and platelets.

While the mechanisms of interaction are still poorly understood, recent studies favour the idea that the PPIase activity of CyPA is crucial to induce downstream signalling by *cis*/*trans* isomerization of prolyl bonds in the extracellular domain of CD147 [17, 18]. Downstream, CyPA-CD147 interaction can lead to activation of p38 MAPK, ERK1/2, NF-κB and JNK and increased Ca^{2+} influx [6]. To add to the complexity, the CyPA-CD147 interaction seems to be dependent on the presence of heparans in the outer membrane of the target cell. It therefore seems plausible that CyPA doesn't primarily bind to CD147, but signalling via CD147 could rather depend on a primary interaction of basic residues located at the C-terminus of CyPA with heparans [19, 20].

Especially on CD4+ leucocytes, studies have shown that activated cells show a higher expression of CD147 on their surface than resting cells. In this context, migration towards extracellular CyPA was increased and was not dependent on the presence of cell surface heparans [21].

Endothelial cells were shown to increase the expression VCAM-1 and E-selectin via stimulation of ERK1/2, JNK and p38 in response to extracellular CyPA, therefore promoting the adhesion and invasion of leucocytes into the inflammatory milieu [22].

CD147 can further facilitate the recruitment of leucocytes by inducing rolling and adhesion on the endothelial surface. Studies have shown that CD147 binds to E-selectin expressed on the surface of endothelial cells and plays a major role in the selectin-mediated invasion into the inflammatory milieu [23].

Migration of leucocytes towards an increasing concentration of extracellular CyPA has been shown in various assays and for many subtypes. Furthermore, inhibition of CD147 and selective inhibition of extracellular but not intracellular CyPA have been shown to ameliorate these effects, underlining the importance of CD147 for extracellular CyPA signalling [18, 24].

Synergistic augmentation of leucocyte recruitment has been shown for many combinations of chemokines. Simultaneous administration of CXCL2 and extracellular CyPA showed an increased CXCR2 receptor internalization, intracellular calcium mobilization and actin polymerization in neutrophils in vitro [25].

Amongst the variety of cellular responses, CyPA-induced matrix metalloproteinase (MMP) induction is an important pathophysiological mechanism in many diseases. Ranging from atherosclerosis to rheumatoid arthritis, MMPs are vital for the invasion and swarming of leucocytes, as well as for degradation and reorganization of extracellular matrix [26, 27]. Especially in tissues with a low cell turnover, such as the heart, the expression of MMPs correlates with an increased percentage of fibrosis, as regeneration is inherently low [28].

Furthermore, CyPA-CD147 interaction, as well as increased expression of MMPs, has been shown to be of great significance in cancer biology. CyPA has been shown to be overexpressed in many cancer types, interfering with apoptosis, proliferation and metastasis [29].

Like CyPA, signalling of extracellular CyPB is largely dependent on interaction with CD147. Similarly to extracellular CyPA, CyPB has been shown to be involved with the recruitment of leucocytes by inducing chemotaxis and adhesion, especially of $CD4^{+}$ $CD45RO^{+}$ T-cells [30–32].

Yet, there are also some interesting differences in the resulting responses. While extracellular CyPA has been shown to stimulate the expression of numerous cytokines, CyPB not only fails to induce them, but macrophages pretreated with CyPB prior to administration of LPS showed reduced expression of pro-inflammatory mediators [33].

12.3 Cyclophilins in Platelet Function

In platelets, being of central interest in many cardiovascular diseases, CyPA plays a pivotal role for activation of platelets and induction of thrombosis in vivo.

Intracellular CyPA has been shown to be a central Ca^{2+} modulator in platelets. Intracellular Ca^{2+} mobilization is a crucial step in the activation of platelets, leading to the release of prothrombotic substances and change of shape and expression of surface molecules. In mice, CyPA deficiency reduces the release of Ca^{2+} from intracellular stores. Furthermore, the recruitment of Ca^{2+} from the extracellular space to the cytoplasm is reduced as well [34]. Inhibition of intracellular CyPA in platelets using CsA reduces the Ca^{2+} reuptake by SERCA2b, by inhibiting the formation of a SERCA2b/CyPA complex [35].

Further studies showed that extracellular CyPA has a significant impact on platelet function as well. In vivo, thrombus formation can be reduced significantly by selectively inhibiting extracellular CyPA. Furthermore, addition of extracellular CyPA to CyPA-deficient platelets enhances thrombus formation in vivo. Especially aggregation, as well as degranulation and therefore expression of P-selectin, SDF-1, active $\alpha_{IIb}\beta_3$, CD41a, CD49b and CD29 ($\alpha_2\beta_1$ integrin) on the surface of the platelet, can be mediated by paracrine CyPA signalling. These effects could also be seen in platelets from CyPA-deficient mice treated with recombinant CyPA. Most of these effects seem to depend on signalling of extracellular CyPA through CD147, as blocking of CD147 abrogates these effects [36].

Interestingly, extracellular CyPB binds to the surface of platelets as well. Studies have shown that addition of CyPB to platelets induces no aggregation or degranulation of any kind. This was also true when platelets were pretreated with low doses of known activators, such as thrombin or ADP. When platelets treated with CyPB were tested for adhesion to collagen, a marked increase was noted, which was diminished by addition of CsA. Further experiments also showed an increase of intracellular free calcium when treated with CyPB [37].

12.4 Mechanisms and Intervention: Cyclophilins and Potential Therapeutic Approaches

Being a universally expressed intracellular protein involved in fundamental cellular functions, CyPA is part of many pathophysiological processes. Amongst these are pathologies that are very commonly seen and account for a vast number of patients seen in daily practice of physicians worldwide.

CsA is the prototypical inhibitor of cyclophilins. With its strong affinity and its ability to penetrate the cell, it is able to inhibit the PPIase activity of CyPA within the cell and the extracellular space. Inhibition of intracellular CyPA by CsA greatly reduces the activation of T lymphocytes and is therefore successfully used as an immunosuppressive drug in clinical practice since decades. Intracellular CsA forms a complex with CyPA, which inhibits the NFAT-dependent activation of T-cells via inhibition of calcineurin [38]. Several derivatives of CsA have been synthesized, differing in their affinity to CyPA, ability to suppress T-cell activation and other properties. The most recent development being CsA derivatives, which cannot penetrate the cell membrane and therefore offer the opportunity to selectively inhibit extracellular cyclophilins, leaving intracellular mechanisms largely intact [39].

Following, we will show examples of recent applications of this new pharmacological concept.

For MM218 and MM284, CsA derivatives that cannot penetrate the cell membrane, in vivo experiments have shown that administration reduces inflammatory response. MM218 was able to drastically reduce inflammation, including recruitment of antigen-specific CD4+ T-cells, in a murine model of allergic lung inflammation [40].

For MM284 studies have shown that administration diminishes inflammation and infiltration of T-cells and macrophages in a model of troponin I-induced autoimmune myocarditis in mice. Furthermore, treatment with MM284 significantly reduced myocardial collagen deposition and expression of MMP-9, being one of the most pro-fibrotic collagenases in the myocardium [24].

Having a distinct inflammatory component, atherosclerosis is a major cause for cardiovascular events. The role of CyPA in atherosclerosis has been studied intensively. The interaction of extracellular CyPA with CD147 has been found to be of great significance for the regulation of MT-1-MMP, MMP-9 and M-CSF during the development of foam cells from CD34+ progenitor cells [26] (Fig. 12.2). MMPs are considered to be drivers of increased vulnerability in the plaque, leading to rupture and subsequent thrombosis by activation of platelets and coagulation factors. In earlier stages of atherosclerosis, CyPA has been shown to be of importance for the uptake of low-density lipoprotein into the vascular wall by a mechanistically still unclear regulation of the expression of scavenger receptors. CyPA-deficient $ApoE^{-/-}$ mice showed a markedly reduced burden of atherosclerosis, reduced VCAM-1 expression and apoptosis while on a high-cholesterol diet. Furthermore, CyPA deficiency led to a reduced recruitment of leucocytes into the lesions [41].

However, recent studies surprisingly showed that pharmacological inhibition of cyclophilins via MM284 increased the burden of atherosclerosis in the aorta. Despite decreasing TNFα levels in the plasma, application of the inhibitor showed no effect on the abundance of different inflammatory cell types; the expression levels of IL-6, IL-10 or MCP-1; or the plasma lipoprotein profile [42].

12.5 Prospective Thoughts

In this very brief overview of the functions and mechanisms of cyclophilins in the context of inflammation and thrombosis, we show that this topic is yet to be discovered as a new therapeutic field, which could be applicable for many diseases with an inflammatory component.

Current studies show conclusively that the inhibition of extracellular CyPA can modulate inflammatory responses in various models of inflammation. From the recruitment of leucocytes and platelets to the expression of pro-fibrotic signals, the CyPA/CyPB-CD147 interaction plays an important role in the complex interactions, which lead to inflammation and its termination.

New pharmacological approaches to selectively decrease the amount of active CyPs in the extracellular space have shown promising results in various settings. With tools to differentiate between extracellular and intracellular functions of CyPs, we have the opportunity to inhibit cyclophilin function leaving T-cell activation intact. There are still many challenges ahead to develop the compounds we use to date into a safe application in humans. But with promising results at hand already, now groups are working on overcoming the pharmacological hurdles for clinical testing. The coming years will show whether this approach will be effective and safe enough.

Compliance with Ethical Standards

Conflict of Interest: The authors declares that they have no conflict of interest.
Ethical Approval: This article does not contain any studies with human participants or animals performed by any of the authors.

References

1. Wang P, Heitman J. The cyclophilins. Genome Biol. 2005;6(7):226.
2. Zenke G, Baumann G, Wenger R, Hiestand P, Quesniaux V, Andersen E, Schreier MH. Molecular mechanisms of immunosuppression by cyclosporins. Ann NY Acad Sci. 1993;685:330–5.
3. Le Hir M, Su Q, Weber L, Woerly G, Granelli-Piperno A, Ryffel B. In situ detection of cyclosporin A: evidence for nuclear localization of cyclosporine and cyclophilins. Lab Invest. 1995;73(5):727–33.
4. Kozlov G, Bastos-Aristizabal S, Maattanen P, Rosenauer A, Zheng F, Killikelly A, Trempe JF, Thomas DY, Gehring K. Structural basis of cyclophilin B binding by the calnexin/calreticulin P-domain. J Biol Chem. 2010;285(46):35551–7.
5. Matzinger P. Tolerance, danger, and the extended family. Annu Rev Immunol. 1994;12:991–1045.
6. Hoffmann H, Schiene-Fischer C. Functional aspects of extracellular cyclophilins. Biol Chem. 2014;395(7–8):721–35.
7. Ryffel B, Woerly G, Greiner B, Haendler B, Mihatsch MJ, Foxwell BM. Distribution of the cyclosporine binding protein cyclophilin in human tissues. Immunology. 1991;72(3):399–404.
8. Satoh K, Matoba T, Suzuki J, O'Dell MR, Nigro P, Cui Z, Mohan A, Pan S, Li L, Jin ZG, Yan C, Abe J, Berk BC. Cyclophilin A mediates vascular remodeling by promoting inflammation and vascular smooth muscle cell proliferation. Circulation. 2008;117(24):3088–98.
9. Billich A, Winkler G, Aschauer H, Rot A, Peichl P. Presence of cyclophilin A in synovial fluids of patients with rheumatoid arthritis. J Exp Med. 1997;185(5):975–80.
10. Seizer P, Geisler T, Bigalke B, Schneider M, Klingel K, Kandolf R, Stellos K, Schreieck J, Gawaz M, May AE. EMMPRIN and its ligand Cyclophilin A as novel diagnostic markers in inflammatory cardiomyopathy. Int J Cardiol. 2013;163(3):299–304.
11. Tegeder I, Schumacher A, John S, Geiger H, Geisslinger G, Bang H, Brune K. Elevated serum cyclophilin levels in patients with severe sepsis. J Clin Immunol. 1997;17(5):380–6.
12. Spik G, Haendler B, Delmas O, Mariller C, Chamoux M, Maes P, Tartar A, Montreuil J, Stedman K, Kocher HP, et al. A novel secreted Cyclophilin-like protein (SCYLP). J Biol Chem. 1991;266(17):10735–8.
13. Price ER, Jin M, Lim D, Pati S, Walsh CT, McKeon FD. Cyclophilin B trafficking through the secretory pathway is altered by binding of cyclosporin A. Proc Natl Acad Sci USA. 1994;91 (9):3931–5.
14. De Ceuninck F, Allain F, Caliez A, Spik G, Vanhoutte PM. High binding capacity of Cyclophilin B to chondrocyte heparan sulfate proteoglycans and its release from the cell surface by matrix metalloproteinases: possible role as a proinflammatory mediator in arthritis. Arthritis Rheum. 2003;48(8):2197–206.

15. Fearon P, Lonsdale-Eccles AA, Ross OK, Todd C, Sinha A, Allain F, Reynolds NJ. Keratinocyte secretion of Cyclophilin B via the constitutive pathway is regulated through its cyclosporin-binding site. J Invest Dermatol. 2011;131(5):1085–94.
16. Liao DF, Jin ZG, Baas AS, Daum G, Gygi SP, Aebersold R, Berk BC. Purification and identification of secreted oxidative stress-induced factors from vascular smooth muscle cells. J Biol Chem. 2000;275(1):189–96.
17. Schlegel J, Redzic JS, Porter CC, Yurchenko V, Bukrinsky M, Labeikovsky W, Armstrong GS, Zhang F, Isern NG, DeGregori J, Hodges R, Eisenmesser EZ. Solution characterization of the extracellular region of CD147 and its interaction with its enzyme ligand Cyclophilin A. J Mol Biol. 2009;391(3):518–35.
18. Malesevic M, Gutknecht D, Prell E, Klein C, Schumann M, Nowak RA, Simon JC, Schiene-Fischer C, Saalbach A. Anti-inflammatory effects of extracellular cyclosporins are exclusively mediated by CD147. J Med Chem. 2013;56(18):7302–11.
19. Yurchenko V, Zybarth G, O'Connor M, Dai WW, Franchin G, Hao T, Guo H, Hung HC, Toole B, Gallay P, Sherry B, Bukrinsky M. Active site residues of Cyclophilin A are crucial for its signaling activity via CD147. J Biol Chem. 2002;277(25):22959–65.
20. Saphire AC, Bobardt MD, Gallay PA. Host Cyclophilin A mediates HIV-1 attachment to target cells via heparans. EMBO J. 1999;18(23):6771–85.
21. Damsker JM, Bukrinsky MI, Constant SL. Preferential chemotaxis of activated human CD4+ T cells by extracellular Cyclophilin A. J Leukoc Biol. 2007;82(3):613–8.
22. Jin ZG, Lungu AO, Xie L, Wang M, Wong C, Berk BC. Cyclophilin A is a proinflammatory cytokine that activates endothelial cells. Arterioscler Thromb Vasc Biol. 2004;24(7):1186–91.
23. Kato N, Yuzawa Y, Kosugi T, Hobo A, Sato W, Miwa Y, Sakamoto K, Matsuo S, Kadomatsu K. The E-selectin ligand basigin/CD147 is responsible for neutrophil recruitment in renal ischemia/reperfusion. J Am Soc Nephrol. 2009;20(7):1565–76.
24. Heinzmann D, Bangert A, Muller AM, von Ungern-Sternberg SN, Emschermann F, Schonberger T, Chatterjee M, Mack AF, Klingel K, Kandolf R, Malesevic M, Borst O, Gawaz M, Langer HF, Katus H, Fischer G, May AE, Kaya Z, Seizer P. The novel extracellular Cyclophilin A (CyPA)—inhibitor MM284 reduces myocardial inflammation and remodeling in a mouse model of troponin I -induced myocarditis. PLoS One. 2015;10(4):e0124606.
25. Heine SJ, Olive D, Gao JL, Murphy PM, Bukrinsky MI, Constant SL. Cyclophilin A cooperates with MIP-2 to augment neutrophil migration. J Inflamm Res. 2011;4:93–104.
26. Seizer P, Schonberger T, Schott M, Lang MR, Langer HF, Bigalke B, Kramer BF, Borst O, Daub K, Heidenreich O, Schmidt R, Lindemann S, Herouy Y, Gawaz M, May AE. EMMPRIN and its ligand Cyclophilin A regulate MT1-MMP, MMP-9 and M-CSF during foam cell formation. Atherosclerosis. 2010;209(1):51–7.
27. Wang L, Wang CH, Jia JF, Ma XK, Li Y, Zhu HB, Tang H, Chen ZN, Zhu P. Contribution of Cyclophilin A to the regulation of inflammatory processes in rheumatoid arthritis. J Clin Immunol. 2010;30(1):24–33.
28. Li YY, McTiernan CF, Feldman AM. Interplay of matrix metalloproteinases, tissue inhibitors of metalloproteinases and their regulators in cardiac matrix remodeling. Cardiovasc Res. 2000;46(2):214–24.
29. Lee J, Kim SS. An overview of cyclophilins in human cancers. J Int Med Res. 2010;38 (5):1561–74.
30. Yurchenko V, O'Connor M, Dai WW, Guo H, Toole B, Sherry B, Bukrinsky M. CD147 is a signaling receptor for Cyclophilin B. Biochem Biophys Res Commun. 2001;288(4):786–8.
31. Allain F, Vanpouille C, Carpentier M, Slomianny MC, Durieux S, Spik G. Interaction with glycosaminoglycans is required for Cyclophilin B to trigger integrin-mediated adhesion of peripheral blood T lymphocytes to extracellular matrix. Proc Natl Acad Sci USA. 2002;99 (5):2714–9.
32. Melchior A, Denys A, Deligny A, Mazurier J, Allain F. Cyclophilin B induces integrin-mediated cell adhesion by a mechanism involving CD98-dependent activation of protein kinase C-delta and p44/42 mitogen-activated protein kinases. Exp Cell Res. 2008;314 (3):616–28.

33. Marcant A, Denys A, Melchior A, Martinez P, Deligny A, Carpentier M, Allain F. Cyclophilin B attenuates the expression of TNF-alpha in lipopolysaccharide-stimulated macrophages through the induction of B cell lymphoma-3. J Immunol. 2012;189(4):2023–32.
34. Elvers M, Herrmann A, Seizer P, Munzer P, Beck S, Schonberger T, Borst O, Martin-Romero FJ, Lang F, May AE, Gawaz M. Intracellular Cyclophilin A is an important Ca(2+) regulator in platelets and critically involved in arterial thrombus formation. Blood. 2012;120 (6):1317–26.
35. Rosado JA, Pariente JA, Salido GM, Redondo PC. SERCA2b activity is regulated by cyclophilins in human platelets. Arterioscler Thromb Vasc Biol. 2010;30(3):419–25.
36. Seizer P, Ungern-Sternberg SN, Schonberger T, Borst O, Munzer P, Schmidt EM, Mack AF, Heinzmann D, Chatterjee M, Langer H, Malesevic M, Lang F, Gawaz M, Fischer G, May AE. Extracellular Cyclophilin A activates platelets via EMMPRIN (CD147) and PI3K/Akt signaling, which promotes platelet adhesion and thrombus formation in vitro and in vivo. Arterioscler Thromb Vasc Biol. 2015;35(3):655–63.
37. Allain F, Durieux S, Denys A, Carpentier M, Spik G. Cyclophilin B binding to platelets supports calcium-dependent adhesion to collagen. Blood. 1999;94(3):976–83.
38. Seizer P, Gawaz M, May AE. Cyclophilin A and EMMPRIN (CD147) in cardiovascular diseases. Cardiovasc Res. 2014;102(1):17–23.
39. Malesevic M, Kuhling J, Erdmann F, Balsley MA, Bukrinsky MI, Constant SL, Fischer G. A cyclosporin derivative discriminates between extracellular and intracellular cyclophilins. Angew Chem Int Ed Engl. 2010;49(1):213–5.
40. Balsley MA, Malesevic M, Stemmy EJ, Gigley J, Jurjus RA, Herzog D, Bukrinsky MI, Fischer G, Constant SL. A cell-impermeable Cyclosporine A derivative reduces pathology in a mouse model of allergic lung inflammation. J Immunol. 2010;185(12):7663–70.
41. Nigro P, Satoh K, O'Dell MR, Soe NN, Cui Z, Mohan A, Abe J, Alexis JD, Sparks JD, Berk BC. Cyclophilin A is an inflammatory mediator that promotes atherosclerosis in apolipoprotein E-deficient mice. J Exp Med. 2011;208(1):53–66.
42. Ditiatkovski M, Neelisetti VN, Cui HL, Malesevic M, Fischer G, Bukrinsky M, Sviridov D. Inhibition of extracellular cyclophilins with cyclosporine analog and development of atherosclerosis in apolipoprotein E-deficient mice. J Pharmacol Exp Ther. 2015;353(3):490–5.

Platelets and Innate Immunity in Atherosclerosis

13

Johannes Patzelt and Harald F. Langer

Abstract

Platelets are classically considered initiators of hemostasis and—in pathology—intravascular thrombosis causing diseases such as myocardial infarction or stroke. However, platelets are also mediators of innate immunity, secrete inflammatory proteins, mediate leukocyte recruitment, and contribute to tissue remodeling. Inflammation and innate immunity have common intersection points with the hemostatic system at various levels. With the complement system being part of the innate immune system, this chapter focuses on the role of platelets and the complement system in the context of atherosclerosis.

Contents

J. Patzelt
University Clinic for Cardiovascular Medicine, University of Tuebingen, Tuebingen, Germany

H.F. Langer (✉)
University Clinic for Cardiovascular Medicine, University of Tuebingen, Tuebingen, Germany

Section for Cardioimmunology, Department of Cardiovascular Medicine, University of Tuebingen, Tuebingen, Germany
e-mail: harald.langer@med.uni-tuebingen.de

A. Zirlik et al. (eds.), *Platelets, Haemostasis and Inflammation*,
Cardiac and Vascular Biology 5, https://doi.org/10.1007/978-3-319-66224-4_13

13.1 Introduction

Beyond their classical role in hemostasis, platelets do also have important functions in microbial defense. The persistence of microbes in platelet thrombi in septic thrombotic diseases such as endocarditis brings up the importance of the interaction of immunity and hemostasis. Besides cellular components such as leukocytes, macrophages, and dendritic cells, the complement system is an integral part of our innate immune system. Atherosclerosis is recognized as an inflammatory disease [1], and accumulating evidence shows that platelets and the complement system do not only have intersections in microbial defense but also in atherosclerotic disease. With atherosclerosis still being the leading cause of death in the Western world, a closer review of the contributions of platelets and the complement system in its genesis is warranted. After injury, platelets cover and close an endothelial wound, and the contact of platelets with the subendothelial matrix triggers their activation and subsequent thrombus formation [2]. Platelet contact with the subendothelial structures of the vascular wall is also an important pathophysiologic principle in early atherosclerotic plaque formation [3]. Additionally, it has been shown that platelets interact with intact endothelium, too, and recruit leukocytes even before an atherosclerotic plaque has formed [4]. The complement system, besides having important protective functions in immune defense, can be a driving force in chronic inflammatory disease [5].

13.2 The Complement System and Atherosclerosis

Many cellular and molecular mediators of the immune system have been identified to modulate the development of atherosclerosis. Complement being part of the innate immune system encompasses a broad range of immune-modulatory effects, including the opsonization of microbial intruders with C1a or mannose-binding lectin (MBL), followed by activation products of the complement cascade C2–C4 (with the opsonins C3b and C4b). Mast cell degranulation is induced by the soluble anaphylatoxins C3a and C5a and inflammatory cells are attracted [6]. The lysis of target cells is mediated by the membrane attack complex (MAC), which is formed by the components C5b–C9. In addition to immune defense, the complement system influences central homeostatic and pathophysiological processes in tissue remodeling and the removal of immune complexes, apoptotic cells, and cellular debris [7]. It is well recognized that the classical pathway with C1q, C2, and C4 is associated with the homeostatic control of such immune complexes, as the deficiency of these components predisposes to diseases characterized by an impaired clearance of cellular debris, for example, in systemic lupus erythematosus (SLE) [8]. Interestingly, cellular debris also accumulates in atherosclerotic plaques. Jonsson et al. identified a significant association between genetic C2 deficiency and atherosclerosis with increased rates of myocardial infarctions and stroke in a cohort of 40 patients [9]. C4 deficiency is a disease with premature atherosclerotic peripheral vascular lesions. A significant part of those patients exhibits circulating immune complexes, which correlate with atherosclerotic lesions [10]. An SNP of the complement receptor C1qRp (CD 93) affects the risk for coronary artery disease as demonstrated in a genome-wide analysis in a cohort with familial hypercholesterolemia [11]. Similarly, polymorphisms with decreased levels of mannose-binding lectin (MBL) are associated with an elevated incidence of coronary heart disease and increased carotid plaque formation [12–14].

The generation of anaphylatoxins in the wake of complement activation promotes inflammation. Elevated levels of the anaphylatoxin C5a have been measured in patients with advanced atherosclerosis and were predictive for major cardiovascular events independently of known risk markers such as C reactive protein (CRP) or fibrinogen [15].

Complement factors are accumulating in atherosclerotic plaques [16, 17]. Under physiological conditions, activated complement components are quickly cleared from the circulation. In cholesterol-fed rabbits, however, activated complement and the MAC were identified within early stages of plaque formation even before the arrival of inflammatory cells and the formation of fatty streaks [18]. Subsequent studies could confirm this finding by demonstrating the presence of the terminal complement complex C5b-9 in human atherosclerotic arteries [19–21].

Two sources for those complement components have been identified: They may derive from blood circulation [22, 23] or they may be produced locally within the plaque as is indicated by the presence of mRNA for several complement components (C1r, C1s, C4, C7, and C8) [23, 24]. Indeed, the power of local complement production in immune processes being the causing factor of disease has been documented by various studies [25–28]. As a matter of fact, there are differences in complement activation between superficial and deeper lavers of the atherosclerotic plaque. While, in the luminal layer, there are signs of classical and alternative but not terminal complement activation (consistent with the presence of complement regulators C4bp and fH) [29, 30], terminal complement complex deposition associated with smooth muscle cells, cell debris, and extracellular lipids is noted in deeper layers of the plaque [29, 30]. In the necrotic core of advanced atherosclerotic lesions, C1q and the receptor for its globular domain (gC1a-R) are found [31]. Furthermore, C3b can be detected as well, and there is a higher concentration in ruptured compared to intact plaques in the same patients [17]. Accumulation of C5a in lipid-rich inflammatory lesions containing exposed cholesterol and necrotic cell debris compared to stable plaques containing collagen and elastin is another sign of increased complement activation within ruptured plaques [32]. These observations are supported by epidemiological data showing increased C5a levels in patients with increased cardiovascular risk independently of nonspecific inflammatory markers [15]. While elevated plasmatic C4 levels are associated with severe atherosclerosis [33], no causality can be derived from mere detection of complement and we do not know whether it has a protective or deleterious role for atherosclerosis. Actually, there are studies suggesting protective effects of complement, while others find pro-atherosclerotic effects. Data from the 1970s show that C6 deficiency protects cholesterol-fed rabbits from atherosclerosis [34]. Similarly, inhibition of C5a or its receptor C5aR1 (CD88) reduces atherosclerotic lesions in murine models [35, 36]. However, C1q deficiency has been associated with significantly larger lesions in atherosclerotic low-density lipoprotein receptor-deficient (LDLR−/−) mice compared to C1q-sufficient controls [37, 38]. This observation could be explained by the fact that C1q binds apoptotic or necrotic cells in plaques directly or indirectly via IgM, thereby promoting classical pathway deposition of C3 activation products and, in consequence, their removal by macrophages. Indeed, C3 deficiency enhances abdominal and thoracic aortic lesions in atherosclerotic LDLR−/− mice compared to C3-sufficient controls [39]. This finding is supported by the fact that mice lacking ApoE and LDLR exhibit a strong increase of aortic lesion load (+84%) when C3 is absent too [40]. Considering the ambiguous results for the role of complement activation in atherogenesis, several clinical trials assessed the efficacy of targeting complement in coronary artery disease (CAD). Treatment with an anti-C5 antibody (Pexelizumab) resulted in significantly reduced mortality in patients with ST-elevation myocardial infarction (STEMI) [41]. Complement inhibition in patients undergoing coronary artery bypass graft surgery led to reduced morbidity and mortality [42].

In conclusion, the function of complement in the development of atherosclerosis is not completely understood and unanswered questions will have to be addressed in future basic and clinical studies.

13.3 Platelets and Atherosclerosis

Platelets play a key role in the late thrombotic complications associated with atherosclerosis and are, thus, an important target for the development of diagnostic and therapeutic tools. Moreover, platelets can also interact with intact endothelium even before an atherosclerotic plaque has formed and therefore may play a role in the genesis of atherosclerosis [43]. Indeed, platelets adhere to intact endothelium via the von Willebrand factor receptor GPIbα and the fibrinogen receptor $GP_{IIbIIIa}$ before a plaque has formed in atherosclerotic apolipoprotein E-deficient (ApoE−/−) mice [4, 43, 44]. However, atherosclerosis may also emerge in the absence of $GP_{IIbIIIa}$ as is known from studies with patients whose platelets lack functional $GP_{IIbIIIa}$ (Glanzmann thrombasthenia). 4 out of 7 of those patients showed atherosclerotic plaques as revealed by ultrasound imaging of the carotid bifurcation [45]. Bringing those findings together, platelet–vessel wall interactions via $GP_{IIbIIIa}$ seem to contribute to but do not seem to be a prerequisite in the genesis of atherosclerosis and may be functionally substituted by other platelet receptors. Deficiency of GPIbα shows no protection from atherosclerotic plaque formation in mice [46]. On the other hand, platelet depletion with a GPIbα-specific antibody in ApoE−/− mice leads to reduced leukocyte accumulation in the arterial intima and attenuated atherosclerotic plaque formation. Importantly, this indicates that adhering platelets form a focal point for the immune cell-driven inflammation that is central to atherosclerosis [4]. Platelets are able to interact with P-selectin expressed on intact endothelium via GPIbα and PSGL-1. This initial rolling of platelets on endothelium

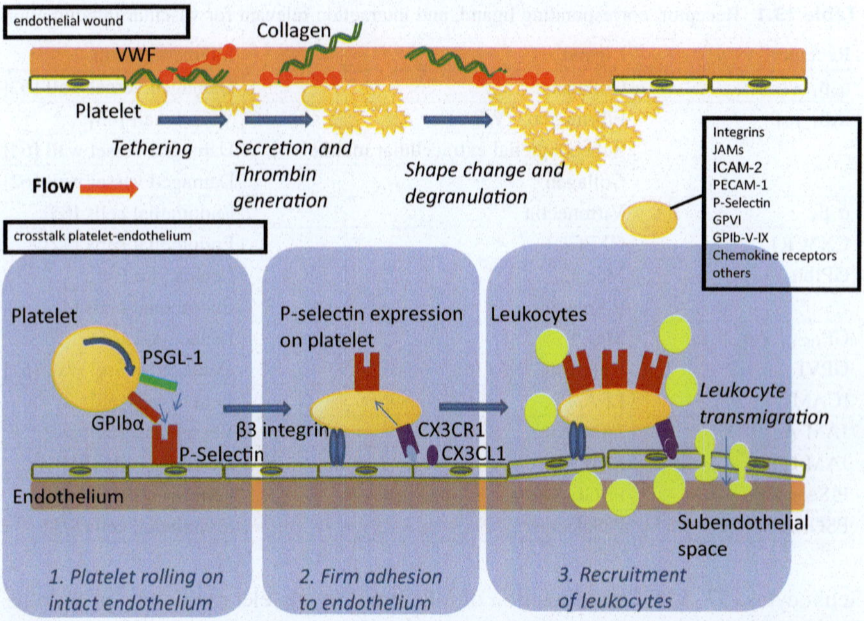

Fig. 13.1 (adapted from Patzelt et al.) [48]. *Upper part*: Model of platelet interaction with the damaged vessel wall: exposure of subendothelial matrix after endothelial injury leads to platelet tethering, activation, and accumulation to provide sealing of the endothelial wound. *Lower part*: Model of platelet interaction with the endothelium. Platelets interact with the endothelium via adhesion receptors such as GPIbα and PSGL-1 promoting rolling and subsequent firm adhesion via β3 integrins. Interaction with endothelial bound chemokines such as Cx3CL1 (fractalkine) or CCL5 (Rantes) results in P-selectin-mediated recruitment of leukocytes to the vessel wall and subsequent transmigration. For interaction with both the subendothelial matrix and the endothelium, major adhesion receptors expressed on platelets are listed

is followed by β3 integrin-mediated firm adhesion of platelets to the vessel wall [43]. For the initiation of atherosclerosis, these aforementioned steps are considered crucial. Various inflammatory receptors are expressed on the platelets' surface including the fractalkine receptor (CX3CR1), which induces P-selectin expression on platelets after binding to fractalkine (CX3CL1) expressed on inflamed endothelial cells [47]. Platelet P-selectin exposure driven by arterial shear forces in turn initiates local accumulation of leukocytes [47] (Fig. 13.1). P-selectin seems to have a central role in atherogenesis, which is corroborated by the positive correlation of intima-media thickness with levels of platelet P-selectin [49]. Both platelet and endothelial P-selectin contributed to lesion formation in a mouse model of atherosclerosis that was based on the adoptive transfer of P-selectin positive or negative platelets [50].

It is unclear whether platelet adhesion to the intima causes direct damage to the endothelium. However, several studies found how platelets contribute to vascular inflammation via their interaction with leukocytes [51–56]. In ApoE-deficient mice, atherosclerosis was exacerbated by activated platelets mediating the recruitment of

Table 13.1 Receptor, corresponding ligand, and interaction relevant for vascular inflammation

Receptor	Ligand	Interaction with
$\alpha_2\beta_1$	Collagen	Damaged vessel wall [62]
$\alpha_5\beta_1$, $\alpha_6\beta_1$	Fibrinogen, vWF	Vessel wall [63]
	Subendothelial extracellular matrix	Damaged vessel wall [64]
	Collagen	Damaged vessel wall [62]
$\alpha_v\beta_3$	Vitronectin	Endothelial cells [65]
CX3CR1	CX3CL1	Endothelial cells [47]
GPIbα	Mac 1	Leukocytes [65]
	P-Selectin	Endothelial cells [4]
$GP_{IIb-IIIa}$ ($\alpha_{IIb}\beta_3$)	Mac 1	Leukocytes [66]
GPVI	Collagen	Damaged vessel wall [67]
ICAM-2	LFA-1	Leukocytes [68]
JAM-A	JAM-A	Vessel wall [69]
JAM-C	MAC-1	Dendritic cells [53]
P-Selectin	PSGL-1	Leukocytes [70]
PSGL-1	P-Selectin	Endothelial cells [71]

leukocytes [57, 58]. The deposition of inflammatory platelet mediators on endothelial cells was facilitated by the formation of platelet–leukocyte aggregates (PLA) [44, 52]. Moreover, platelet activation increased the number of circulating PLAs [59, 60]. Several receptor/ligand pairs have been identified that potentially support platelet–leukocyte cross talk, including integrins or members of the JAM family of proteins [57, 58, 61]. A list of platelet surface receptors with potential relevance for atherosclerosis is given in Table 13.1. Additionally, platelets may also influence vascular inflammation via release of factors from their granules [72]. For example, the release of the chemokines CCL5 or CXCL4 contributes to atherosclerosis in a P-selectin-dependent manner [73]. Fatty acids play a major role in the genesis of atherosclerosis. Importantly, platelets can bind oxidized LDL—playing a key role in atherogenesis—and interaction with lipoproteins can change platelet function [74, 75]. Not surprisingly, platelets of hypercholesterolemic patients show enhanced activity in vivo and hyperaggregability in vitro [76, 77].

Together, on the one hand, platelet activation seems to exert pro-atherosclerotic effects; on the other hand, it can also provide effects of atheromodulation and tissue/vascular remodeling.

13.4 Platelets and the Complement System

Platelets as well as the complement system are associated with atherogenesis and its late complications. Considering this, a closer review of their interactions is warranted. In platelet isolates, a wide variety of complement factors and receptor can be detected [78]. A potentially self-reinforcing cycle with complement activating platelets and, in turn, (thrombin) activated platelets initiating the complement cascade has been described [79]. Platelets express CR4 (a receptor for iC3b), although its function remains elusive, so far [80]. Besides this C3-fragment receptor, platelets also express the C1q receptors gC1qR/p33 and cC1qR that were shown to mediate platelet activation and aggregation [81–83]. The anaphylatoxins C3a and C5a are generated further downstream in the complement cascade. Receptors for those are also found on platelets. In vitro, C3a and its derivative C3a-des-Arg mediate platelet aggregation and activation [84, 85]. Recently, we found that in patients with CAD the expression of these anaphylatoxin receptors (C3aR and C5aR) and the expression of activation markers such as P-selectin are correlated [86].

Platelets actively counteract complement deposition onto their surface, because of their propensity to become activated by the complement cascade. In fact, platelets bind and express many complement control proteins (CCPs) [87]. Absence or dysfunction of such CCPs is associated with platelet malfunction and activation-induced thrombocytopenia [87]. In line with this notion, in atypical hemolytic uremic syndrome (aHUS) excessive complement activation on platelets leads to thrombocytopenia and prothrombotic complications, which is caused by deficiencies or mutations in CCPs (frequently factor H) [88]. Moreover, in paroxysmal nocturnal hemoglobinuria (PNH) platelets are affected by overwhelming complement activation. In this disease, effective anchoring of CCPs decay-accelerating factor (DAF, CD55) and protectin (DC59) on the platelet surface is prevented by a mutation in the phosphatidylinositol glycan A (PIGA) [89]. Both conditions can nowadays be treated with Eculizumab, a humanized version of the anti-C5 antibody h5G1.1, first described in 1996. The mAb Eculizumab prevents the cleavage of C5-C5b by binding directly to C5 and, thus, blocks the formation of the membrane damaging MAC [90].

Interaction between platelets and the complement system can also occur via proteins that are not classically seen as complement receptors such as P-selectin [91] or GPIbα [92]. P-selectin was observed to bind C3b and mediate generation of C3a as well as MAC formation [91]. Platelet activation is enhanced by MAC formation causing a prothrombotic state in aHUS and PNH [93]. Studies with bacterial infection in mice revealed that upon systemic infection, C3b-opsonized bacteria form complexes with platelets in the bloodstream. Such complexes are built in the presence of the alpha chain of GPIbα on the platelet surface, indicating that GPIb interacts directly or indirectly with activated complement C3 [92]. These molecules may synergize in physiological hemostatic processes as suggested by the fact that both C3- and GPIb-deficient mice show prolonged bleeding times [94, 95].

13.5 Conclusions

Various experimental and clinical studies indicate that the complement system and platelets modulate the genesis of atherosclerosis and are able to modulate each other's function.

The heterogeneous results highlight the complexity of pathological mechanisms underlying atherosclerosis and the contribution of platelet activation, complement activation, and their cross talk for this disease. Thus, there is a clear need for further intensified clinical and experimental studies to characterize the interaction of platelets and the complement system in the context of atherosclerosis.

Compliance with Ethical Standards

Conflict of Interest: Johannes Patzelt and Harald F. Langer declares that they have no conflict of interest.

Ethical Approval: This article does not contain any studies with human participants or animals performed by any of the authors.

References

1. Ross R. Atherosclerosis—an inflammatory disease. N Engl J Med. 1999;340:115–26.
2. Kuijper PH, Gallardo Torres HI, Lammers JW, Sixma JJ, Koenderman L, Zwaginga JJ. Platelet and fibrin deposition at the damaged vessel wall: cooperative substrates for neutrophil adhesion under flow conditions. Blood. 1997;89:166–75.
3. Langer HF, Bigalke B, Seizer P, Stellos K, Fateh-Moghadam S, Gawaz M. Interaction of platelets and inflammatory endothelium in the development and progression of coronary artery disease. Semin Thromb Hemost. 2010;36:131–8.
4. Massberg S, Brand K, Gruner S, Page S, Muller E, Muller I, Bergmeier W, Richter T, Lorenz M, Konrad I, Nieswandt B, Gawaz M. A critical role of platelet adhesion in the initiation of atherosclerotic lesion formation. J Exp Med. 2002;196:887–96.
5. Walport MJ. Complement. First of two parts. N Engl J Med. 2001;344:1058–66.
6. Verschoor A, Carroll MC. Complement and its receptors in infection. In: Kaufmann SHE, Medzhitov R, Gordon S, editors. The innate immune response to infection. Washington, DC: American Society for Microbiology Press; 2004.
7. Ricklin D, Hajishengallis G, Yang K, Lambris JD. Complement: a key system for immune surveillance and homeostasis. Nat Immunol. 2010;11:785–97.
8. Aggarwal R, Sestak AL, D'Sousa A, Dillon SP, Namjou B, Scofield RH. Complete complement deficiency in a large cohort of familial systemic lupus erythematosus. Lupus. 2010;19:52–7.
9. Jonsson G, Truedsson L, Sturfelt G, Oxelius VA, Braconier JH, Sjoholm AG. Hereditary c2 deficiency in sweden: frequent occurrence of invasive infection, atherosclerosis, and rheumatic disease. Medicine. 2005;84:23–34.
10. Nityanand S, Truedsson L, Mustafa A, Bergmark C, Lefvert AK. Circulating immune complexes and complement c4 null alleles in patients in patients operated on for premature atherosclerotic peripheral vascular disease. J Clin Immunol. 1999;19:406–13.
11. van der Net JB, Oosterveer DM, Versmissen J, Defesche JC, Yazdanpanah M, Aouizerat BE, Steyerberg EW, Malloy MJ, Pullinger CR, Kastelein JJ, Kane JP, Sijbrands EJ. Replication study of 10 genetic polymorphisms associated with coronary heart disease in a specific high-risk population with familial hypercholesterolemia. Eur Heart J. 2008;29:2195–201.
12. Madsen HO, Videm V, Svejgaard A, Svennevig JL, Garred P. Association of mannose-binding-lectin deficiency with severe atherosclerosis. Lancet. 1998;352:959–60.

13. Hegele RA, Ban MR, Anderson CM, Spence JD. Infection-susceptibility alleles of mannose-binding lectin are associated with increased carotid plaque area. J Invest Med. 2000;48:198–202.
14. Best LG, Davidson M, North KE, MacCluer JW, Zhang Y, Lee ET, Howard BV, DeCroo S, Ferrell RE. Prospective analysis of mannose-binding lectin genotypes and coronary artery disease in american indians: the strong heart study. Circulation. 2004;109:471–5.
15. Speidl WS, Exner M, Amighi J, Kastl SP, Zorn G, Maurer G, Wagner O, Huber K, Minar E, Wojta J, Schillinger M. Complement component c5a predicts future cardiovascular events in patients with advanced atherosclerosis. Eur Heart J. 2005;26:2294–9.
16. Speidl WS, Kastl SP, Huber K, Wojta J. Complement in atherosclerosis: friend or foe? J Thromb Haemost. 2011;9:428–40.
17. Laine P, Pentikainen MO, Wurzner R, Penttila A, Paavonen T, Meri S, Kovanen PT. Evidence for complement activation in ruptured coronary plaques in acute myocardial infarction. Am J Cardiol. 2002;90:404–8.
18. Seifert PS, Hugo F, Hansson GK, Bhakdi S. Prelesional complement activation in experimental atherosclerosis. Terminal c5b-9 complement deposition coincides with cholesterol accumulation in the aortic intima of hypercholesterolemic rabbits. Lab Invest. 1989;60:747–54.
19. Vlaicu R, Niculescu F, Rus HG, Cristea A. Immunohistochemical localization of the terminal c5b-9 complement complex in human aortic fibrous plaque. Atherosclerosis. 1985;57:163–77.
20. Niculescu F, Rus HG, Vlaicu R. Immunohistochemical localization of c5b-9, s-protein, c3d and apolipoprotein b in human arterial tissues with atherosclerosis. Atherosclerosis. 1987;65:1–11.
21. Torzewski M, Klouche M, Hock J, Messner M, Dorweiler B, Torzewski J, Gabbert HE, Bhakdi S. Immunohistochemical demonstration of enzymatically modified human ldl and its colocalization with the terminal complement complex in the early atherosclerotic lesion. Arterioscler Thromb Vasc Biol. 1998;18:369–78.
22. Vlaicu R, Rus HG, Niculescu F, Cristea A. Quantitative determinations of immunoglobulins and complement components in human aortic atherosclerotic wall. Med Interne. 1985;23:29–35.
23. Niculescu F, Rus H. The role of complement activation in atherosclerosis. Immunol Res. 2004;30:73–80.
24. Yasojima K, Schwab C, McGeer EG, McGeer PL. Complement components, but not complement inhibitors, are upregulated in atherosclerotic plaques. Arterioscler Thromb Vasc Biol. 2001;21:1214–9.
25. Verschoor A, Brockman MA, Knipe DM, Carroll MC. Cutting edge: myeloid complement c3 enhances the humoral response to peripheral viral infection. J Immunol. 2001;167:2446–51.
26. Verschoor A, Brockman MA, Gadjeva M, Knipe DM, Carroll MC. Myeloid c3 determines induction of humoral responses to peripheral herpes simplex virus infection. J Immunol. 2003;171:5363–71.
27. Gadjeva M, Verschoor A, Brockman MA, Jezak H, Shen LM, Knipe DM, Carroll MC. Macrophage-derived complement component c4 can restore humoral immunity in c4-deficient mice. J Immunol. 2002;169:5489–95.
28. Li K, Sacks SH, Zhou W. The relative importance of local and systemic complement production in ischaemia, transplantation and other pathologies. Mol Immunol. 2007;44:3866–74.
29. Oksjoki R, Jarva H, Kovanen PT, Laine P, Meri S, Pentikainen MO. Association between complement factor h and proteoglycans in early human coronary atherosclerotic lesions: implications for local regulation of complement activation. Arterioscler Thromb Vasc Biol. 2003;23:630–6.
30. Oksjoki R, Kovanen PT, Mayranpaa MI, Laine P, Blom AM, Meri S, Pentikainen MO. Complement regulation in human atherosclerotic coronary lesions. Immunohistochemical evidence that c4b-binding protein negatively regulates the classical complement pathway, and that c5b-9 is formed via the alternative complement pathway. Atherosclerosis. 2007;192:40–8.

31. Peerschke EI, Minta JO, Zhou SZ, Bini A, Gotlieb A, Colman RW, Ghebrehiwet B. Expression of gc1q-r/p33 and its major ligands in human atherosclerotic lesions. Mol Immunol. 2004;41:759–66.
32. Speidl WS, Kastl SP, Hutter R, Katsaros KM, Kaun C, Bauriedel G, Maurer G, Huber K, Badimon JJ, Wojta J. The complement component c5a is present in human coronary lesions in vivo and induces the expression of mmp-1 and mmp-9 in human macrophages in vitro. FASEB J. 2011;25:35–44.
33. Muscari A, Bozzoli C, Gerratana C, Zaca F, Rovinetti C, Zauli D, La Placa M, Puddu P. Association of serum iga and c4 with severe atherosclerosis. Atherosclerosis. 1988;74:179–86.
34. Geertinger P, Sorensen H. Complement as a factor in arteriosclerosis. Acta Pathol Microbiol Scand A Pathol. 1970;78:284–8.
35. Shagdarsuren E, Bidzhekov K, Mause SF, Simsekyilmaz S, Polakowski T, Hawlisch H, Gessner JE, Zernecke A, Weber C. C5a receptor targeting in neointima formation after arterial injury in atherosclerosis-prone mice. Circulation. 2010;122:1026–36.
36. Manthey HD, Thomas AC, Shiels IA, Zernecke A, Woodruff TM, Rolfe B, Taylor SM. Complement c5a inhibition reduces atherosclerosis in apoe-/- mice. FASEB J. 2011;25:2447–55.
37. Bhatia VK, Yun S, Leung V, Grimsditch DC, Benson GM, Botto MB, Boyle JJ, Haskard DO. Complement c1q reduces early atherosclerosis in low-density lipoprotein receptor-deficient mice. Am J Pathol. 2007;170:416–26.
38. Lewis MJ, Malik TH, Ehrenstein MR, Boyle JJ, Botto M, Haskard DO. Immunoglobulin m is required for protection against atherosclerosis in low-density lipoprotein receptor-deficient mice. Circulation. 2009;120:417–26.
39. Buono C, Come CE, Witztum JL, Maguire GF, Connelly PW, Carroll M, Lichtman AH. Influence of c3 deficiency on atherosclerosis. Circulation. 2002;105:3025–31.
40. Persson L, Boren J, Robertson AK, Wallenius V, Hansson GK, Pekna M. Lack of complement factor c3, but not factor b, increases hyperlipidemia and atherosclerosis in apolipoprotein e-/- low-density lipoprotein receptor-/- mice. Arterioscler Thromb Vascu Biol. 2004;24:1062–7.
41. Granger CB, Mahaffey KW, Weaver WD, Theroux P, Hochman JS, Filloon TG, Rollins S, Todaro TG, Nicolau JC, Ruzyllo W, Armstrong PW, Investigators C. Pexelizumab, an anti-c5 complement antibody, as adjunctive therapy to primary percutaneous coronary intervention in acute myocardial infarction: the complement inhibition in myocardial infarction treated with angioplasty (comma) trial. Circulation. 2003;108:1184–90.
42. Testa L, Meco M, Cirri S, Bedogni F. Pexelizumab and survival in cardiac surgery. HSR Proc Intensive Care Cardiovasc Anesth. 2011;3:23–4.
43. Gawaz M, Langer H, May AE. Platelets in inflammation and atherogenesis. J Clin Invest. 2005;115:3378–84.
44. Huo Y, Schober A, Forlow SB, Smith DF, Hyman MC, Jung S, Littman DR, Weber C, Ley K. Circulating activated platelets exacerbate atherosclerosis in mice deficient in apolipoprotein e. Nat Med. 2003;9:61–7.
45. Shpilberg O, Rabi I, Schiller K, Walden R, Harats D, Tyrrell KS, Coller B, Seligsohn U. Patients with glanzmann thrombasthenia lacking platelet glycoprotein alpha(iib)beta(3) (gpiib/iiia) and alpha(v)beta(3) receptors are not protected from atherosclerosis. Circulation. 2002;105:1044–8.
46. Strassel C, Hechler B, Bull A, Gachet C, Lanza F. Studies of mice lacking the gpib-v-ix complex question the role of this receptor in atherosclerosis. J Thromb Haemost. 2009;7:1935–8.
47. Schulz C, Schafer A, Stolla M, Kerstan S, Lorenz M, von Bruhl ML, Schiemann M, Bauersachs J, Gloe T, Busch DH, Gawaz M, Massberg S. Chemokine fractalkine mediates leukocyte recruitment to inflammatory endothelial cells in flowing whole blood: a critical role for p-selectin expressed on activated platelets. Circulation. 2007;116:764–73.
48. Patzelt J, Verschoor A, Langer HF. Platelets and the complement cascade in atherosclerosis. Front Physiol. 2015;6:49.

49. Koyama H, Maeno T, Fukumoto S, Shoji T, Yamane T, Yokoyama H, Emoto M, Shoji T, Tahara H, Inaba M, Hino M, Shioi A, Miki T, Nishizawa Y. Platelet p-selectin expression is associated with atherosclerotic wall thickness in carotid artery in humans. Circulation. 2003;108:524–9.
50. Burger PC, Wagner DD. Platelet p-selectin facilitates atherosclerotic lesion development. Blood. 2003;101:2661–6.
51. Santoso S, Sachs UJ, Kroll H, Linder M, Ruf A, Preissner KT, Chavakis T. The junctional adhesion molecule 3 (jam-3) on human platelets is a counterreceptor for the leukocyte integrin mac-1. J Exp Med. 2002;196:679–91.
52. Schober A, Manka D, von Hundelshausen P, Huo Y, Hanrath P, Sarembock IJ, Ley K, Weber C. Deposition of platelet rantes triggering monocyte recruitment requires p-selectin and is involved in neointima formation after arterial injury. Circulation. 2002;106:1523–9.
53. Langer HF, Daub K, Braun G, Schonberger T, May AE, Schaller M, Stein GM, Stellos K, Bueltmann A, Siegel-Axel D, Wendel HP, Aebert H, Roecken M, Seizer P, Santoso S, Wesselborg S, Brossart P, Gawaz M. Platelets recruit human dendritic cells via mac-1/jam-c interaction and modulate dendritic cell function in vitro. Arterioscler Thromb Vasc Biol. 2007;27:1463–70.
54. Langer HF, Choi EY, Zhou H, Schleicher R, Chung KJ, Tang Z, Gobel K, Bdeir K, Chatzigeorgiou A, Wong C, Bhatia S, Kruhlak MJ, Rose JW, Burns JB, Hill KE, Qu H, Zhang Y, Lehrmann E, Becker KG, Wang Y, Simon DI, Nieswandt B, Lambris JD, Li X, Meuth SG, Kubes P, Chavakis T. Platelets contribute to the pathogenesis of experimental autoimmune encephalomyelitis. Circ Res. 2012;110:1202–10.
55. Ley K, Laudanna C, Cybulsky MI, Nourshargh S. Getting to the site of inflammation: the leukocyte adhesion cascade updated. Nat Rev Immunol. 2007;7:678–89.
56. Langer HF, Chavakis T. Leukocyte-endothelial interactions in inflammation. J Cell Mol Med. 2009;13:1211–20.
57. Wagner DD, Frenette PS. The vessel wall and its interactions. Blood. 2008;111:5271–81.
58. von Hundelshausen P, Koenen RR, Weber C. Platelet-mediated enhancement of leukocyte adhesion. Microcirculation. 2009;16:84–96.
59. van Gils JM, Zwaginga JJ, Hordijk PL. Molecular and functional interactions among monocytes, platelets, and endothelial cells and their relevance for cardiovascular diseases. J Leukoc Biol. 2009;85:195–204.
60. Totani L, Evangelista V. Platelet-leukocyte interactions in cardiovascular disease and beyond. Arterioscler Thromb Vasc Biol. 2010;30:2357–61.
61. von Hundelshausen P, Weber C. Platelets as immune cells: bridging inflammation and cardiovascular disease. Circ Res. 2007;100:27–40.
62. Inoue O, Suzuki-Inoue K, Dean WL, Frampton J, Watson SP. Integrin alpha2beta1 mediates outside-in regulation of platelet spreading on collagen through activation of src kinases and plcgamma2. J Cell Biol. 2003;160:769–80.
63. Bombeli T, Schwartz BR, Harlan JM. Adhesion of activated platelets to endothelial cells: Evidence for a gpiibiiia-dependent bridging mechanism and novel roles for endothelial intercellular adhesion molecule 1 (icam-1), alphavbeta3 integrin, and gpibalpha. J Exp Med. 1998;187:329–39.
64. Gruner S, Prostredna M, Schulte V, Krieg T, Eckes B, Brakebusch C, Nieswandt B. Multiple integrin-ligand interactions synergize in shear-resistant platelet adhesion at sites of arterial injury in vivo. Blood. 2003;102:4021–7.
65. Gawaz M, Neumann FJ, Dickfeld T, Reininger A, Adelsberger H, Gebhardt A, Schomig A. Vitronectin receptor (alpha(v)beta3) mediates platelet adhesion to the luminal aspect of endothelial cells: Implications for reperfusion in acute myocardial infarction. Circulation. 1997;96:1809–18.
66. Weber C, Springer TA. Neutrophil accumulation on activated, surface-adherent platelets in flow is mediated by interaction of mac-1 with fibrinogen bound to alphaiibbeta3 and stimulated by platelet-activating factor. J Clin Invest. 1997;100:2085–93.

67. Massberg S, Gawaz M, Gruner S, Schulte V, Konrad I, Zohlnhofer D, Heinzmann U, Nieswandt B. A crucial role of glycoprotein vi for platelet recruitment to the injured arterial wall in vivo. J Exp Med. 2003;197:41–9.
68. Weber KS, Alon R, Klickstein LB. Sialylation of icam-2 on platelets impairs adhesion of leukocytes via lfa-1 and dc-sign. Inflammation. 2004;28:177–88.
69. Karshovska E, Zhao Z, Blanchet X, Schmitt MM, Bidzhekov K, Soehnlein O, von Hundelshausen P, Mattheij NJ, Cosemans JM, Megens RT, Koeppel TA, Schober A, Hackeng TM, Weber C, Koenen RR. Hyperreactivity of junctional adhesion molecule a-deficient platelets accelerates atherosclerosis in hyperlipidemic mice. Circ Res. 2015;116:587–99.
70. Dole VS, Bergmeier W, Patten IS, Hirahashi J, Mayadas TN, Wagner DD. Psgl-1 regulates platelet p-selectin-mediated endothelial activation and shedding of p-selectin from activated platelets. Thromb Haemost. 2007;98:806–12.
71. Frenette PS, Denis CV, Weiss L, Jurk K, Subbarao S, Kehrel B, Hartwig JH, Vestweber D, Wagner DD. P-selectin glycoprotein ligand 1 (psgl-1) is expressed on platelets and can mediate platelet-endothelial interactions in vivo. J Exp Med. 2000;191:1413–22.
72. Langer HF, Gawaz M. Platelet-vessel wall interactions in atherosclerotic disease. Thromb Haemost. 2008;99:480–6.
73. von Hundelshausen P, Schmitt MM. Platelets and their chemokines in atherosclerosis-clinical applications. Front Physiol. 2014;5:294.
74. Siegel-Axel D, Daub K, Seizer P, Lindemann S, Gawaz M. Platelet lipoprotein interplay: trigger of foam cell formation and driver of atherosclerosis. Cardiovasc Res. 2008;78:8–17.
75. Stellos K, Sauter R, Fahrleitner M, Grimm J, Stakos D, Emschermann F, Panagiota V, Gnerlich S, Perk A, Schonberger T, Bigalke B, Langer HF, Gawaz M. Binding of oxidized low-density lipoprotein on circulating platelets is increased in patients with acute coronary syndromes and induces platelet adhesion to vascular wall in vivo—brief report. Arterioscler Thromb Vasc Biol. 2012;32:2017–20.
76. Cipollone F, Mezzetti A, Porreca E, Di Febbo C, Nutini M, Fazia M, Falco A, Cuccurullo F, Davi G. Association between enhanced soluble cd40l and prothrombotic state in hypercholesterolemia: Effects of statin therapy. Circulation. 2002;106:399–402.
77. Ferroni P, Basili S, Santilli F, Davi G. Low-density lipoprotein-lowering medication and platelet function. Pathophysiol Haemost Thromb. 2006;35:346–54.
78. Hamad OA, Nilsson PH, Wouters D, Lambris JD, Ekdahl KN, Nilsson B. Complement component c3 binds to activated normal platelets without preceding proteolytic activation and promotes binding to complement receptor 1. J Immunol. 2010;184:2686–92.
79. Hamad OA, Ekdahl KN, Nilsson PH, Andersson J, Magotti P, Lambris JD, Nilsson B. Complement activation triggered by chondroitin sulfate released by thrombin receptor-activated platelets. J Thromb Haemost. 2008;6:1413–21.
80. Vik DP, Fearon DT. Cellular distribution of complement receptor type 4 (cr4): Expression on human platelets. J Immunol. 1987;138:254–8.
81. Wautier JL, Souchon H, Reid KB, Peltier AP, Caen JP. Studies on the mode of reaction of the first component of complement with platelets: Interaction between the collagen-like portion of c1q and platelets. Immunochemistry. 1977;14:763–6.
82. Peerschke EI, Ghebrehiwet B. C1q augments platelet activation in response to aggregated ig. J Immunol. 1997;159:5594–8.
83. Peerschke EI, Ghebrehiwet B. Human blood platelet gc1qr/p33. Immunol Rev. 2001;180:56–64.
84. Polley MJ, Nachman RL. Human platelet activation by c3a and c3a des-arg. J Exp Med. 1983;158:603–15.
85. Martel C, Cointe S, Maurice P, Matar S, Ghitescu M, Theroux P, Bonnefoy A. Requirements for membrane attack complex formation and anaphylatoxins binding to collagen-activated platelets. PloS One. 2011;6:e18812.

86. Patzelt J, Mueller KA, Breuning S, Karathanos A, Schleicher R, Seizer P, Gawaz M, Langer HF, Geisler T. Expression of anaphylatoxin receptors on platelets in patients with coronary heart disease. Atherosclerosis. 2014;238:289–95.
87. Langer H, Verschoor A. Crosstalk between platelets and the complement system in immune protection and disease. Thromb Haemost. 2013;110:910–9.
88. Stahl AL, Vaziri-Sani F, Heinen S, Kristoffersson AC, Gydell KH, Raafat R, Gutierrez A, Beringer O, Zipfel PF, Karpman D. Factor h dysfunction in patients with atypical hemolytic uremic syndrome contributes to complement deposition on platelets and their activation. Blood. 2008;111:5307–15.
89. Nicholson-Weller A, Burge J, Fearon DT, Weller PF, Austen KF. Isolation of a human erythrocyte membrane glycoprotein with decay-accelerating activity for c3 convertases of the complement system. J Immunol. 1982;129:184–9.
90. Thomas TC, Rollins SA, Rother RP, Giannoni MA, Hartman SL, Elliott EA, Nye SH, Matis LA, Squinto SP, Evans MJ. Inhibition of complement activity by humanized anti-c5 antibody and single-chain fv. Mol Immunol. 1996;33:1389–401.
91. Del Conde I, Cruz MA, Zhang H, Lopez JA, Afshar-Kharghan V. Platelet activation leads to activation and propagation of the complement system. J Exp Med. 2005;201:871–9.
92. Verschoor A, Neuenhahn M, Navarini AA, Graef P, Plaumann A, Seidlmeier A, Nieswandt B, Massberg S, Zinkernagel RM, Hengartner H, Busch DH. A platelet-mediated system for shuttling blood-borne bacteria to cd8alpha+ dendritic cells depends on glycoprotein gpib and complement c3. Nat Immunol. 2011;12:1194–201.
93. Sims PJ, Wiedmer T. The response of human platelets to activated components of the complement system. Immunol Today. 1991;12:338–42.
94. Strassel C, Nonne C, Eckly A, David T, Leon C, Freund M, Cazenave JP, Gachet C, Lanza F. Decreased thrombotic tendency in mouse models of the bernard-soulier syndrome. Arterioscler Thromb Vasc Biol. 2007;27:241–7.
95. Gushiken FC, Han H, Li J, Rumbaut RE, Afshar-Kharghan V. Abnormal platelet function in c3-deficient mice. J Thromb Haemost. 2009;7:865–70.

Platelets and HMGB1 in Sterile and Non-sterile Inflammation

14

Sebastian Vogel and Meinrad Gawaz

Abstract

Platelets play a critical role in hemostasis, thrombosis, wound healing, and inflammation. We have recently shown that the damage-associated molecular pattern molecule (DAMP) high-mobility group box 1 (HMGB1) derived from platelets plays a critical role in mediating thrombosis and inflammation. The specific role of platelet-derived HMGB1 in various platelet-relevant disease states and events and the underlying pathophysiological triggers are still poorly understood. Here, we give an overview of HMGB1 and platelets in the context of sterile and non-sterile inflammation, with a focus on ischemia/reperfusion injury and dengue virus infection.

Contents

14.1 Introduction

Platelets are cellular fragments that accumulate at sites of vascular and tissue lesions and control hemostasis, thrombosis, wound healing, and inflammation [1–7]. Beyond their role as cellular mediators of hemostasis, platelets act as sentinel innate

S. Vogel • M. Gawaz (✉)
University Hospital Tübingen, Otfried-Müller-Straße 10, 72076 Tübingen, Germany
e-mail: Meinrad.Gawaz@med.uni-tuebingen.de

A. Zirlik et al. (eds.), *Platelets, Haemostasis and Inflammation*,
Cardiac and Vascular Biology 5, https://doi.org/10.1007/978-3-319-66224-4_14

immune cells that exert a unique link between coagulation and immune responses. High-mobility group box 1 (HMGB1) is a highly conserved DNA-binding protein, which is abundant in the nucleus of almost all mammalian cells and acts as a damage-associated molecular pattern (DAMP) molecule when present in the extracellular space [8–11]. Although lacking a nucleus, platelets also contain HMGB1 [5, 6, 12]. The research field around HMGB1 derived from platelets has recently experienced an unprecedented renaissance. However, the role of platelet-derived HMGB1 in mediating platelet-relevant diseases is still poorly understood. In innate immune cells, secretion and sensing of HMGB1 involves Toll-like receptor 4 (TLR4) and the NOD-like receptor NLRP3. The NLRP3 inflammasome is a key inflammatory process not only employed by innate immune cells [13] but also platelets [14]. We will summarize important findings on platelet-derived HMGB1 and sterile and non-sterile inflammation and will highlight ischemia/reperfusion (I/R) injury and dengue virus infection as two highly relevant disease states.

14.2 Platelets and HMGB1

The lifespan of platelets is 7–10 days and a platelet diameter is one third of the diameter of an erythrocyte. Platelets originate from hematopoietic progenitor cells, which develop to megakaryocytes, which in turn undergo the process of fragmentation [15]. Besides their function as mediators of primary hemostasis, platelets have been shown to exert multiple other effects at the intersection between thrombosis, inflammation, and wound healing [1, 2, 5, 6]. Upon endothelial disruption, platelets are recruited at sites of injury and interact with various extracellular matrix proteins in vessel walls, which propagate platelet activation, aggregation, and thrombosis [16]. This process is tightly regulated by cytokines and chemokines released from dying cells or stressed immune cells [2, 4]. Moreover, platelets themselves also release a whole range of cytokines and chemokines and thereby control recruitment and adhesion of immune cells and regenerative progenitor cells at sites of vascular and tissue injury [5, 6, 17–19]. HMGB1 is a DAMP when released by dying cells or actively secreted by stressed immune cells and promotes inflammation [9–11, 20]. HMGB1 activates pattern recognition receptors, including Toll-like receptors (TLRs) and the receptor of advanced glycation end products (RAGE), a transmembrane multiligand receptor of the immunoglobulin superfamily [21, 22]. HMGB1 plays a critical role in regulating inflammation and tissue repair in the setting of both sterile and infectious damage [20]. Elevated HMGB1 plasma levels are associated with abnormal coagulation, sepsis, acute pancreatitis, myocardial infarction, cancer, and other disease states [23–28]. Platelets contain significant amounts of HMGB1 as well [12]. We and others have recently shown that platelet-derived HMGB1 is critically involved in thrombus formation [5], the regulation of cellular tropisms [6], and formation of neutrophil extracellular traps (NETs) [5, 29]. Neutrophil-platelet interactions are discussed further in a separate paragraph.

14.3 Platelet Interactions with Neutrophils

Platelets have been recognized to induce the formation of NETs [30–32]. NETosis is a distinct type of cell death that results in the release of neutrophil nuclear DNA, which forms web-like structures with granular proteins, i.e., NETs [33]. NETs have been shown to exert antimicrobial activity during infections [31, 33]. Moreover, NETs play a critical role in innate immunity and microvascular thrombus formation during sterile inflammation, such as small-vessel vasculitis [34], transfusion-related acute lung injury [30], trauma/hemorrhagic shock [5], and diseases associated with cancer, such as pancreatic cancer [35] and the occurrence of liver metastases [36]. In coronary thrombi, it has been shown recently that activated platelets co-localize with neutrophils and induce NET formation mediated by platelet-derived HMGB1 and induction of neutrophil autophagy [29]. NETs promote thrombosis [37], and the activation of the canonical MAP kinase pathway (Raf/MEK/ERK) appears to be critical during NETosis [30]. Further studies are needed to understand the significance of platelet-induced NETosis in platelet-relevant diseases and contribution of platelet-derived HMGB1 and other DAMPs to the regulation of this process.

Release of mitochondria and mitochondrial DNA from activated platelets is another critical determinant in neutrophil-mediated inflammation and potentially in cardiovascular disease states [38]. Similar to NETs, the presence of platelet-derived mitochondrial DNA in the extracellular matrix may regulate sterile and non-sterile inflammation. It has been shown that the catapult-like release of mitochondrial DNA by eosinophils is crucial for maintaining the intestinal barrier function after inflammation-associated epithelial cell damage [39]. It is tempting to speculate if the release of mitochondrial DNA by platelets is involved in host defense as well. Controversial data has been reported on the influence of HMGB1 in mitochondrial quality control [40, 41]. We are currently investigating the effect of platelet-derived HMGB1 and other DAMPs on mitochondrial metabolism and DNA in platelets and contribution to inflammation and host defense.

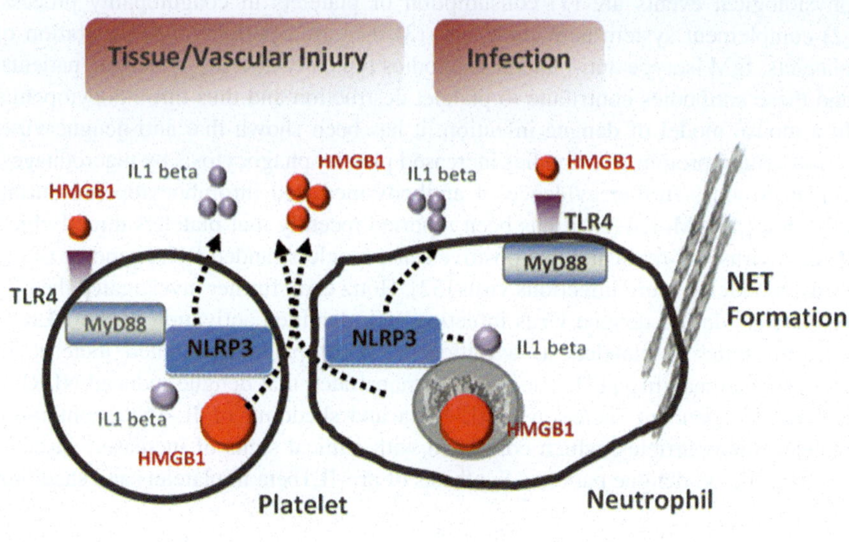

14.4 Ischemia/Reperfusion Injury

I/R injury typically leads to sterile inflammation, and platelets and platelet-derived DAMPs have been shown to play a critical role during I/R damage [4, 7, 8, 20, 42]. In a murine model of experimental I/R damage of the myocardium, HMGB1 serum levels were elevated 30 min after induction of infarction, and administration of recombinant HMGB1 worsened I/R damage, indicating a critical role of circulating HMGB1 [42]. Increased HMGB1 serum levels were also reported in patients with acute myocardial infarction, which were associated with adverse clinical outcomes including pump failure, cardiac rupture, and in-hospital deaths [43]. Conflicting results exist regarding the effect of HMGB1 on the damaged myocardium, as detrimental as well as salutary effects have been reported [42, 44, 45]. In a rat model of experimental I/R damage of the myocardium, the administration of a neutralizing anti-HMGB1 antibody significantly worsened the infarction [45]. Moreover, administration of recombinant HMGB1 into the myocardium during experimental myocardial infarction upregulated tissue healing through the activation of c-kit positive cells to form new myocytes [44]. During hepatic I/R injury, however, inhibition of HMGB1 activity with a neutralizing antibody significantly decreased liver damage [46]. Platelets mediate thrombus formation during I/R damage [47]. The role of platelet-derived HMGB1 and other DAMPs in I/R damage is currently under investigation.

14.5 Dengue Virus Infection

Dengue virus is a positive-stranded RNA virus that infects more than 50 million individuals per year and typically causes hemorrhage and coagulopathy, which may result in a life-threatening syndrome associated with increased vascular permeability, hypovolemia, hypotension, and hemorrhagic shock [48–50]. The major pathophysiological events are (1) consumption of platelets in coagulopathy process, (2) complement system activation, and (3) increased peripheral sequestration of platelets. IgM isotype anti-platelet antibodies typically develop in dengue patients, and these antibodies contribute to platelet destruction and thus thrombocytopenia. In a mouse model of dengue infection, it has been shown that anti-dengue virus nonstructural protein 1 antibodies increased platelet phagocytosis by macrophages, which provides further evidence of antibody-mediated thrombocytopenia during infection [51]. Moreover, it has been reported recently that platelets directly bind dengue virus and replicated the positive sense single-stranded RNA genome of the virus, producing more infectious virus [52]. Hottz et al. further investigated the role of platelets during dengue virus infection and identified activation of the NLRP3 inflammasome in platelets as a critical event of platelet-mediated increase in vascular permeability [14]. The group demonstrated that dengue induced NLRP3/caspase-1-dependent secretion of IL-1beta and shedding of IL-1beta-containing platelet microparticles, which correlated with clinical signs of increased vascular permeability in dengue patients. Synthesis of pro-IL1beta in platelets and shedding in its mature form in membrane microvesicles have been shown 15 years ago [19]. The NLRP3 inflammasome is a critical inflammatory process not only employed by innate immune cells [13] but also platelets [14], making platelet-derived NLRP3 a potential target in treating dengue and possibly other infectious diseases associated with abnormal coagulation. Indeed, knockout mice lacking inflammasome components NLRP3 and caspase-1 exhibited markedly decreased pathological alterations in a "two-hit" model of secondary dengue virus infection [53]. Moreover, various other pattern recognition receptors including TLR2 and TLR4 have been shown to play a critical role in dengue [54, 55]. A potential role of platelet-derived HMGB1 in dengue has not been reported yet.

14.6 Conclusion

Platelet-derived HMGB1 may signal via pattern recognition receptors, including membrane-bound Toll-like receptors and cytosolic NOD-like receptors. We have recently identified platelet-derived HMGB1 as a critical mediator of thrombosis and NET formation [5]. Further studies are needed to investigate the role of platelet-derived HMGB1 in other diseases associated with sterile and non-sterile inflammation and identify the underlying molecular mechanisms.

Compliance with Ethical Standards

Conflict of Interest: Sebastian Vogel and Meinrad Gawaz declares that they have no conflict of interest.

Ethical Approval: This article does not contain any studies with human participants or animals performed by any of the authors.

References

1. Gawaz M, Langer H, May AE. Platelets in inflammation and atherogenesis. J Clin Invest. 2005;115(12):3378–84.
2. Gawaz M, Vogel S. Platelets in tissue repair: control of apoptosis and interactions with regenerative cells. Blood. 2013;122(15):2550–4.
3. Morrell CN, Aggrey AA, Chapman LM, Modjeski KL. Emerging roles for platelets as immune and inflammatory cells. Blood. 2014;123(18):2759–67.
4. Nurden AT. Platelets, inflammation and tissue regeneration. Thromb Haemost. 2011;105 (Suppl 1):S13–33.
5. Vogel S, et al. Platelet-derived HMGB1 is a critical mediator of thrombosis. J Clin Invest. 2015;125(12):4638–54.
6. Vogel S, Chatterjee M, Metzger K, Borst O, Geisler T, Seizer P, Muller I, Mack A, Schumann S, Buhring HJ, et al. Activated platelets interfere with recruitment of mesenchymal stem cells to apoptotic cardiac cells via high mobility group box 1/toll-like receptor 4-mediated down-regulation of hepatocyte growth factor receptor MET. J Biol Chem. 2014;289(16):11068–82.

7. von Hundelshausen P, Weber C. Platelets as immune cells: bridging inflammation and cardiovascular disease. Circ Res. 2007;100(1):27–40.
8. Andersson U, Tracey KJ. HMGB1 Is a therapeutic target for sterile inflammation and infection. Annu Rev Immunol. 2011;29(139–162):24.
9. Fiuza C, Bustin M, Talwar S, Tropea M, Gerstenberger E, Shelhamer JH, Suffredini AF. Inflammation-promoting activity of HMGB1 on human microvascular endothelial cells. Blood. 2003;101(7):2652–60.
10. Scaffidi P, Misteli T, Bianchi ME. Release of chromatin protein HMGB1 by necrotic cells triggers inflammation. Nature. 2002;418(6894):191–5.
11. Vogel S, Borger V, Peters C, Forster M, Liebfried P, Metzger K, Meisel R, Daubener W, Trapp T, Fischer JC, et al. Necrotic cell-derived high mobility group box 1 attracts antigen-presenting cells but inhibits hepatocyte growth factor-mediated tropism of mesenchymal stem cells for apoptotic cell death. Cell Death Differ. 2015;22(7):1219–30.
12. Rouhiainen A, Imai S, Rauvala H, Parkkinen J. Occurrence of amphoterin (HMG1) as an endogenous protein of human platelets that is exported to the cell surface upon platelet activation. Thromb Haemost. 2000;84(6):1087–94.
13. Willingham SB, Allen IC, Bergstralh DT, Brickey WJ, Huang MT, Taxman DJ, Duncan JA, Ting JP. NLRP3 (NALP3, Cryopyrin) facilitates in vivo caspase-1 activation, necrosis, and HMGB1 release via inflammasome-dependent and -independent pathways. J Immunol. 2009;183(3):2008–15.
14. Hottz ED, Lopes JF, Freitas C, Valls-de-Souza R, Oliveira MF, Bozza MT, Da Poian AT, Weyrich AS, Zimmerman GA, Bozza FA, et al. Platelets mediate increased endothelium permeability in dengue through NLRP3-inflammasome activation. Blood. 2013;122 (20):3405–14.
15. Machlus KR, Italiano JE Jr. The incredible journey: from megakaryocyte development to platelet formation. J Cell Biol. 2013;201(6):785–96.
16. Nieswandt B, Pleines I, Bender M. Platelet adhesion and activation mechanisms in arterial thrombosis and ischaemic stroke. J Thromb Haemost. 2011;9(Suppl 1):92–104.
17. Chatterjee M, von Ungern-Sternberg SN, Seizer P, Schlegel F, Buttcher M, Sindhu NA, Muller S, Mack A, Gawaz M. Platelet-derived CXCL12 regulates monocyte function, survival, differentiation into macrophages and foam cells through differential involvement of CXCR4-CXCR7. Cell Death Dis. 2015;6:e1989.
18. Lev EI, Estrov Z, Aboulfatova K, Harris D, Granada JF, Alviar C, Kleiman NS, Dong JF. Potential role of activated platelets in homing of human endothelial progenitor cells to subendothelial matrix. Thromb Haemost. 2006;96(4):498–504.
19. Lindemann S, Tolley ND, Dixon DA, McIntyre TM, Prescott SM, Zimmerman GA, Weyrich AS. Activated platelets mediate inflammatory signaling by regulated interleukin 1beta synthesis. J Cell Biol. 2001;154(3):485–90.
20. Andersson U, Rauvala H. Introduction: HMGB1 in inflammation and innate immunity. J Intern Med. 2011;270(4):296–300.
21. Park JS, Gamboni-Robertson F, He Q, Svetkauskaite D, Kim JY, Strassheim D, Sohn JW, Yamada S, Maruyama I, Banerjee A, et al. High mobility group box 1 protein interacts with multiple toll-like receptors. Am J Physiol Cell Physiol. 2006;290(3):C917–24.
22. Park JS, Svetkauskaite D, He Q, Kim JY, Strassheim D, Ishizaka A, Abraham E. Involvement of toll-like receptors 2 and 4 in cellular activation by high mobility group box 1 protein. J Biol Chem. 2004;279(9):7370–7.
23. Angus DC, Yang L, Kong L, Kellum JA, Delude RL, Tracey KJ, Weissfeld L. Circulating high-mobility group box 1 (HMGB1) concentrations are elevated in both uncomplicated pneumonia and pneumonia with severe sepsis. Crit Care Med. 2007;35(4):1061–7.
24. Chung HW, Lee SG, Kim H, Hong DJ, Chung JB, Stroncek D, Lim JB. Serum high mobility group box-1 (HMGB1) is closely associated with the clinical and pathologic features of gastric cancer. J Transl Med. 2009;7:38.

25. Hashimoto T, Ishii J, Kitagawa F, Yamada S, Hattori K, Okumura M, Naruse H, Motoyama S, Matsui S, Tanaka I, et al. Circulating high-mobility group box 1 and cardiovascular mortality in unstable angina and non-ST-segment elevation myocardial infarction. Atherosclerosis. 2012;221(2):490–5.
26. Hatada T, Wada H, Nobori T, Okabayashi K, Maruyama K, Abe Y, Uemoto S, Yamada S, Maruyama I. Plasma concentrations and importance of high mobility group box protein in the prognosis of organ failure in patients with disseminated intravascular coagulation. Thromb Haemost. 2005;94(5):975–9.
27. Kargi A, Demirpence O, Gunduz S, Goktas S, Alikanoglu AS, Yildirim M. Serum levels of HMGB1 have a diagnostic role in metastatic renal cell cancer. Cancer Biomark. 2016;17 (1):17–20.
28. Kocsis AK, Szabolcs A, Hofner P, Takacs T, Farkas G, Boda K, Mandi Y. Plasma concentrations of high-mobility group box protein 1, soluble receptor for advanced glycation end-products and circulating DNA in patients with acute pancreatitis. Pancreatology. 2009;9 (4):383–91.
29. Maugeri N, Campana L, Gavina M, Covino C, De Metrio M, Panciroli C, Maiuri L, Maseri A, D'Angelo A, Bianchi ME, et al. Activated platelets present high mobility group box 1 to neutrophils inducing autophagy and promoting the extrusion of neutrophil extracellular traps. J Thromb Haemost. 2014;12(12):2074–88.
30. Caudrillier A, Kessenbrock K, Gilliss BM, Nguyen JX, Marques MB, Monestier M, Toy P, Werb Z, Looney MR. Platelets induce neutrophil extracellular traps in transfusion-related acute lung injury. J Clin Invest. 2012;122(7):2661–71.
31. Clark SR, Ma AC, Tavener SA, McDonald B, Goodarzi Z, Kelly MM, Patel KD, Chakrabarti S, McAvoy E, Sinclair GD, et al. Platelet TLR4 activates neutrophil extracellular traps to ensnare bacteria in septic blood. Nat Med. 2007;13(4):463–9.
32. Maugeri N, Franchini S, Campana L, Baldini M, Ramirez GA, Sabbadini MG, Rovere-Querini P, Manfredi AA. Circulating platelets as a source of the damage-associated molecular pattern HMGB1 in patients with systemic sclerosis. Autoimmunity. 2012;45(8):584–7.
33. Brinkmann V, Reichard U, Goosmann C, Fauler B, Uhlemann Y, Weiss DS, Weinrauch Y, Zychlinsky A. Neutrophil extracellular traps kill bacteria. Science. 2004;303(5663):1532–5.
34. Kessenbrock K, Krumbholz M, Schonermarck U, Back W, Gross WL, Werb Z, Grone HJ, Brinkmann V, Jenne DE. Netting neutrophils in autoimmune small-vessel vasculitis. Nat Med. 2009;15(6):623–5.
35. Boone BA, Orlichenko L, Schapiro NE, Loughran P, Gianfrate GC, Ellis JT, Singhi AD, Kang R, Tang D, Lotze MT, et al. The receptor for advanced glycation end products (RAGE) enhances autophagy and neutrophil extracellular traps in pancreatic cancer. Cancer Gene Ther. 2015;22(6):326–34.
36. Tohme S, Yazdani HO, Al-Khafaji AB, Chidi AP, Loughran P, Mowen K, Wang Y, Simmons RL, Huang H, Tsung A. Neutrophil extracellular traps promote the development and progression of liver metastases after surgical stress. Cancer Res. 2016;76(6):1367–80.
37. Fuchs TA, Brill A, Duerschmied D, Schatzberg D, Monestier M, Myers DD Jr, Wrobleski SK, Wakefield TW, Hartwig JH, Wagner DD. Extracellular DNA traps promote thrombosis. Proc Natl Acad Sci USA. 2010;107(36):15880–5.
38. Boudreau LH, Duchez AC, Cloutier N, Soulet D, Martin N, Bollinger J, Paré A, Rousseau M, Naik GS, Lévesque T, et al. Platelets release mitochondria serving as substrate for bactericidal group IIA-secreted phospholipase A2 to promote inflammation. Blood. 2014;124 (14):2173–83.
39. Yousefi S, Gold JA, Andina N, Lee JJ, Kelly AM, Kozlowski E, Schmid I, Straumann A, Reichenbach J, Gleich GJ, et al. Catapult-like release of mitochondrial DNA by eosinophils contributes to antibacterial defense. Nat Med. 2008;14(9):949–53.
40. Huebener P, Gwak GY, Pradere JP, Quinzii CM, Friedman R, Lin CS, Trent CM, Mederacke I, Zhao E, Dapito DH, et al. High-mobility group box 1 is dispensable for autophagy, mitochondrial quality control, and organ function in vivo. Cell Metab. 2014;19(3):539–47.

41. Tang D, Kang R, Livesey KM, Kroemer G, Billiar TR, Van Houten B, Zeh HJ, Lotze MT. High-mobility group box 1 is essential for mitochondrial quality control. Cell Metabolism. 2011;13(6):701–11.
42. Andrassy M, Volz HC, Igwe JC, Funke B, Eichberger SN, Kaya Z, Buss S, Autschbach F, Pleger ST, Lukic IK, et al. High-mobility group box 1 in ischemia-reperfusion injury of the heart. Circulation. 2008;117:3216–26.
43. Goldstein RS, Gallowitsch-Puerta M, Yang L, Rosas-Ballina M, Huston JM, Czura CJ, Lee DC, Ward MF, Bruchfeld AN, Wang H, et al. Elevated high-mobility group box 1 levels in patients with cerebral and myocardial ischemia. Shock. 2006;25(6):571–4.
44. Kitahara T, Takeishi Y, Harada M, Niizeki T, Suzuki S, Sasaki T, Ishino M, Bilim O, Nakajima O, Kubota I. High-mobility group box 1 restores cardiac function after myocardial infarction in transgenic mice. Cardiovasc Res. 2008;80(1):40–6.
45. Oozawa S, Mori S, Kanke T, Takahashi H, Liu K, Tomono Y, Asanuma M, Miyazaki I, Nishibori M, Sano S. Effects of HMGB1 on ischemia-reperfusion injury in the rat heart. Circ J. 2008;72(7):1178–84.
46. Tsung A, Sahai R, Tanaka H, Nakao A, Fink MP, Lotze MT, Yang H, Li J, Tracey KJ, Geller DA, et al. The nuclear factor HMGB1 mediates hepatic injury after murine liver ischemia-reperfusion. J Exp Med. 2005;201(7):1135–43.
47. Choudhri TF, Hoh BL, Zerwes HG, Prestigiacomo CJ, Kim SC, Connolly ES Jr, Kottirsch G, Pinsky DJ. Reduced microvascular thrombosis and improved outcome in acute murine stroke by inhibiting GP IIb/IIIa receptor-mediated platelet aggregation. J Clin Invest. 1998;102 (7):1301–10.
48. Bhatt S, Gething PW, Brady OJ, Messina JP, Farlow AW, Moyes CL, Drake JM, Brownstein JS, Hoen AG, Sankoh O, et al. The global distribution and burden of dengue. Nature. 2013;496 (7446):504–7.
49. de Azeredo EL, Monteiro RQ, de Oliveira Pinto LM. Thrombocytopenia in dengue: interrelationship between virus and the imbalance between coagulation and fibrinolysis and inflammatory mediators. Mediators Inflamm. 2015;2015:313842.
50. Schmidt AC. Response to dengue fever—the good, the bad, and the ugly? N Engl J Med. 2010;363(5):484–7.
51. Wan SW, Yang YW, Chu YT, Lin CF, Chang CP, Yeh TM, Anderson R, Lin YS. Anti-dengue virus nonstructural protein 1 antibodies contribute to platelet phagocytosis by macrophages. Thromb Haemost. 2016;115(3):646–56.
52. Simon AY, Sutherland MR, Pryzdial EL. Dengue virus binding and replication by platelets. Blood. 2015;126(3):378–85.
53. Lien TS, Sun DS, Chang CM, Wu CY, Dai MS, Chan H, Wu WS, Su SH, Lin YY, Chang HH. Dengue virus and antiplatelet autoantibodies synergistically induce haemorrhage through Nlrp3-inflammasome and FcgammaRIII. Thromb Haemost. 2015;113(5):1060–70.
54. Azeredo EL, Neves-Souza PC, Alvarenga AR, Reis SR, Torrentes-Carvalho A, Zagne SM, Nogueira RM, Oliveira-Pinto LM, Kubelka CF. Differential regulation of toll-like receptor-2, toll-like receptor-4, CD16 and human leucocyte antigen-DR on peripheral blood monocytes during mild and severe dengue fever. Immunology. 2010;130(2):202–16.
55. Modhiran N, Watterson D, Muller DA, Panetta AK, Sester DP, Liu L, Hume DA, Stacey KJ, Young PR. Dengue virus NS1 protein activates cells via toll-like receptor 4 and disrupts endothelial cell monolayer integrity. Sci Transl Med. 2015;7(304):304ra142.

cGMP Signaling in Platelets

15

Lai Wen, Susanne Feil, and Robert Feil

Abstract

The cyclic nucleotide cGMP is a key intracellular signaling molecule in mammals. It mediates many effects of nitric oxide (NO) including the regulation of vascular tone and platelet activity. Pharmacological and genetic studies have indicated that the NO-cGMP pathway could be an attractive target for antithrombotic drugs. Here, we summarize the biochemistry and (patho-) physiology of cGMP signaling in platelets. These cells generate and degrade cGMP by the NO-activated soluble guanylate cyclase and several phosphodiesterases (PDE2, PDE3, and PDE5), respectively. An increase of the cGMP concentration activates cGMP-dependent protein kinase type I (cGKI), which phosphorylates several platelet proteins. Among the cGKI substrates are small G-proteins (e.g., Rap1B), regulators of G-protein signaling (e.g., RGS18) and intracellular Ca^{2+} release (e.g., IRAG), and actin-binding proteins (e.g., VASP). According to the prevalent view, cGKI-dependent substrate phosphorylation limits platelet activation and thrombus formation through the inhibition of intracellular Ca^{2+} release, integrin activation, cytoskeletal remodeling, and granule secretion. Interestingly, several studies suggest that cGMP also promotes specific aspects of platelet activation. We discuss these seemingly contradictory findings and propose a new model of cGMP-regulated hemostasis that leads to optimal platelet activation in response to vascular injury. This model integrates both platelet stimulation and inhibition by dynamic shear stress-regulated cGMP signals that are generated during different phases of thrombus formation under flow in vivo.

L. Wen • S. Feil • R. Feil (✉)
Interfakultäres Institut für Biochemie, University of Tübingen, Hoppe-Seyler-Str. 4, 72076 Tübingen, Germany
e-mail: robert.feil@uni-tuebingen.de

A. Zirlik et al. (eds.), *Platelets, Haemostasis and Inflammation*,
Cardiac and Vascular Biology 5, https://doi.org/10.1007/978-3-319-66224-4_15

Contents

15.1 Introduction

Platelets are key players in hemostasis and thrombosis [1–4]. After vascular injury, platelets rapidly adhere to the subendothelial matrix exposed at the site of injury and become activated. Initially, platelets adhere to von Willebrand factor (vWF) via the glycoprotein Ib/IX/V (GPIb/IX/V) complex and to collagen via GPVI. This results in platelet activation and transformation of integrin $\alpha_{IIb}\beta_3$ (fibrinogen receptor, also called GPIIb/IIIa) and integrin $\alpha_2\beta_1$ (collagen receptor), which firmly bind to their respective extracellular matrix ligands. Then, platelets spread and form a surface for the recruitment of additional platelets via fibrinogen bridges between $\alpha_{IIb}\beta_3$ receptors leading to platelet aggregation and formation of a hemostatic plug that stops bleeding. Arterial thrombosis is primarily an exaggerated hemostatic response at the site of vascular injury. Thrombosis plays a causative role in myocardial infarction, and antithrombotic therapy takes center stage in the management of acute coronary syndromes. To avoid vessel occlusion, thrombus growth is limited by two endogenous inhibitors supplied by the endothelium, prostacyclin, and nitric oxide (NO). Prostacycylin and NO increase the levels of cAMP and cGMP in platelets, respectively. Here, we focus on the biochemistry, pharmacology, and (patho-)physiological role of the NO-cGMP signaling cascade in platelets. We will also discuss recent developments and controversies in this field, in particular whether an increase in platelet cGMP inhibits and/or promotes hemostasis and thrombosis.

NO is generated from L-arginine by NO synthases (NOS). There are three known NOS enzymes encoded by distinct genes: the neuronal NOS (nNOS or NOS1), the inducible NOS (iNOS or NOS2), and the endothelial NOS (eNOS or NOS3) [5]. nNOS and eNOS are constitutively expressed in various tissues and both enzymes are activated by Ca^{2+}/calmodulin. In contrast, expression of iNOS is inducible and the enzyme is constitutively active. Its expression is induced in several cell types including macrophages, vascular smooth muscle cells (VSMCs), and endothelial cells after exposure to lipopolysaccharide (LPS) or cytokines. eNOS is the major enzyme responsible for NO production in the vascular endothelium. NO generated in endothelial cells rapidly diffuses across cell membranes into

VSMCs, where it activates the NO receptor, soluble guanylate cyclase (sGC). sGC produces cyclic guanosine monophosphate (cGMP) in VSMCs resulting in vasodilation. Endothelium-derived NO also diffuses into the vessel lumen, where it interacts with several blood cell types including platelets.

cGMP was isolated from rat urine in the 1960s [6]. Later, it was found that cGMP mediates many effects of NO in platelets and other cell types. cGMP is formed by guanylate cyclases via cyclization of guanosine triphosphate (GTP) (Fig. 15.1). The NO-activated sGC is mainly located in the cytosol [7, 8], while

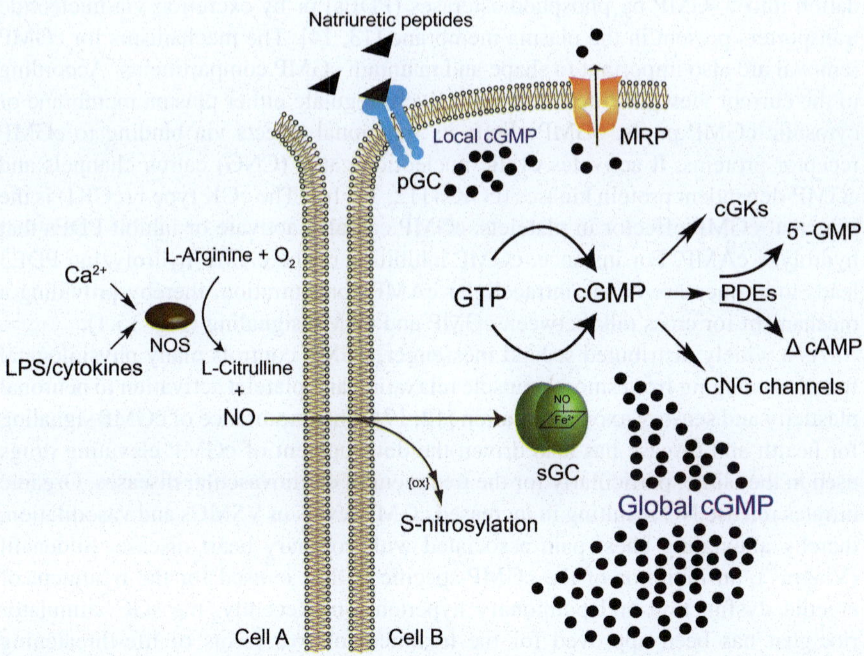

Fig. 15.1 Basic principles of cGMP signaling. NO is synthesized by a NOS in cell A (e.g., Ca^{2+}-activated eNOS in an endothelial cell, or iNOS, whose expression is induced by LPS/cytokines, in a macrophage) and diffuses across cell membranes to a nearby target cell B (e.g., a VSMC or platelet). NO binds to the Fe^{2+} of the heme group of sGC and induces the generation of a global cGMP pool (*black balls*) in the cytosol. Binding of natriuretic peptides to pGCs triggers the formation of local cGMP microdomains at the plasma membrane. cGMP exerts its functions via several effector proteins, mainly cGKs, PDEs, and CNG channels. Removal of cGMP is accomplished via hydrolysis to 5′-GMP by PDEs (e.g., PDE5) and via excretion by transporters in the plasma membrane including members of the MRP family (e.g., MRP4). cGMP can also modulate cAMP levels through stimulation or inhibition of cAMP-degrading PDEs (e.g., PDE2 or PDE3, respectively). NO can also undergo oxidation and react with thiols of cysteine side chains leading to S-nitrosylation of proteins. The major elements of cGMP signaling in platelets are NO, sGC (predominantly the $\alpha_1\beta_1$ isoform), cGKI (predominantly the cGKIβ isoform), and PDEs (predominantly PDE2, PDE3, and PDE5). Note that pGCs have not been detected in platelets. *cGKs*, cGMP-dependent protein kinases; *CNG* channels, cyclic nucleotide-gated cation channels; *LPS*, lipopolysaccharide; *MRP*, multidrug resistance protein; *NOS*, nitric oxide synthase; PDEs, phosphodiesterases; *pGC*, particulate guanylate cyclase; *sGC*, soluble guanylate cyclase

particulate guanylate cyclases (pGCs) are transmembrane receptors activated by peptides such as atrial, brain, and C-type natriuretic peptide (ANP, BNP, and CNP, respectively) [9, 10]. Accumulating evidence suggests that cGMP produced by different guanylate cyclases has different functional outcomes, even in the same cell, thus indicating the existence of subcellular cGMP signaling compartments [11, 12]. One of the simplest models predicts that the NO-sGC system generates global cGMP signals in the cytosol, whereas the natriuretic peptide-pGC system produces localized cGMP microdomains at the plasma membrane (Fig. 15.1). In order to terminate cGMP signaling, cGMP is efficiently removed, either by degradation into 5′-GMP by phosphodiesterases (PDEs) or by excretion via nucleotide transporters present in the plasma membrane [13, 14]. The mechanisms for cGMP removal are also important to shape and maintain cGMP compartments. According to the current view, distinct PDEs selectively regulate either plasma membrane or cytosolic cGMP pools. cGMP elicits its functional effects via binding to cGMP receptor proteins. It activates cyclic nucleotide-gated (CNG) cation channels and cGMP-dependent protein kinases (cGKs) [12, 15, 16]. The cGK type I (cGKI) is the principal cGMP effector in platelets. cGMP can also activate or inhibit PDEs that hydrolyze cAMP. For instance, cGMP inhibition of the cAMP-hydrolyzing PDE3 leads to an increase of the intracellular cAMP concentration, thereby providing a mechanism for cross talk between cGMP and cAMP signaling (Fig. 15.1).

As a widely distributed second messenger, cGMP controls many physiological functions ranging from smooth muscle relaxation and platelet activation to neuronal plasticity and sensory axon bifurcation [12, 17]. The importance of cGMP signaling for health and disease has also driven the development of cGMP-elevating drugs used in the clinic, particularly for the treatment of cardiovascular diseases. Organic nitrates release NO resulting in increased cGMP levels in VSMCs and vasodilation, thereby alleviating chest pain associated with coronary heart disease. Sildenafil (Viagra®), an inhibitor of the cGMP-specific PDE5, is used for the treatment of erectile dysfunction and pulmonary hypertension. Recently, the sGC stimulator riociguat has been approved for the treatment of two forms of life-threatening pulmonary hypertension. Interestingly, the physiological relevance and therapeutic potential of platelet NO-cGMP signaling for the regulation of platelet activity, hemostasis, and thrombosis is still elusive.

15.2 Shaping cGMP Signals in Platelets

15.2.1 Generation of cGMP

The intracellular cGMP concentration is determined by the balance of its synthesis and removal. To date, pGCs have not been detected in platelets. However, platelets do express sGC, the only definitive receptor for NO. Thus, cGMP synthesis in platelets is exclusively triggered by NO-activated sGC. sGC is a heterodimer consisting of two homologous subunits: an α-subunit of 73–82 kDa and a heme-binding β-subunit of ≈70 kDa [8, 18, 19]. For each subunit, two isoforms have been identified (α_1 and α_2, β_1 and β_2), but the β_2 isoform, which is preferentially expressed in kidney, does not form functional heterodimers [20]. The $\alpha_1\beta_1$ heterodimer is expressed in platelets, VSMCs, and many other cell types, while expression of the $\alpha_2\beta_1$ isoform is more restricted and most abundant in brain, lung, and placenta [21–24]. The catalytic activity of sGC requires both an α- and a β-subunit. Each subunit consists of four distinct regions that are conserved among eukaryotes. The β_1-subunit contains an N-terminal heme-binding domain, a Per/Arnt/Sim (PAS) domain, a putative amphipathic helix/coiled-coil, and a C-terminal catalytic domain. The α_1-subunit shares 30% sequence identity with the β_1-subunit and has a similar structural organization, except that its N-terminus does not bind heme and is of unknown function. The heme-binding domain of the β_1-subunit is also termed H-NOX (heme nitric oxide/oxygen) domain based on its ligand binding properties, which are conserved among similar proteins found in prokaryotes and eukaryotes [8, 25]. NO binds to the Fe^{2+} of the heme moiety forming a NO-Fe^{2+}-His complex and converting sGC into an active enzyme with over 200-fold increase in activity compared to the basal state [26]. A reduced Fe^{2+}-bound heme is essential for the activation of sGC by NO. Oxidation of the heme Fe^{2+} to Fe^{3+} strongly attenuates the enzyme's sensitivity to NO [27, 28]. The commonly used sGC inhibitor ODQ (1H-[1, 2, 4]oxadiazolo[4,3-a]quinoxalin-1-one) oxidizes the sGC heme Fe^{2+} to Fe^{3+} resulting in irreversible desensitization of sGC to NO [27].

Platelets predominantly express the $\alpha_1\beta_1$ isoform of sGC [29]. In line with this expression profile, platelets of sGC α_2-subunit knockout mice showed no functional difference compared to wild-type platelets [29]. However, genetic inactivation of the β_1-subunit resulted in impaired inhibition of platelet aggregation by NO and decreased tail bleeding times in whole body knockout animals [30, 31]. Consistent with the results from sGC β_1-deficient mice, mouse mutants expressing heme-deficient NO-unresponsive sGC also exhibited loss of NO-mediated platelet inhibition and shorter tail bleeding times than wild-type littermates [32]. These results indicated an inhibitory role of the NO-cGMP pathway in hemostasis and thrombosis. How sGC activity is regulated in intact platelets under native conditions during platelet activation and thrombus formation is not well understood. Besides the dramatic stimulation of sGC by NO, it is likely that sGC activity is also affected by alternative mechanisms. Interestingly, elevation of platelet cGMP has been detected after exposure of platelets to platelet-activating agonists including vWF, collagen, and thrombin [33–35]. Binding of vWF to platelet GPIb leads to phosphorylation of the sGC β_1-subunit and cGMP generation in an NO-independent manner [34]. It has also been suggested that several proteins including CCTη [36], Hsp70 [37], PSD95 [38], LGN [39], and PDI [40] are able to interact with sGC and, thereby, modulate cGMP synthesis in intact cells.

15.2.2 Removal of cGMP

Cyclic nucleotide PDEs catalyze the hydrolysis of one of the phosphoester bonds of cAMP or cGMP, producing inactive 5′-AMP or 5′-GMP, respectively. There are

21 genes known to encode PDEs, and they are grouped into 11 gene families based on their amino acid sequence, regulatory properties, and catalytic characteristics. PDE 4, 7, and 8 are highly specific for cAMP, while PDE 5, 6, and 9 are highly specific for cGMP, and the remaining five families (PDE 1, 2, 3, 10, and 11) hydrolyze both cGMP and cAMP, although with different affinities and efficiencies [13].

Platelets express three cGMP-hydrolyzing PDEs: PDE2, PDE3, and PDE5 [41, 42]. PDE2 is a cGMP-activated cGMP/cAMP PDE. Binding of cGMP to an allosteric site of PDE2 promotes the hydrolysis of both cGMP and cAMP. In contrast, PDE3 is a "cGMP-inhibited" cGMP/cAMP PDE. In fact, PDE3 has similar affinities for cGMP and cAMP, but the reaction rate for cGMP is only ≈10% of that for cAMP. Thus, cGMP exerts competitive inhibition of cAMP hydrolysis and PDE3 is usually referred to as "cGMP-inhibited" PDE. The cGMP-specific PDE5 is abundantly expressed in platelets [43]. By binding to an allosteric site, cGMP increases the catalytic activity of the enzyme and, thus, promotes its own degradation [44]. Activation of PDE5 by cGMP is further augmented via phosphorylation of a specific serine residue (Ser92 or Ser102 of bovine or human PDE5, respectively) by cGKI [45].

Intracellular cGMP can also be removed through cGMP efflux. Several cGMP transporters located in the plasma membrane have been identified including members of the multidrug resistance protein (MRP) family such as MRP4/5/8 [46]. MRP4-mediated removal of cGMP in VSMCs contributes to the control of muscle tonus to an extent similar to PDE5-mediated cGMP degradation [47]. Efflux of cGMP from activated human platelets was reported more than 20 years ago [48]. While MRP5 and MRP8 were not detected in platelets [49], MRP4 is highly abundant in dense granules and also at lower levels in the plasma membrane of human platelets [50]. MRP4-mediated cGMP efflux reduces the inhibitory effects of cGMP in human platelets [50, 51], but the relative importance of this process in comparison to cGMP removal via PDEs remains to be established.

15.3 cGMP Effector Mechanisms in Platelets

cGKs are central mediators of cGMP signaling [12]. They belong to the AGC subfamily of serine/threonine protein kinases and are activated by cGMP concentrations in the range of ≈0.1–1 μM [52, 53]. Mammals have two cGK genes, *prkg1* and *prkg2*, encoding cGKI and cGK type II, respectively. In the cardiovascular system, cGKI is more commonly expressed than cGK type II, in particular in VSMCs, cardiomyocytes, and platelets. The *prkg1* gene encodes two cGKI isoforms, termed cGKIα and cGKIβ, which differ in their N-terminal domains (≈100 amino acids). Human platelets only express cGKIβ, whereas both cGKIβ and a small amount of cGKIα were detected in mouse platelets [54, 55]. In an ischemia-induced thrombosis model, cGKI-deficient mouse platelets showed increased adhesion and aggregation compared to wild-type platelets indicating an inhibitory role of cGKI-mediated signaling in platelets [56]. cGKI substrates such as vasodilator-stimulated phosphoprotein (VASP) and inositol-1,4,5-trisphosphate (IP_3) receptor-associated cGMP kinase substrate (IRAG) are also abundantly expressed in platelets. Consistent with the analysis of cGKI mouse mutants, mice deficient in VASP or IRAG showed impaired NO/cGMP-dependent inhibition of platelet aggregation in vivo [55, 57]. Taken together, the data from genetic manipulation of sGC, cGKI, and cGKI substrates have clearly established the importance of the NO-sGC-cGMP-cGKI pathway for platelet inhibition in vivo. What are the molecular mechanisms of cGMP inhibition of platelet activity? Many studies have shown that activation of NO-cGMP signaling interferes with several key events during platelet aggregation under in vitro conditions including Ca^{2+} release from intracellular stores, granule secretion, and activation of small G-proteins and integrins.

Ca^{2+} plays a central role in platelet aggregation by promoting platelet adhesion, granule release, soluble agonist-induced platelet activation, cytoskeleton reorganization, and integrin activation [2]. Soluble platelet agonists that activate $G\alpha_q$-coupled receptors (ADP, thrombin, thromboxane A_2) as well as collagen induce activation of phospholipase C, which generates IP_3. IP_3 binds to and activates the IP_3 receptor, a Ca^{2+} channel in the endoplasmic reticulum membrane mediating Ca^{2+} release from intracellular Ca^{2+} stores followed by store-operated Ca^{2+} entry from the extracellular space [58, 59]. Sustained cytosolic Ca^{2+} elevation is required for both platelet aggregation and blood coagulation. Platelets provide a suitable surface for the assembly of coagulation protein complexes by expressing phosphatidylserine. The exteriorization of phosphatidylserine is mediated by the Ca^{2+}-dependent regulation of a phospholipid scramblase called TMEM16F. Phosphatidylserine exposure accelerates thrombin formation via the coagulation cascade, through which fibrinogen is converted into fibrin, thereby stabilizing the thrombus [4]. It is well known that activation of cGKI suppresses agonist-induced intracellular Ca^{2+} release in VSMCs [60–62], at least in part, by forming a complex containing cGKIβ, its substrate IRAG, and the IP_3 receptor type I (IP_3RI) [63, 64]. A similar mechanism is also active in platelets. It has been shown that cGKI phosphorylates IRAG in intact platelets and that thrombin-induced Ca^{2+} release was attenuated by NO or cGMP in wild-type platelets but not in IRAG-deficient platelets [55]. Thus, the interaction of cGKI-phosphorylated IRAG and the IP_3RI appears to suppress IP_3-induced Ca^{2+} release from the endoplasmic reticulum of platelets.

Following platelet activation, platelets undergo a dramatic shape change mediated by cytoskeletal remodeling and accompanied by the release of platelet granule contents. The release of soluble agonists such as thromboxane A_2 and ADP is crucial for irreversible platelet activation at the site of vascular injury. These agonists activate specific G protein-coupled receptors and act in an autocrine or paracrine manner to enhance platelet activation and thrombus formation [4]. The platelet cytoskeleton binds and positions specific signaling molecules. In particular, reorganization of the actin cytoskeleton promotes the formation of active integrin signaling complexes. An increase in cGMP also affects the cytoskeleton through cGKI-mediated phosphorylation of cytoskeleton-associated proteins. The actin-binding protein VASP is a major cGKI substrate in platelets [65]. cGKI phosphorylates VASP at Ser157 and Ser239. VASP participates in actin fiber formation and VASP phosphorylation affects its own intracellular localization [66]. Phosphorylation of VASP at Ser157 correlates with inhibition of the fibrinogen receptor integrin $\alpha_{IIb}\beta_3$ [67]. VASP-deficient platelets show enhanced binding of integrin $\alpha_{IIb}\beta_3$ to fibrinogen revealing an inhibitory function of VASP in integrin activation [68, 69]. Deletion of VASP in mice also leads to enhanced adhesion of platelets to the vascular wall in an ischemia-induced thrombosis model [57]. These findings suggest that cGMP inhibits platelet aggregation, at least in part, via cGKI-mediated VASP phosphorylation and interference with cytoskeletal remodeling, which is normally associated with platelet activation. However, the exact functional role of VASP phosphorylation at Ser157 and/or Ser239 in hemostasis and thrombosis in vivo is unknown.

In addition to IRAG and VASP, several other cGKI substrates have been identified that might contribute to the inhibitory actions of cGMP in platelets [70, 71]. Many of these proteins represent G-proteins or modulators of G-protein activity known to control platelet activation via their effects on integrins, intracellular Ca^{2+} release, and the cytoskeleton. For instance, the small G-protein Rap1B is involved in the activation of integrin $\alpha_{IIb}\beta_3$ [72]. Inhibition of integrin $\alpha_{IIb}\beta_3$ by NO/cGMP in human platelets is, in part, attributed to cGKI-mediated phosphorylation of Rap1B [73] or Rap1GAP2 [74], a GTPase-activating protein for Rap1B. Moreover, cGKI has been shown to phosphorylate regulator of G-protein signaling 18 (RGS18) [75]. RGS18 is a GTPase-activating protein for the $G\alpha_q$-subunits of heterotrimeric G-proteins that interact with G protein-coupled receptors at the platelet plasma membrane. Phosphorylation of RGS18 by cGKI potentiates RGS18 function and, consequently, attenuates $G\alpha_q$ signaling leading to reduced receptor-mediated Ca^{2+} release from intracellular stores triggered by platelet agonists like thrombin [75]. Recently, two novel cGKI substrates in platelets have been identified, ARHGAP17, a GTPase-activating protein, and ARHGEF6, a guanine nucleotide exchange factor, which regulate the activity of the small G-protein Rac1 [76]. Phosphorylation of ARHGAP17 and ARHGEF6 by cGKI induces a rearrangement of their associated protein complexes, resulting in a reduced level of active Rac1, a key player in cytoskeletal remodeling and platelet activation [77].

In addition to cGMP, cAMP is another important cyclic nucleotide messenger playing an inhibitory role in platelet activation [71]. cGMP and cAMP activate cGMP- and cAMP-dependent protein kinases, respectively, and most of the abovementioned cGKI substrate proteins can also be phosphorylated by cAMP-dependent protein kinase. Thus, it appears that in many cases cGMP and cAMP signals converge at the level of downstream mechanisms. There are also other types of interplay between cGMP and cAMP signaling in platelets. Perhaps the most significant cGMP-cAMP cross talk is mediated through the stimulatory or inhibitory effect of cGMP on cAMP hydrolysis by PDE2 or PDE3, respectively. A rise in the intracellular cGMP concentration can decrease the cAMP level via cGMP-activated PDE2, and it can increase the cAMP level via cGMP-inhibited PDE3 [41, 78]. Thus, cGMP might affect platelet activity not only via activation of cGKI but also via modulation of cAMP signaling (Fig. 15.1). This mode of cGMP-cAMP cross talk has been best studied in cardiomyocytes, where it might contribute to the spatiotemporal control of cyclic nucleotide signaling compartments and cardiac (patho-)physiology [79]. Indeed, cGMP-mediated inhibition of PDE3 has also been implicated in elevation of cAMP levels and activation of cAMP-dependent protein kinase in human platelets, which could mediate at least some aspects of NO-induced platelet inhibition [80, 81]. Another study has described a compartment-specific signaling complex in human platelets containing PDE5, cGKIβ, and the IP_3RI and proposed a model in which cGKI selectively activates PDE5 within a defined microdomain allowing spatial and temporal control of cGMP signaling in platelets [82]. However, the in vivo relevance of cGMP-cAMP cross talk and compartmentalized cyclic nucleotide signaling in platelets remains to be established.

15.4 Relevance of cGMP in Hemostasis, Thrombosis, and Antithrombotic Therapy

The biochemical, genetic, and pharmacological studies described above demonstrate that platelets express a robust NO-sGC-cGMP-cGKI pathway and that activation of this signaling cascade results in platelet inhibition. It is widely accepted that a cGMP increase works as a brake to limit thrombosis. Several clinically used antiplatelet drugs elevate cGMP and/or cAMP levels. For instance, dipyridamole and cilostazol inhibit cyclic nucleotide PDEs. Moreover, recent evidence indicates that activation of the NO-cGMP axis contributes to the antithrombotic effects of certain β-blockers like nebivolol [83] as well as to the action of $P2Y_{12}$ receptor blockers that are commonly used in antiplatelet therapy (e.g., clopidogrel, prasugrel, ticagrelor) [84]. On the other hand, disabling the cGMP brake is predicted to promote platelet activation and aggregation. Indeed, a recent report suggests that hyperlipidemia and oxidized low-density lipoprotein promote platelet hyperactivity by inducing reactive oxygen species that desensitize cGKI-mediated platelet inhibition [85].

It is important to note that most of our knowledge about cGMP signaling in platelets has been derived from in vitro analysis of wild-type and mutant platelets isolated from mice and humans, for instance, by biochemical or aggregometry assays. In vitro experiments with isolated platelets do probably not completely mimic the in vivo situation, in particular with regard to platelet interactions with immobilized substrates, other cell-types, blood flow, and shear forces that influence platelet activity under in vivo conditions [3, 4, 86]. Moreover, the interpretation of experimental results is often complicated by technical and biological issues. (1) The selectivity and efficacy of several cGMP analogs commonly used to study cGMP signaling in platelets is questionable [87–89]. (2) NO, especially if used at high concentrations, can also exert cGMP-independent effects [90], for instance, via protein S-nitrosylation [91]. (3) Conventional methods for detection of cGMP (e.g., RIA or ELISA) are performed with platelet extracts. These cGMP assays have low temporal resolution and lack spatial resolution. They are not able to monitor dynamic cGMP signals in real time in living platelets during activation and aggregation. The transferability of cGMP levels determined in platelet extracts to cGMP signaling in living platelets appears limited if one considers that platelet cGMP is very rapidly degraded by PDEs [43] and that changes of the cGMP concentration in subcellular compartments might remain undetected when cGMP levels are determined in cell extracts [82]. Moreover, it has been noted that some commercially available ELISA kits for cGMP can give false-positive results [92]. (4) Last but not least, the interpretation of experiments with gene-targeted mice or with platelets isolated from these mice might be complicated due to additional, probably unknown, phenotypes of the mutant mice. For instance, cGKI-deficient mice show significantly elevated IL-6 serum levels [93]. The high IL-6 concentration is not due to dysfunctional cGMP signaling in platelets and leads to thrombocytosis [94]. The high platelet count and potential alterations of platelet functions secondary to the high IL-6 levels might lead to thrombosis and bleeding phenotypes in global cGKI knockout mice that are not caused by a lack of cGMP-cGKI signaling in platelets. Furthermore, it should be considered that vascular endothelial-generated NO and the smooth muscle sGC-cGMP-cGKI pathway is important in vasodilation. Thus, the whole body deficiency of sGC or cGKI may cause vasoconstriction and consequently reduce bleeding time. To specifically address the role of cGMP signaling proteins in platelets, it is recommended to analyze platelet-specific knockout mice.

To date, several studies have investigated platelet cGMP signaling by using in vivo models of hemostasis and thrombosis. One study applied intravital microscopy in mice after intestinal ischemia/reperfusion injury and found that platelet cGKI but not endothelial or smooth muscle cGKI is required to prevent intravascular adhesion and aggregation of platelets after ischemia in vivo [56]. Surprisingly, another study reported that cGKI-deficient platelets showed impaired activation in response to vWF and low-dose thrombin and that global cGKI knockout mice had prolonged tail bleeding times [33]. A follow-up study of the same group reported that aggregation of sGC-deficient platelets was reduced at low concentrations of collagen and thrombin, and tail bleeding times and thrombus formation were increased in platelet-specific sGC knockout mice [35]. These findings indicated a stimulatory role for the sGC-cGMP-cGKI pathway in platelet activation and provoked an ongoing controversy about cGMP's function(s) in platelets [95–98]. It is well known that platelet cGMP levels increase in response to various platelet agonists (e.g., vWF, collagen, thrombin, ADP) [33–35, 99, 100]. Unfortunately, the functional effect of agonist-induced cGMP elevation was not always investigated in these studies, thus leaving the question open whether the cGMP increase associated with a specific agonist leads to platelet activation or inhibition. In general, the cGMP level reached during the initial phase of agonist-induced platelet activation appears to be much (≈tenfold) smaller than the cGMP concentration reached during NO-induced platelet inhibition. Mechanistic details as to how platelet agonists increase cGMP and how cGMP might promote the platelet activation process are not well understood. It is a matter of debate whether platelet agonists activate a platelet NOS to produce NO or whether platelet sGC is activated in an NOS/NO-independent manner [92]. A recent study indicated that platelets produce NO during adhesion to immobilized collagen in flowing blood under high shear rates, but under these conditions NO production limited rather than promoted further platelet deposition [101]. Other work indicated that vWF stimulates cGMP production in platelets independent of NOS and NO via Src kinase-mediated Tyr192 phosphorylation and activation of sGC [34]. It was suggested that the increase in cGMP upon stimulation of GPIb by vWF activates cGKI, which in turn leads to the stimulation of mitogen-activated protein kinases (MAPKs) and activation of the platelet integrin $\alpha_{IIb}\beta_3$ [102]. However, other studies could not confirm a stimulatory role of the cGMP-cGKI-MAPK axis in platelet activation by thrombin or low-dose collagen [103, 104].

Interestingly, an increasing number of publications describe a stimulatory role of cGMP in platelet activation elicited by ligands of pattern recognition receptors such as TLR4 and NOD2. These receptors play a key role in innate immunity [105] and might provide a bridge between infection/inflammation and thrombotic events. Bacteria-derived LPS stimulates platelet aggregation and thrombus formation via binding to TLR4 on the platelet surface. TLR4-induced intracellular signaling in platelets required the adaptor protein MyD88 and was associated with elevated cGMP and cGKI activity, which may initiate platelet activation [106]. Very recently, it was shown that the damage-associated molecular pattern molecule HMGB1 is a critical mediator of thrombosis, whose effects are transmitted via TLR4/MyD88-dependent recruitment of sGC toward the platelet plasma membrane, followed by cGMP synthesis and activation of cGKI [107]. Another study found that platelets also express the pattern recognition receptor NOD2 and that activation of platelet NOD2 by bacteria-derived muramyl dipeptide elicits an increase of cGMP-cGKI signaling associated with enhanced platelet activation and thrombosis [108]. In sum, these studies support the notion that activation of the cGMP-cGKI pathway via certain upstream triggers can have a stimulatory effect on platelets, which may be critical for abnormal platelet activation and aggregation in response to bacterial infection and inflammation.

The pathophysiological relevance of platelet cGMP signaling in humans is supported by recent human genetic analysis. One study identified the segregation of heterozygous mutations in two genes functionally related to NO-cGMP signaling in an extended myocardial infarction family [109]. The mutated genes encode the sGC α_1-subunit and the CCTη chaperone, which stabilizes sGC. Platelets from digenic mutation carriers contained less sGC protein and displayed reduced NO-induced cGMP formation, and sGC α_1-deficient mice showed accelerated thrombosis in the microcirculation after local trauma [109]. These findings indicate that dysfunctional NO-cGMP signaling increases the risk of myocardial infarction, perhaps through increased thrombus formation. Unfortunately, the platelet function of mutated carriers was not analyzed in this study. Therefore, it is not clear whether variants in sGC α_1 and CCTη promote myocardial infarction through platelet activation or other mechanisms. Another report described an autosomal-recessive syndrome resulting in severe moyamoya (a cerebrovascular condition leading to stroke) and early-onset achalasia (a rare disease characterized by aperistalsis of the esophagus) that is associated with homozygous mutations in the sGC α_1 gene [110]. Mutated carriers had a complete loss of sGC protein in their platelets. Interestingly, loss of platelet sGC led to a defect in platelet activation, strongly suggesting that sGC-cGMP signaling has a stimulatory role in human platelets [110]. Two recent large-scale genome-wide association studies identified a common variant on chromosome four overlapping with the sGC α_1 gene, and this variant showed a significant association with coronary artery disease and myocardial infarction [111, 112]. Thus, human genetic studies provide substantial evidence for a causal involvement of sGC and cGMP signaling in the pathogenesis of atherothrombotic diseases. cGMP-elevating drugs, such as the PDE5 inhibitor sildenafil or the sGC stimulator riociguat, are increasingly recognized as a treatment option for cardiovascular and cardiopulmonary disorders. However, it is not clear whether an increase of cGMP in platelets specifically would have beneficial and/or detrimental effects on these diseases.

Taken together, the in vitro and in vivo data discussed above strongly suggest a biphasic role of cGMP signaling in both platelet activation and inhibition. We speculate that biphasic cGMP signaling in platelets is indeed highly relevant for an optimal hemostatic platelet response in vivo and can be best explained when we consider the dynamics of platelet cGMP signals and thrombus formation under flow. It is now widely recognized that blood flow and the resulting shear forces are key factors affecting platelet aggregation [113, 114]. Shear rates in arteries and arterioles are in the range of 300–800 s^{-1} and 500–1600 s^{-1}, respectively, whereas those in veins are about ten times lower at 20–200 s^{-1} [115]. During thrombus formation, there is a dramatic increase of shear rates, which can surge above 10,000 s^{-1} in stenotic arteries [116]. It has long been known that shear stress activates eNOS in endothelial cells increasing NO production [117] and that shear stress-induced NO can diffuse into the vessel lumen and stimulate cGMP synthesis in platelets [118]. Thus, it is likely that endothelial NO release and platelet cGMP signals are modulated by the changing shear forces during thrombus formation.

Based on these assumptions and the myriad of data published in the field, we propose a new model that integrates seemingly contradictory findings as well as cGMP compartmentation and blood flow to describe dynamic platelet cGMP signaling during hemostasis and thrombosis in vivo (Fig. 15.2). In this model, a small/compartmentalized increase of platelet cGMP during the initial phase of platelet activation promotes platelet aggregation at the site of injury and is then followed by a stronger cGMP signal that inhibits further recruitment of platelets and limits thrombus growth. How is the spatiotemporal dynamics of platelet cGMP signals regulated, and how can an increase of cGMP both stimulate and block platelet aggregation? To provide answers to these questions, our model incorporates two important features: (1) adjustment of the intraplatelet cGMP concentration by shear-dependent NO release from the endothelium and (2) generation of different cGMP pools during early and late stages of platelet aggregation. In the initial phase of platelet adhesion and activation, shear stress and endothelial NO production at the site of injury are relatively low (Fig. 15.2, *left*). In response to adhesive ligands

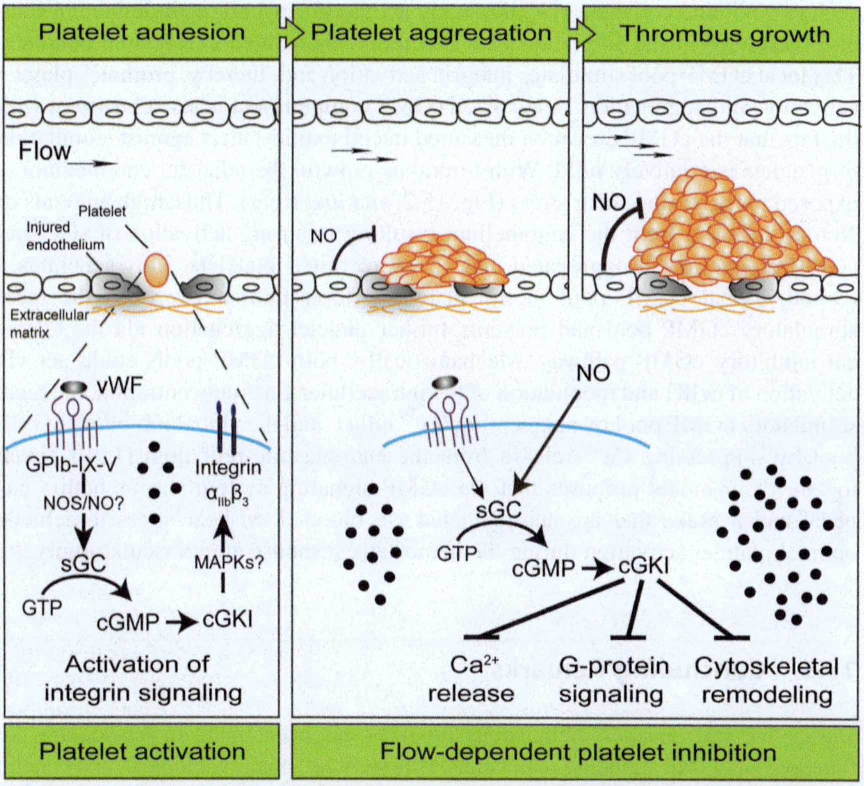

Fig. 15.2 A model for autoregulated hemostasis and thrombosis by dynamic flow-dependent cGMP signaling. The regulation of platelet activity by cGMP (*black balls*) is flow dependent. An important factor that adjusts the intraplatelet cGMP concentration is endothelial NO release, which is augmented with increasing shear stress during thrombus growth. The increasing shear stress exerted by blood flow during thrombosis is indicated by *horizontal arrows* in the vessel lumen. In the early phase of platelet activation, a mild cGMP increase is stimulatory (*left*). At this stage, shear stress and NO production at the site of injury are low. In response to adhesive ligands such as vWF and collagen, sGC is weakly activated via a not well-defined mechanism, perhaps via intraplatelet NOS/NO, resulting in a relatively small and/or compartmentalized cGMP increase at the platelet plasma membrane. This early cGMP signal promotes platelet adhesion via activation of integrin $\alpha_{IIb}\beta_3$ signaling, perhaps via cGKI and MAPKs. However, in later stages of platelet aggregation and thrombus growth, strong inhibitory cGMP signals are generated (*mid* and *right*). Flow-induced shear rates and endothelial NO production increase with thrombus growth. The high local NO concentration results in strong stimulation of sGC activity in aggregated and freshly recruited platelets. The high cGMP concentration in these platelets strongly activates cGKI, which phosphorylates multiple substrates resulting in inhibition of intracellular Ca^{2+} release, G-protein signaling, and cytoskeletal remodeling, so that further recruitment of platelets and overgrowth of hemostatic thrombi is prevented. With decreasing thrombus size, the shear forces acting on the endothelium and the production of cGMP decline, and the intraplatelet cGMP concentration drops due to hydrolysis mainly by PDE5 (not shown). Thus, the shear stress dependency of the sGC-cGMP-cGKI pathway could provide a mechanism for a biphasic role of cGMP, whereby weak/compartmentalized cGMP signals stimulate initial platelet

such as vWF and collagen, sGC is weakly activated via a not well-defined mechanism resulting in the formation of a membrane-associated cGMP microdomain. This local cGMP pool stimulates integrin activation and, thereby, promotes platelet activation and aggregation. A localized cGMP compartment is also consistent with the fact that the cGMP elevation measured in cell extracts after agonist stimulation of platelets is relatively mild. With thrombus growth, the adjacent endothelium is exposed to increasing shear stress (Fig. 15.2, *mid and right*). Thus, high amounts of NO are released from the endothelium resulting in strong activation of sGC and cGMP production in aggregated and newly recruited platelets. This generates a second, global cGMP pool in the platelet cytoplasm that overrides the local stimulatory cGMP pool and prevents further platelet aggregation via the classical inhibitory cGMP pathway. Mechanistically, both cGMP pools could act via activation of cGKI and modulation of the intracellular Ca^{2+} concentration, the local stimulatory cGMP pool by augmenting Ca^{2+} influx, and the global inhibitory cGMP pool by suppressing Ca^{2+} release from the endoplasmic reticulum [119]. Taken together, our model proposes that the cGMP signaling system acts as both a gas pedal and a brake that are autoregulated via blood flow/shear stress to achieve optimal platelet activation during the hemostatic response after vascular injury.

15.5 Concluding Remarks

During the last decades, tremendous progress has been made in elucidating the functions of cGMP signaling in platelets. Biochemical, pharmacological, and genetic studies have clearly established an inhibitory role of the canonical NO-sGC-cGMP-cGKI pathway in platelet aggregation, mainly by inhibiting intracellular Ca^{2+} release and cytoskeletal remodeling. However, accumulating evidence indicates that cGMP can also promote initial platelet activation. The mechanism of this stimulatory action of cGMP is less well understood than inhibitory cGMP signaling. These contradictory findings combined with the fact that many of the previous studies were performed under in vitro conditions, which probably do not completely mimic the in vivo situation, make it difficult to conclude whether an increase in platelet cGMP has beneficial and/or detrimental effects on hemostasis and thrombosis in vivo. Major questions in platelet cGMP signaling remain to be answered, for instance:

- Can platelets produce NO?
- Does NO also exhibit cGMP-independent effects on platelet activity?
- Does cGMP also exhibit cGKI-independent effects on platelet activity?

Fig. 15.2 (continued) activation at the site of vascular injury followed by stronger cGMP signals that limit an exaggerated hemostatic response and occlusive thrombosis. As such, the cGMP signaling system could serve as both a gas pedal and a brake that are autoregulated via blood flow/shear stress to achieve optimal platelet activation during the hemostatic response after vascular injury

- How exactly does cGMP signaling interact with other signaling pathways to control platelet functions? For instance, what is the role of cGMP-cAMP cross talk? What are the in vivo substrates of cGKI?
- In addition to cGMP hydrolysis via PDEs, how important is cGMP efflux via cyclic nucleotide transporters in shaping cGMP signals in platelets?
- Do platelets have subcellular cGMP signaling compartments? If so, are they functionally relevant?
- How does cGMP promote initial platelet activation? How is sGC activated in response to platelet agonists and is the resulting cGMP production compartmentalized, for instance, at the plasma membrane? How does the cGMP increase promote integrin activation and platelet aggregation?
- Are cardiovascular diseases associated with dysfunctional cGMP signaling in platelets? Can we treat these disorders by targeting platelets with cGMP-elevating drugs?

To improve our understanding of cGMP's function in hemostasis and thrombosis, it is important to study cGMP signaling in platelets under in vivo conditions, in the context of platelet interactions with the vessel wall and blood flow. We propose a new model of cGMP-regulated hemostasis that integrates both platelet stimulation and inhibition by dynamic shear stress-regulated cGMP signals during different phases of thrombus formation under flow in vivo (Fig. 15.2). In this model, rapidly changing and compartmentalized cGMP signals are crucial for the appropriate functioning of platelets under native conditions. This hypothesis has to be tested in the future. However, the spatiotemporal dynamics of cGMP signals in living platelets cannot be monitored with conventional cell-destructive cGMP assays such as RIA and ELISA. In recent years, cGMP sensor proteins have been developed for the visualization of cGMP signals in real time in living cells [120, 121]. Transgenic mice expressing such a cGMP biosensor are available [122, 123] and should allow for the imaging of dynamic cGMP signals in platelets under flow conditions. The spatiotemporal cGMP profile within platelets can then be correlated with changes in platelet behavior during platelet aggregation in vitro or even during thrombus formation in vivo. The study of cGMP in hemostasis and thrombosis should benefit from these technical advances.

What is the therapeutic potential of cGMP-elevating drugs for the treatment of thrombotic conditions? An inherent weakness with all currently used antiplatelet agents is their deleterious impact on hemostasis, with the most potent antithrombotic drugs typically conferring the greatest bleeding risk [4]. Optimal platelet activation in response to vascular injury in the context of hemostasis means preventing circulating platelets from activating needlessly, allowing them to respond quickly when necessary and limiting platelet activation to avoid excessive platelet accumulation and thrombus growth [124]. Dynamic flow-regulated NO-cGMP signaling might indeed provide a self-regulating gas and brake for optimal platelet activation. As such, pharmacological stimulation of platelet cGMP signaling, perhaps with innovative compounds that selectively trigger inhibitory cGMP signals in a growing thrombus without affecting stimulatory cGMP signals during initial formation of the hemostatic plug, is an interesting strategy for antithrombotic therapy with a lower risk of bleeding. Indeed, novel sGC-stimulating drugs have been shown to reduce thrombus formation in animal models [125].

Acknowledgements The authors would like to thank Michael Paolillo for reading the manuscript as well as the current and past members of the Feil laboratory for critical discussions. We apologize to all our colleagues, whose work could not be cited due to space limitations. The work in the authors' laboratory is supported by the Fund for Science and Deutsche Forschungsgemeinschaft (FE 438/2-4, FOR 2060 projects FE 438/5-1 and FE 438/6-1, KFO 274 projects FE 438/7-1 and FE 438/8-2).

Compliance with Ethical Standards

Conflict of Interest: Lai Wen, Susanne Feil, and Robert Feil declare that they have no conflict of interest.

Ethical Approval: This article does not contain any studies with human participants or animals performed by any of the authors.

References

1. Gawaz M, Langer H, May AE. Platelets in inflammation and atherogenesis. J Clin Invest. 2005;115:3378–84.
2. Furie B, Furie BC. Mechanisms of thrombus formation. N Engl J Med. 2008;359:938–49.
3. Ruggeri ZM. Platelet adhesion under flow. Microcirculation. 2009;16:58–83.
4. Jackson SP. Arterial thrombosis—insidious, unpredictable and deadly. Nat Med. 2011;17:1423–36.
5. Alderton WK, Cooper CE, Knowles RG. Nitric oxide synthases: structure, function and inhibition. Biochem J. 2001;357:593–615.
6. Ashman DF, Lipton R, Melicow MM, Price TD. Isolation of adenosine 3′, 5′-monophosphate and guanosine 3′, 5′-monophosphate from rat urine. Biochem Biophys Res Commun. 1963;11:330–4.
7. Friebe A, Koesling D. Regulation of nitric oxide-sensitive guanylyl cyclase. Circ Res. 2003;93:96–105.
8. Derbyshire ER, Marletta MA. Structure and regulation of soluble guanylate cyclase. Annu Rev Biochem. 2012;81:533–59.
9. Kuhn M. Function and dysfunction of mammalian membrane guanylyl cyclase receptors: lessons from genetic mouse models and implications for human diseases. Handb Exp Pharmacol. 2009;191:47–69.
10. Potter LR. Regulation and therapeutic targeting of peptide-activated receptor guanylyl cyclases. Pharmacol Ther. 2011;130:71–82.
11. Fischmeister R, Castro LR, Abi-Gerges A, Rochais F, Jurevicius J, Leroy J, Vandecasteele G. Compartmentation of cyclic nucleotide signaling in the heart: the role of cyclic nucleotide phosphodiesterases. Circ Res. 2006;99:816–28.
12. Kemp-Harper B, Feil R. Meeting report: cGMP matters. Sci Signal. 2008;1:pe12.
13. Francis SH, Blount MA, Corbin JD. Mammalian cyclic nucleotide phosphodiesterases: molecular mechanisms and physiological functions. Physiol Rev. 2011;91:651–90.

14. Cheepala S, Hulot JS, Morgan JA, Sassi Y, Zhang W, Naren AP, Schuetz JD. Cyclic nucleotide compartmentalization: contributions of phosphodiesterases and ATP-binding cassette transporters. Annu Rev Pharmacol Toxicol. 2013;53:231–53.
15. Biel M, Zong X, Ludwig A, Sautter A, Hofmann F. Structure and function of cyclic nucleotide-gated channels. Rev Physiol Biochem Pharmacol. 1999;135:151–71.
16. Hofmann F. The biology of cyclic GMP-dependent protein kinases. J Biol Chem. 2005;280:1–4.
17. Schmidt H, Rathjen FG. Signalling mechanisms regulating axonal branching in vivo. Bioessays. 2010;32:977–85.
18. Gerzer R, Hofmann F, Bohme E, Ivanova K, Spies C, Schultz G. Purification of soluble guanylate cyclase without loss of stimulation by sodium nitroprusside. Adv Cyclic Nucleotide Res. 1981;14:255–61.
19. Kamisaki Y, Saheki S, Nakane M, Palmieri JA, Kuno T, Chang BY, Waldman SA, Murad F. Soluble guanylate cyclase from rat lung exists as a heterodimer. J Biol Chem. 1986;261:7236–41.
20. Yuen PS, Potter LR, Garbers DL. A new form of guanylyl cyclase is preferentially expressed in rat kidney. Biochemistry. 1990;29:10872–8.
21. Harteneck C, Wedel B, Koesling D, Malkewitz J, Bohme E, Schultz G. Molecular cloning and expression of a new alpha-subunit of soluble guanylyl cyclase. Interchangeability of the alpha-subunits of the enzyme. FEBS Lett. 1991;292:217–22.
22. Russwurm M, Behrends S, Harteneck C, Koesling D. Functional properties of a naturally occurring isoform of soluble guanylyl cyclase. Biochem J. 1998;335(Pt 1):125–30.
23. Budworth J, Meillerais S, Charles I, Powell K. Tissue distribution of the human soluble guanylate cyclases. Biochem Biophys Res Commun. 1999;263:696–701.
24. Mergia E, Russwurm M, Zoidl G, Koesling D. Major occurrence of the new alpha2beta1 isoform of NO-sensitive guanylyl cyclase in brain. Cell Signal. 2003;15:189–95.
25. Gerzer R, Bohme E, Hofmann F, Schultz G. Soluble guanylate cyclase purified from bovine lung contains heme and copper. FEBS Lett. 1981;132:71–4.
26. Humbert P, Niroomand F, Fischer G, Mayer B, Koesling D, Hinsch KD, Gausepohl H, Frank R, et al. Purification of soluble guanylyl cyclase from bovine lung by a new immunoaffinity chromatographic method. Eur J Biochem. 1990;190:273–8.
27. Schrammel A, Behrends S, Schmidt K, Koesling D, Mayer B. Characterization of 1H-[1,2,4] oxadiazolo[4,3-a]quinoxalin-1-one as a heme-site inhibitor of nitric oxide-sensitive guanylyl cyclase. Mol Pharmacol. 1996;50:1–5.
28. Zhao Y, Brandish PE, Di Valentin M, Schelvis JP, Babcock GT, Marletta MA. Inhibition of soluble guanylate cyclase by ODQ. Biochemistry. 2000;39:10848–54.
29. Mergia E, Friebe A, Dangel O, Russwurm M, Koesling D. Spare guanylyl cyclase NO receptors ensure high NO sensitivity in the vascular system. J Clin Invest. 2006;116:1731–7.
30. Friebe A, Mergia E, Dangel O, Lange A, Koesling D. Fatal gastrointestinal obstruction and hypertension in mice lacking nitric oxide-sensitive guanylyl cyclase. Proc Natl Acad Sci USA. 2007;104:7699–704.
31. Dangel O, Mergia E, Karlisch K, Groneberg D, Koesling D, Friebe A. Nitric oxide-sensitive guanylyl cyclase is the only nitric oxide receptor mediating platelet inhibition. J Thromb Haemost. 2010;8:1343–52.
32. Thoonen R, Cauwels A, Decaluwe K, Geschka S, Tainsh RE, Delanghe J, Hochepied T, De Cauwer L, et al. Cardiovascular and pharmacological implications of haem-deficient NO-unresponsive soluble guanylate cyclase knock-in mice. Nat Commun. 2015;6:8482.
33. Li Z, Xi X, Gu M, Feil R, Ye RD, Eigenthaler M, Hofmann F, Du X. A stimulatory role for cGMP-dependent protein kinase in platelet activation. Cell. 2003;112:77–86.
34. Gambaryan S, Kobsar A, Hartmann S, Birschmann I, Kuhlencordt PJ, Muller-Esterl W, Lohmann SM, Walter U. NO-synthase-/NO-independent regulation of human and murine platelet soluble guanylyl cyclase activity. J Thromb Haemost. 2008;6:1376–84.

35. Zhang G, Xiang B, Dong A, Skoda RC, Daugherty A, Smyth SS, Du X, Li Z. Biphasic roles for soluble guanylyl cyclase (sGC) in platelet activation. Blood. 2011;118:3670–9.
36. Hanafy KA, Martin E, Murad F. CCTeta, a novel soluble guanylyl cyclase-interacting protein. J Biol Chem. 2004;279:46946–53.
37. Balashova N, Chang FJ, Lamothe M, Sun Q, Beuve A. Characterization of a novel type of endogenous activator of soluble guanylyl cyclase. J Biol Chem. 2005;280:2186–96.
38. Russwurm M, Wittau N, Koesling D. Guanylyl cyclase/PSD-95 interaction: targeting of the nitric oxide-sensitive alpha2beta1 guanylyl cyclase to synaptic membranes. J Biol Chem. 2001;276:44647–52.
39. Chauhan S, Jelen F, Sharina I, Martin E. The G-protein regulator LGN modulates the activity of the NO receptor soluble guanylate cyclase. Biochem J. 2012;446:445–53.
40. Heckler EJ, Crassous PA, Baskaran P, Beuve A. Protein disulfide-isomerase interacts with soluble guanylyl cyclase via a redox-based mechanism and modulates its activity. Biochem J. 2013;452:161–9.
41. Haslam RJ, Dickinson NT, Jang EK. Cyclic nucleotides and phosphodiesterases in platelets. Thromb Haemost. 1999;82:412–23.
42. Rondina MT, Weyrich AS. Targeting phosphodiesterases in anti-platelet therapy. Handb Exp Pharmacol. 2012;210:225–38.
43. Mullershausen F, Russwurm M, Thompson WJ, Liu L, Koesling D, Friebe A. Rapid nitric oxide-induced desensitization of the cGMP response is caused by increased activity of phosphodiesterase type 5 paralleled by phosphorylation of the enzyme. J Cell Biol. 2001;155:271–8.
44. Rybalkin SD, Rybalkina IG, Shimizu-Albergine M, Tang XB, Beavo JA. PDE5 is converted to an activated state upon cGMP binding to the GAF A domain. EMBO J. 2003;22:469–78.
45. Mullershausen F, Friebe A, Feil R, Thompson WJ, Hofmann F, Koesling D. Direct activation of PDE5 by cGMP: long-term effects within NO/cGMP signaling. J Cell Biol. 2003;160:719–27.
46. Sager G. Cyclic GMP transporters. Neurochem Int. 2004;45:865–73.
47. Krawutschke C, Koesling D, Russwurm M. Cyclic GMP in vascular relaxation: export is of similar Importance as degradation. Arterioscler Thromb Vasc Biol. 2015;35:2011–9.
48. Wu XB, Brune B, Von Appen F, Ullrich V. Efflux of cyclic GMP from activated human platelets. Mol Pharmacol. 1993;43:564–8.
49. Rius M, Hummel-Eisenbeiss J, Keppler D. ATP-dependent transport of leukotrienes B4 and C4 by the multidrug resistance protein ABCC4 (MRP4). J Pharmacol Exp Ther. 2008;324:86–94.
50. Jedlitschky G, Tirschmann K, Lubenow LE, Nieuwenhuis HK, Akkerman JW, Greinacher A, Kroemer HK. The nucleotide transporter MRP4 (ABCC4) is highly expressed in human platelets and present in dense granules, indicating a role in mediator storage. Blood. 2004;104:3603–10.
51. Borgognone A, Pulcinelli FM. Reduction of cAMP and cGMP inhibitory effects in human platelets by MRP4-mediated transport. Thromb Haemost. 2012;108:955–62.
52. Hofmann F, Feil R, Kleppisch T, Schlossmann J. Function of cGMP-dependent protein kinases as revealed by gene deletion. Physiol Rev. 2006;86:1–23.
53. Francis SH, Busch JL, Corbin JD, Sibley D. cGMP-dependent protein kinases and cGMP phosphodiesterases in nitric oxide and cGMP action. Pharmacol Rev. 2010;62:525–63.
54. Geiselhoringer A, Gaisa M, Hofmann F, Schlossmann J. Distribution of IRAG and cGKI-isoforms in murine tissues. FEBS Lett. 2004;575:19–22.
55. Antl M, von Bruhl ML, Eiglsperger C, Werner M, Konrad I, Kocher T, Wilm M, Hofmann F, et al. IRAG mediates NO/cGMP-dependent inhibition of platelet aggregation and thrombus formation. Blood. 2007;109:552–9.
56. Massberg S, Sausbier M, Klatt P, Bauer M, Pfeifer A, Siess W, Fassler R, Ruth P, et al. Increased adhesion and aggregation of platelets lacking cyclic guanosine 3′,5′-monophosphate kinase I. J Exp Med. 1999;189:1255–64.

57. Massberg S, Gruner S, Konrad I, Garcia Arguinzonis MI, Eigenthaler M, Hemler K, Kersting J, Schulz C, et al. Enhanced in vivo platelet adhesion in vasodilator-stimulated phosphoprotein (VASP)-deficient mice. Blood. 2004;103:136–42.
58. Bergmeier W, Stefanini L. Novel molecules in calcium signaling in platelets. J Thromb Haemost. 2009;7(Suppl 1):187–90.
59. Li Z, Delaney MK, O'Brien KA, Du X. Signaling during platelet adhesion and activation. Arterioscler Thromb Vasc Biol. 2010;30:2341–9.
60. Pfeifer A, Klatt P, Massberg S, Ny L, Sausbier M, Hirneiss C, Wang GX, Korth M, et al. Defective smooth muscle regulation in cGMP kinase I-deficient mice. EMBO J. 1998;17:3045–51.
61. Feil R, Gappa N, Rutz M, Schlossmann J, Rose CR, Konnerth A, Brummer S, Kuhbandner S, et al. Functional reconstitution of vascular smooth muscle cells with cGMP-dependent protein kinase I isoforms. Circ Res. 2002;90:1080–6.
62. Muller PM, Gnugge R, Dhayade S, Thunemann M, Krippeit-Drews P, Drews G, Feil R. H(2) O(2) lowers the cytosolic Ca(2)(+) concentration via activation of cGMP-dependent protein kinase Ialpha. Free Radic Biol Med. 2012;53:1574–83.
63. Schlossmann J, Ammendola A, Ashman K, Zong X, Huber A, Neubauer G, Wang GX, Allescher HD, et al. Regulation of intracellular calcium by a signalling complex of IRAG, IP3 receptor and cGMP kinase Ibeta. Nature. 2000;404:197–201.
64. Geiselhoringer A, Werner M, Sigl K, Smital P, Worner R, Acheo L, Stieber J, Weinmeister P, et al. IRAG is essential for relaxation of receptor-triggered smooth muscle contraction by cGMP kinase. EMBO J. 2004;23:4222–31.
65. Halbrugge M, Friedrich C, Eigenthaler M, Schanzenbacher P, Walter U. Stoichiometric and reversible phosphorylation of a 46-kDa protein in human platelets in response to cGMP- and cAMP-elevating vasodilators. J Biol Chem. 1990;265:3088–93.
66. Benz PM, Blume C, Moebius J, Oschatz C, Schuh K, Sickmann A, Walter U, Feller SM, et al. Cytoskeleton assembly at endothelial cell-cell contacts is regulated by alphaII-spectrin-VASP complexes. J Cell Biol. 2008;180:205–19.
67. Horstrup K, Jablonka B, Honig-Liedl P, Just M, Kochsiek K, Walter U. Phosphorylation of focal adhesion vasodilator-stimulated phosphoprotein at Ser157 in intact human platelets correlates with fibrinogen receptor inhibition. Eur J Biochem. 1994;225:21–7.
68. Aszodi A, Pfeifer A, Ahmad M, Glauner M, Zhou XH, Ny L, Andersson KE, Kehrel B, et al. The vasodilator-stimulated phosphoprotein (VASP) is involved in cGMP- and cAMP-mediated inhibition of agonist-induced platelet aggregation, but is dispensable for smooth muscle function. EMBO J. 1999;18:37–48.
69. Reinhard M, Jarchau T, Walter U. Actin-based motility: stop and go with Ena/VASP proteins. Trends Biochem Sci. 2001;26:243–9.
70. Walter U, Gambaryan S. cGMP and cGMP-dependent protein kinase in platelets and blood cells. Handb Exp Pharmacol. 2009;191:533–48.
71. Smolenski A. Novel roles of cAMP/cGMP-dependent signaling in platelets. J Thromb Haemost. 2012;10:167–76.
72. Chrzanowska-Wodnicka M, Smyth SS, Schoenwaelder SM, Fischer TH, White GC 2nd. Rap1b is required for normal platelet function and hemostasis in mice. J Clin Invest. 2005;115:680–7.
73. Danielewski O, Schultess J, Smolenski A. The NO/cGMP pathway inhibits Rap 1 activation in human platelets via cGMP-dependent protein kinase I. Thromb Haemost. 2005;93:319–25.
74. Hoffmeister M, Riha P, Neumuller O, Danielewski O, Schultess J, Smolenski AP. Cyclic nucleotide-dependent protein kinases inhibit binding of 14-3-3 to the GTPase-activating protein Rap1GAP2 in platelets. J Biol Chem. 2008;283:2297–306.
75. Gegenbauer K, Elia G, Blanco-Fernandez A, Smolenski A. Regulator of G-protein signaling 18 integrates activating and inhibitory signaling in platelets. Blood. 2012;119:3799–807.

76. Nagy Z, Wynne K, von Kriegsheim A, Gambaryan S, Smolenski A. Cyclic nucleotide-dependent protein kinases target ARHGAP17 and ARHGEF6 complexes in platelets. J Biol Chem. 2015;290:29974–83.
77. Aslan JE, McCarty OJ. Rho GTPases in platelet function. J Thromb Haemost. 2013;11:35–46.
78. Schwarz UR, Walter U, Eigenthaler M. Taming platelets with cyclic nucleotides. Biochem Pharmacol. 2001;62:1153–61.
79. Zaccolo M, Movsesian MA. cAMP and cGMP signaling cross-talk: role of phosphodiesterases and implications for cardiac pathophysiology. Circ Res. 2007;100:1569–78.
80. Li Z, Ajdic J, Eigenthaler M, Du X. A predominant role for cAMP-dependent protein kinase in the cGMP-induced phosphorylation of vasodilator-stimulated phosphoprotein and platelet inhibition in humans. Blood. 2003;101:4423–9.
81. Jensen BO, Selheim F, Doskeland SO, Gear AR, Holmsen H. Protein kinase A mediates inhibition of the thrombin-induced platelet shape change by nitric oxide. Blood. 2004;104:2775–82.
82. Wilson LS, Elbatarny HS, Crawley SW, Bennett BM, Maurice DH. Compartmentation and compartment-specific regulation of PDE5 by protein kinase G allows selective cGMP-mediated regulation of platelet functions. Proc Natl Acad Sci USA. 2008;105:13650–5.
83. Momi S, Caracchini R, Falcinelli E, Evangelista S, Gresele P. Stimulation of platelet nitric oxide production by nebivolol prevents thrombosis. Arterioscler Thromb Vasc Biol. 2014;34:820–9.
84. Kirkby NS, Lundberg MH, Chan MV, Vojnovic I, Solomon AB, Emerson M, Mitchell JA, Warner TD. Blockade of the purinergic P2Y12 receptor greatly increases the platelet inhibitory actions of nitric oxide. Proc Natl Acad Sci USA. 2013;110:15782–7.
85. Magwenzi S, Woodward C, Wraith KS, Aburima A, Raslan Z, Jones H, McNeil C, Wheatcroft S, et al. Oxidized LDL activates blood platelets through CD36/NOX2-mediated inhibition of the cGMP/protein kinase G signaling cascade. Blood. 2015;125:2693–703.
86. Nesbitt WS, Westein E, Tovar-Lopez FJ, Tolouei E, Mitchell A, Fu J, Carberry J, Fouras A, et al. A shear gradient-dependent platelet aggregation mechanism drives thrombus formation. Nat Med. 2009;15:665–73.
87. Gambaryan S, Geiger J, Schwarz UR, Butt E, Begonja A, Obergfell A, Walter U. Potent inhibition of human platelets by cGMP analogs independent of cGMP-dependent protein kinase. Blood. 2004;103:2593–600.
88. Marshall SJ, Senis YA, Auger JM, Feil R, Hofmann F, Salmon G, Peterson JT, Burslem F, et al. GPIb-dependent platelet activation is dependent on Src kinases but not MAP kinase or cGMP-dependent kinase. Blood. 2004;103:2601–9.
89. Valtcheva N, Nestorov P, Beck A, Russwurm M, Hillenbrand M, Weinmeister P, Feil R. The commonly used cGMP-dependent protein kinase type I (cGKI) inhibitor Rp-8-Br-PET-cGMPS can activate cGKI in vitro and in intact cells. J Biol Chem. 2009;284:556–62.
90. Wanstall JC, Homer KL, Doggrell SA. Evidence for, and importance of, cGMP-independent mechanisms with NO and NO donors on blood vessels and platelets. Curr Vasc Pharmacol. 2005;3:41–53.
91. Hess DT, Stamler JS. Regulation by S-nitrosylation of protein post-translational modification. J Biol Chem. 2012;287:4411–8.
92. Gambaryan S, Tsikas D. A review and discussion of platelet nitric oxide and nitric oxide synthase: do blood platelets produce nitric oxide from L-arginine or nitrite? Amino Acids. 2015;47:1779–93.
93. Lutz SZ, Hennige AM, Feil S, Peter A, Gerling A, Machann J, Krober SM, Rath M, et al. Genetic ablation of cGMP-dependent protein kinase type I causes liver inflammation and fasting hyperglycemia. Diabetes. 2011;60:1566–76.
94. Zhang L, Lukowski R, Gaertner F, Lorenz M, Legate KR, Domes K, Angermeier E, Hofmann F, et al. Thrombocytosis as a response to high interleukin-6 levels in cGMP-dependent protein kinase I mutant mice. Arterioscler Thromb Vasc Biol. 2013;33:1820–8.

95. Du X, Marjanovic JA, Li Z. On the roles of cGMP and glycoprotein Ib in platelet activation. Blood. 2004;103:4371–2. author reply 4372-3.
96. Walter U, Gambaryan S. Roles of cGMP/cGMP-dependent protein kinase in platelet activation. Blood. 2004;104:2609.
97. Gambaryan S, Friebe A, Walter U. Does the NO/sGC/cGMP/PKG pathway play a stimulatory role in platelets? Blood. 2012;119:5335–6. author reply 5336-7.
98. Li Z, Du X. Response: yes, cGMP plays a stimulatory role in platelet activation. Blood. 2012;119:5336–7.
99. Riba R, Sharifi M, Farndale RW, Naseem KM. Regulation of platelet guanylyl cyclase by collagen: evidence that glycoprotein VI mediates platelet nitric oxide synthesis in response to collagen. Thromb Haemost. 2005;94:395–403.
100. Riba R, Oberprieler NG, Roberts W, Naseem KM. Von Willebrand factor activates endothelial nitric oxide synthase in blood platelets by a glycoprotein Ib-dependent mechanism. J Thromb Haemost. 2006;4:2636–44.
101. Cozzi MR, Guglielmini G, Battiston M, Momi S, Lombardi E, Miller EC, De Zanet D, Mazzucato M, et al. Visualization of nitric oxide production by individual platelets during adhesion in flowing blood. Blood. 2015;125:697–705.
102. Li Z, Zhang G, Feil R, Han J, Du X. Sequential activation of p38 and ERK pathways by cGMP-dependent protein kinase leading to activation of the platelet integrin alphaIIb beta3. Blood. 2006;107:965–72.
103. Begonja AJ, Geiger J, Rukoyatkina N, Rauchfuss S, Gambaryan S, Walter U. Thrombin stimulation of p38 MAP kinase in human platelets is mediated by ADP and thromboxane A2 and inhibited by cGMP/cGMP-dependent protein kinase. Blood. 2007;109:616–8.
104. Jackson EC, McNicol A. Cyclic nucleotides inhibit MAP kinase activity in low-dose collagen-stimulated platelets. Thromb Res. 2010;125:147–51.
105. Thaiss CA, Levy M, Itav S, Elinav E. Integration of innate immune signaling. Trends Immunol. 2016. https://doi.org/10.1016/j.it.2015.12.003.
106. Zhang G, Han J, Welch EJ, Ye RD, Voyno-Yasenetskaya TA, Malik AB, Du X, Li Z. Lipopolysaccharide stimulates platelet secretion and potentiates platelet aggregation via TLR4/MyD88 and the cGMP-dependent protein kinase pathway. J Immunol. 2009;182:7997–8004.
107. Vogel S, Bodenstein R, Chen Q, Feil S, Feil R, Rheinlaender J, Schaffer TE, Bohn E, et al. Platelet-derived HMGB1 is a critical mediator of thrombosis. J Clin Invest. 2015;125:4638–54.
108. Zhang S, Zhang S, Hu L, Zhai L, Xue R, Ye J, Chen L, Cheng G, et al. Nucleotide-binding oligomerization domain 2 receptor is expressed in platelets and enhances platelet activation and thrombosis. Circulation. 2015;131:1160–70.
109. Erdmann J, Stark K, Esslinger UB, Rumpf PM, Koesling D, de Wit C, Kaiser FJ, Braunholz D, et al. Dysfunctional nitric oxide signalling increases risk of myocardial infarction. Nature. 2013;504:432–6.
110. Herve D, Philippi A, Belbouab R, Zerah M, Chabrier S, Collardeau-Frachon S, Bergametti F, Essongue A, et al. Loss of alpha1beta1 soluble guanylate cyclase, the major nitric oxide receptor, leads to moyamoya and achalasia. Am J Hum Genet. 2014;94:385–94.
111. Lu X, Wang L, Chen S, He L, Yang X, Shi Y, Cheng J, Zhang L, et al. Genome-wide association study in Han Chinese identifies four new susceptibility loci for coronary artery disease. Nat Genet. 2012;44:890–4.
112. Deloukas P, Kanoni S, Willenborg C, Farrall M, Assimes TL, Thompson JR, Ingelsson E, Saleheen D, et al. Large-scale association analysis identifies new risk loci for coronary artery disease. Nat Genet. 2013;45:25–33.
113. Goto S, Ikeda Y, Saldivar E, Ruggeri ZM. Distinct mechanisms of platelet aggregation as a consequence of different shearing flow conditions. J Clin Invest. 1998;101:479–86.
114. Jackson SP. The growing complexity of platelet aggregation. Blood. 2007;109:5087–95.

115. Kroll MH, Hellums JD, McIntire LV, Schafer AI, Moake JL. Platelets and shear stress. Blood. 1996;88:1525–41.
116. Sakariassen KS. Thrombus formation on apex of arterial stenoses: the need for a fluid high shear stenosis diagnostic device. Future Cardiol. 2007;3:193–201.
117. Corson MA, James NL, Latta SE, Nerem RM, Berk BC, Harrison DG. Phosphorylation of endothelial nitric oxide synthase in response to fluid shear stress. Circ Res. 1996;79:984–91.
118. Lamontagne D, Pohl U, Busse R. Mechanical deformation of vessel wall and shear stress determine the basal release of endothelium-derived relaxing factor in the intact rabbit coronary vascular bed. Circ Res. 1992;70:123–30.
119. Blackmore PF. Biphasic effects of nitric oxide on calcium influx in human platelets. Thromb Res. 2011;127:e8–14.
120. Nikolaev VO, Lohse MJ. Novel techniques for real-time monitoring of cGMP in living cells. Handb Exp Pharmacol. 2009;191:229–43.
121. Thunemann M, Fomin N, Krawutschke C, Russwurm M, Feil R. Visualization of cGMP with cGi biosensors. Methods Mol Biol. 2013;1020:89–120.
122. Thunemann M, Wen L, Hillenbrand M, Vachaviolos A, Feil S, Ott T, Han X, Fukumura D, et al. Transgenic mice for cGMP imaging. Circ Res. 2013;113:365–71.
123. Thunemann M, Schmidt K, de Wit C, Han X, Jain RK, Fukumura D, Feil R. Correlative intravital imaging of cGMP signals and vasodilation in mice. Front Physiol. 2014;5:394.
124. Brass LF, Ma P. Applying the brakes to platelet activation. Blood. 2012;119:3651–2.
125. Stasch JP, Pacher P, Evgenov OV. Soluble guanylate cyclase as an emerging therapeutic target in cardiopulmonary disease. Circulation. 2011;123:2263–73.

Platelets and Stroke

16

Felix Fluri, Bernhard Nieswandt, Guido Stoll,
and Christoph Kleinschnitz

Abstract

Ischemic stroke is the third most common cause of disability worldwide. Since ischemic stroke results from an occlusion of a brain-supplying artery, rapid recanalization either by intravenous thrombolysis or by endovascular thrombectomy is the primary clinical objective. Following restoration of blood flow, reperfusion injury often occurs encompassing a large number of detrimental biochemical processes. Based on animal models of stroke, tethering of platelets on endothelial cells is mediated by the glycoprotein (GP) Ib–V–IX receptor complex on the surface of platelets. This complex facilitates the binding of von Willebrand factor (vWF) to the damaged sub-endothelium and thus the first critical step in platelet adhesion. Further glycoprotein receptors, i.e., glycoprotein (PG) VI, and GPIIb/IIIa are also involved in aggregation and adhesion of platelets. Aggregates of adherent, activated platelets express adhesive molecules, namely, P-selectin, which binds to P-selectin glycoprotein ligand-1 (PSGL-1), the main receptor for P-selectin on leukocytes. Adherence of leukocytes and the subsequent release of pro-inflammatory factors are particular mechanisms that show how platelets induce inflammatory processes. Platelets are also connected in further ways to cellular and humoral components of the

F. Fluri • G. Stoll
Neurologische Klinik und Poliklinik, Universitätsklinikum Würzburg, Josef-Schneider-Straße 11, 97080 Würzburg, Germany

B. Nieswandt (✉)
Rudolf-Virchow-Zentrum für Experimentelle Biomedizin, Universität Würzburg, Josef-Schneider-Straße 2, 97080 Würzburg, Germany
e-mail: bernhard.nieswandt@virchow.uni-wuerzburg.de

C. Kleinschnitz
Universitätsklinikum Essen (AöR), Klinik für Neurologie, Hufelandstraße 55, 45147 Essen, Germany

A. Zirlik et al. (eds.), *Platelets, Haemostasis and Inflammation*,
Cardiac and Vascular Biology 5, https://doi.org/10.1007/978-3-319-66224-4_16

immune system, such as T cells, macrophages, and the complement system. This has led us to develop the term "thrombo-inflammation." Both thrombotic and inflammatory mechanisms are highly intertwined in the pathophysiology of ischemic stroke. Thrombus formation and inflammation are therefore promising targets for the development of novel therapeutic strategies. The growing insights into thrombo-inflammation after cerebral ischemia might provide a platform for further exploration of the critical interface between inflammation and thrombosis after ischemic stroke. These interesting findings in the field of cerebral thrombo-inflammation should encourage stroke researchers to seek further treatments that target the reduction of thrombo-inflammation in stroke and other cardiovascular diseases.

Contents

16.1 Introduction

Estimates from the Global Burden of Disease Study (2010) ranked stroke as the second most common cause of death [1] and the third most common cause of disability-adjusted life years worldwide [2]. Approximately 15–30% of patients with recurrent stroke become permanently disabled, often requiring institutionalized care [3]. Therefore, understanding safe and effective ways to prevent ischemic stroke from occurring (or recurring) is crucial. There are many causes for ischemic stroke; thromboembolism causes about 30% of all stroke cases [4]. Thus, current treatments for the prevention of secondary stroke are frequently based on platelet inhibitors, namely, aspirin and clopidogrel. These drugs attenuate platelet activation and aggregation. Prospective, randomized studies have demonstrated that the use of platelet inhibitors is associated with a reduced incidence of secondary stroke [2]. Aspirin inhibits cyclooxygenase-1 (COX-1)-induced production of thromboxane A_2 and is most often used for the prevention of recurrent stroke. In patients with a history of a cerebrovascular ischemic event, aspirin reduced the risk of a subsequent stroke by 15–18% compared with placebo [5]. Regarding its efficacy, its low cost, and an acceptable side-effect profile, aspirin is recommended for the prevention of ischemic stroke in patients with a history of ischemic cerebrovascular events

[6]. When analyzed by intention-to-treat analysis encompassing more than 19,000 patients, the annual risk of combined vascular events in patients treated with clopidogrel was lower compared to patients receiving aspirin (5.32% vs. 5.83%, respectively; relative risk reduction of 8.7% in favor of clopidogrel, $p = 0.043$) [7]. The combination of clopidogrel with aspirin for stroke prevention has been investigated in the *M*anagement of *At*herothrombosis with *C*lopidogrel in *H*igh-Risk Patients with Recent Transient Ischemic Attacks or Ischemic Stroke (MATCH) trial, which yielded no additional clinical benefit in patients receiving the combined treatment [8] (for more details concerning the study design, see Table 16.1). A subanalysis of the *C*lopidogrel for *H*igh *A*therothrombotic *R*isk and *I*schemic *S*tabilization, *M*anagement, and *A*voidance (CHARISMA) trial suggested that early addition of clopidogrel to aspirin in patients with transient ischemic attack (TIA) and ischemic stroke of arterial origin may be more effective and may provide acceptable safety levels compared with aspirin alone [9]. Findings of a recently published trial, the *C*lopidogrel in *H*igh-risk Patients with *A*cute *N*ondisabling *C*erebrovascular *E*vents (CHANCE) study, revealed that the combination of aspirin and clopidogrel given for 21 days, followed by clopidogrel alone up to Day 90, was more effective than aspirin alone in preventing recurrent strokes in Chinese patients who had had a minor stroke or TIA [10]. The effect of a dual antiplatelet therapy to prevent vascular events in patients with atrial fibrillation has been investigated in the *A*trial Fibrillation *C*lopidogrel *T*rial With *I*rbesartan for Prevention of *V*ascular *E*vents (ACTIVE) program, in which two randomized controlled trials were designed to test the efficacy of clopidogrel plus aspirin versus oral anticoagulant therapy (ACTIVE W) [11] and the effect of clopidogrel plus aspirin versus aspirin alone in patients with contraindications to oral anticoagulants or in patients unwilling to take oral anticoagulants (ACTIVE A) [12]. The former trial showed that anticoagulant therapy remains the superior choice for preventing vascular events such as stroke and systemic embolus in patients with atrial fibrillation [11]. In the ACTIVE A trial, a reduction in major vascular events, particularly stroke, was found compared with the control arm of the study, whereas the incidence of major bleedings increased [12]. Thus, the antithrombotic properties of these agents play an important role in reducing the risk of recurrent stroke. By reducing platelet activation, these agents probably not only prevent thrombosis but in addition decrease the interactions of platelets with leukocytes and endothelial cells and thus influence inflammatory mechanisms [13]. The interplay between platelets, leukocytes, and inflammatory factors is particularly relevant in stroke pathogenesis and contributes to the development of plaques and arterial stenosis [14]. In this chapter, we will discuss potential interactions between platelets, endothelial cells, leukocytes, and other factors of the immune system, such as thrombo-inflammation, and their influence on inflammatory and thrombotic processes during the course of an ischemic stroke.

Table 16.1 Characteristics of trials

Trial	Treatment daily dose	Comparator daily dose	Participants, n	Age, years	Female, %	Endpoint	Onset-to-treatment interval	Duration of dual treatment
ACTIVE A 2009 [12]	A 75–100 mg + C 75 mg	A 75–100 mg	7554	71	42.0	Stroke	na	3.6 years
ACTIVE W 2006 [11]	OAC	A 75–100 mg + C 75 mg	6706	70.2	33.0	Stroke	na	1.3 years
CHANCE 2013 [10]	A 75 mg + C 75 mg	A 75 mg	5170	62	33.8	Minor stroke, TIA	<24 h	21 days
CHARISMA substudy 2011 [9]	A 75–100 mg + C 75 mg	A 75–100 mg	1331	65	36.8	Stroke, TIA	<5 years	2.1 years
MATCH 2004 [8]	A 75 mg + C 75 mg	C 75 mg	7599	66	37.1	Stroke, TIA	<3 months	3.4 years

A aspirin, *C* clopidogrel, *na* not available, *OAC* oral anticoagulant, TIA transient ischemic attack

16.2 Mechanisms of Platelet Activation in Acute Ischemic Stroke

After vessel wall injury (e.g., due to ischemia or rupture of an atherosclerotic plaque), platelets readily adhere to the damaged endothelium in a stepwise process [15] (for an overview, see Fig. 16.1). Within this process, tethering of platelets is the first event of platelet activation [15]. At this stage, platelet adhesion is not firm, and tethering is still completely reversible. Tethering of platelets on endothelial cells is mediated by the glycoprotein (GP) Ib–V–IX receptor complex, which is expressed on the surface of platelets [16]. The GPIb–V–IX receptor complex initiates the first critical step in platelet adhesion by facilitating the binding of von Willebrand factor (vWF) to the damaged sub-endothelium. This protein originates from the Weibel–Palade bodies of endothelial cells or the α-granules of platelets [17]. vWF bears different binding domains; the A1-binding domain binds to the glycoprotein Ibα (GPIbα), the major subunit of the GPIb–V–IX receptor complex [16]. Blockade of the vWF-binding domain on GPIbα using the monoclonal GPIbα antibody (p0p/B) in wild-type mice abolished platelet tethering and adhesion to the injured endothelial cells after mechanically induced arterial thrombosis [18]. The significance of GPIbα in platelet adhesion was also characterized in transgenic mice, in which the extracellular domain of GPIbα has been replaced by the α-subunit of the human interleukin (IL)-4 receptor. This study demonstrated that platelet adhesion to the exposed extracellular matrix is completely inhibited in arterioles of IL-4Rα/GPIbα-transgenic mice [19]. Additionally, the interaction of GPIbα with the A1 domain of vWF is important in the context of strong hydrodynamic forces [19]—that is, binding of GPIbα to the A1 domain decelerates the fast-flowing platelets at high shear stress. The interaction of vWF and its platelet receptor GPIbα is also crucial in brain infarct evolution. Complete blockade of the vWF-binding domain of GPIbα using Fab fragments of the antibody p0p/B resulted in significantly smaller infarct volumes in a murine stroke model [20]. This effect was still observed when the blocking antibodies were administered 1 h after transient occlusion of the middle cerebral artery (tMCAo) of mice [20]. It is of note that blocking the GPIbα receptors did not result in an increased risk of developing intracerebral hemorrhage (ICH). The aforementioned findings were also supported by another study showing that $vWF^{-/-}$ mice undergoing tMCAo had significantly smaller infarct volumes 1 day after the induction of stroke [21]. Recently, an analysis of a national (US) discharge register showed that the prevalence of cardiovascular events (including ischemic cerebrovascular disease) in patients with a vWF deficiency (VWD) was 15% lower than in non-VWD patients, when adjusted for vascular risk factors [22]. This finding might further underline that VWD is protective against cardiovascular diseases.

Another membrane glycoprotein receptor is glycoprotein VI (GPVI), which plays a crucial role in the collagen-induced activation and aggregation of platelets [23]. GPVI has been detected exclusively on the surface of platelets and megakaryocytes [24]. GPVI fosters the procoagulant activation of platelets and thrombin in response to fibrillar and non-fibrillar collagens by inducing binding

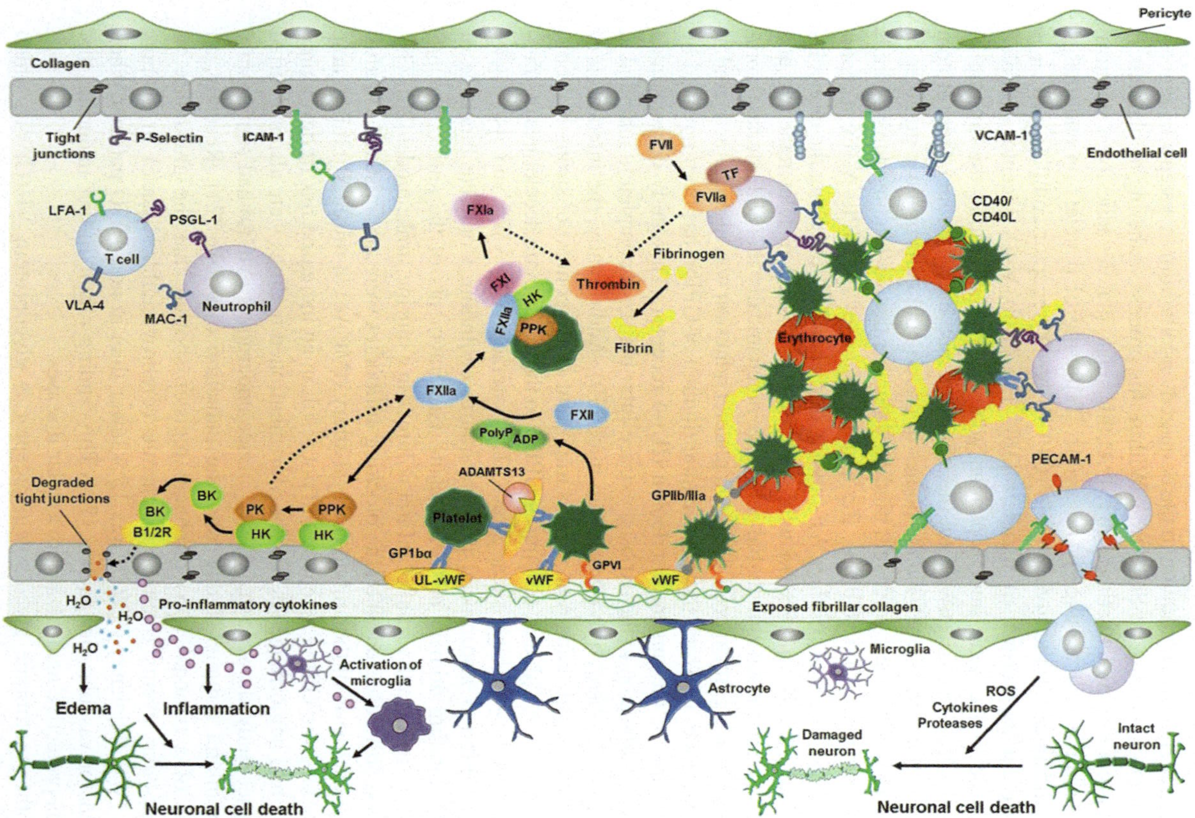

Fig. 16.1 Mechanisms of thrombo-inflammation in brain damage caused by stroke. At the site of ischemic vascular injury, blood platelets adhere to released or exposed vWF/ULvWF via their GPIbα receptor. Upon this initial adhesion step, binding of exposed fibrillar collagen to GPVI induces (1) platelet activation with the functional upregulation of GPIIb/IIIa and (2) the release of polyphosphates from the activated platelets themselves. Polyphosphates and negatively

sites of integrin α2β1 for collagen [25]. Platelets in which GPVI has been depleted by in vivo administration of antibodies against the receptor do not respond to collagen [24]. Furthermore, deficiency of GPVI in mice resulted in abnormal phosphatidylserine exposure after stimulation with thrombin [26]. Recently, another mechanism of GPVI-mediated platelet adhesion has been described by Bültmann and co-workers, which investigated the interaction of GPVI and fibronectin. In flow chamber experiments, platelet adhesion to fibronectin was significantly inhibited by GPVI-Fc and the anti-GPVI antibody, indicating an interplay between GPVI and fibronectin regarding platelet adhesion [27]. GPVI is also linked to another component of platelets, the glycoprotein vitronectin. Recombinant GPVI-Fc has been shown to bind to activated endothelium, mainly via vitronectin, which in turn has been shown to prevent platelet/endothelial interactions [28]. GPVI is also a receptor for laminin, which supports platelet adhesion via binding to integrin $\alpha_6\beta_1$ [29]. The aforementioned studies suggest that the role of GPVI in thrombus formation is not only the consequence of its interaction with collagen. Whether GPVI binds to other adhesive molecules is not known and remains to be investigated. Interestingly, GPVI is also involved in cerebral infarct growth. After tMCAo, depletion of GPVI with the JAQ1 antibody [20] or blocking of collagen-binding domains using revacept, a recombinant soluble dimeric GPVI-Fc [30], significantly decreased infarct size. There is some evidence that GPVI is also a promising biomarker for acute ischemic stroke in humans. Patients with TIA

Fig. 16.1 (continued) charged surfaces interact with and activate FXII, the initiator protease of the intrinsic coagulation pathway. As well as triggering thrombus via fibrin generation, FXIIa also promotes the activation of the contact–kinin system: FXIIa cleaves PPK to form the active serine protease PK, which in turn cleavages HMWK to release the inflammatory peptide hormone BK. Binding of BK to its endothelial receptor initiates signaling cascades that induce (1) endothelial cell damage, leading to vascular edema, and (2) the expression of pro-inflammatory cytokines that induce glial activation, inflammation, and, finally, neuronal cell death. Simultaneously, circulating leukocytes (T cells, neutrophils) become activated by the ischemic insult, resulting in a sterile inflammatory reaction involving the upregulation of chemoattractants, chemokines, and adhesion molecules both on endothelial cells and immune cells. After recruitment (via P-selectin/PSGL-1) and stable tethering (via ICAM-1/LFA-1 and VCAM-1/VLA-4) to the vasculature, T lymphocytes interact with activated platelets via CD40/CD40L to form a solid thrombus. Neutrophils also contribute to thrombus formation as they (1) interact with platelets (via MAC-1/GP1bα and P-selectin/PSGL-1), (2) participate in fibrin cross-linkage (via MAC-1/fibrin interaction), and (3) trigger thrombin activation by inducing the extrinsic TF/FVIIa pathway. The successive infiltration of immune cells into the brain parenchyma triggers further tissue inflammation and liberation of ROS, cytokines, and proteases and consequently induces neuronal cell damage. *ADAMTS13* A disintegrin-like and metalloprotease with thrombospondin type I repeats-13, *BK* Bradykinin, *B1R* Bradykinin receptor B1, *B2R* Bradykinin receptor B2, *FVII/FXI/FXII(a)* (Activated) coagulation factor VII/XI/XII, *GP* Glycoprotein, *HMWK* High-molecular-weight kininogen, *ICAM-1* Intercellular adhesion molecule-1, *LFA-1* Lymphocyte function-associated antigen-1, *MAC-1* Macrophage-1 antigen, *PECAM-1* Platelet endothelial cell adhesion molecule, *PK* Plasma kallikrein, *PPK* plasma prekallikrein, *PSGL-1* P-selectin glycoprotein ligand-1, *ROS* Reactive oxygen species, *TF* Tissue factor, *VCAM-1* Vascular cell adhesion molecule-1, *VLA-4* Leukocyte very late antigen-4, *vWF/UL-vWF* von Willebrand factor/ultra-large vWF. Reprinted with kind permission of AHA Journals

and ischemic stroke showed significantly elevated levels of GPVI expression on platelets on admission compared to patients without cerebrovascular events [31]. On the other hand, patients with stroke had significantly decreased plasma levels of soluble GPVI compared to patients without cerebrovascular events [32].

Another platelet receptor, GPIIb/IIIa, is also involved in platelet adhesion [33]. Platelet activation induces the expression of GPIIb/IIIa, the most abundant receptor on the surface of platelets. After activation of platelets, the GPIIb/IIIa receptor shifts from an inactive state to an active ligand-binding conformation. Activated GPIIb/IIIa mediates further platelet aggregation by binding to its primary ligands fibrinogen and the C1 domain of vWF, thereby cross-linking activated platelets and stabilizing the growing platelet formation. GPIIb/IIIa is involved in infarct evolution, as shown in several animal experiments, in which partial blockade of GPIIb/IIIa receptors resulted in a reduction of infarct size up to 70% [34–36]. Interestingly, when GPIIb/IIIa receptors are partially blocked either in "healthy" [20] or in aged and comorbid mice [37] using $(Fab)_2$ fragments of the mouse GPIIb/IIIa-blocking monoclonal antibody, no neuroprotective effect has been observed regarding infarct evolution [20, 37]; however, this study revealed an increased rate of ICH, which is in accordance with a phase III trial investigating the GPIIb/IIIa antagonist abciximab in patients with ischemic stroke [38]. This trial was stopped prematurely due to an increased rate of symptomatic or fatal ICH [38]. The unfavorable risk/benefit ratio of GPIIb/IIIa antagonists in the setting of ischemic stroke might be explained by their narrow therapeutic window.

In contrast, $GPIIb^{-/-}ApoE^{-/-}$ mice revealed a smaller infarct size 1 day after tMCAO compared to wild-type mice [39]. These findings corroborate a dual role of GPIIb/IIIa antagonists: At peak concentrations, these agents act as platelet antagonists and thus cause bleedings. However, at lower concentrations, they might act as partial platelet agonists, which results in platelet activation and thrombus formation [40]. The pro-inflammatory nature of subtherapeutic levels of GPIIb/IIIa receptor inhibitors might also be an important factor regarding the adverse events of these agents. Abciximab has been shown to increase the expression of P-selectin in response to adenosine diphosphate, finally resulting in the formation of platelet–leukocyte aggregates [41]. On the other hand, inhibition of GPIIb/IIIa receptors seems to affect inflammatory processes in a positive way [42]. Treatment with abciximab is accompanied by a reduction in the serum levels of C-reactive protein (CRP), IL-6, and tumor necrosis factor-α (TNF-α) [42] after coronary angioplasty. Finally, there might be direct toxic effects via apoptosis; GPIIb/IIIa inhibitors (i.e., orbofiban and xemilofiban) have been shown to activate procaspase-3 and cause dose-dependent apoptosis, especially under hypoxemic conditions [43]. Thus, GPIb or GPVI receptor inhibitors have revealed neuroprotective properties in experimental stroke models without showing an increase in bleeding complications, whereas agents that target the GPIIb/IIIa receptor showed no positive effect on stroke outcome and raised the incidence of ICH and mortality [20].

16.3 The Concept of Thrombo-inflammation in Acute Ischemic Stroke

As outlined above, thrombus formation followed by occlusion of cerebral vessels is one of the main mechanisms leading to cerebral ischemia. However, there is growing evidence that—as well as thrombotic vessel occlusion—additional mechanisms are involved in the development of stroke. This hypothesis is corroborated by a recently published study of our group showing that the blockade of the final common pathway of platelet aggregation with anti-GPIIb/IIIa $F(ab)_2$ fragments has no positive effect on stroke size and functional outcome in a murine stroke model, but it does increase the frequency of ICH and the rate of mortality [20]. Additionally, animals that survived the treatment revealed infarct sizes that were comparable to control animals [20]. Platelets are connected in many ways to cellular and humoral components of the immune system, such as T cells, macrophages, and the complement system. This has led us to develop the term "thrombo-inflammation." In this context, certain immune cells (e.g., T cells, macrophages, and neutrophils), as well as platelets, play an important role. Investigations of small vessels in the ischemic cerebral tissue have shown the existence of aggregates of degranulated platelets, as well as leukocytes and fibrin [44]. The presence of this thrombotic material results in the occlusion of small cerebral vessels and thus contributes to the ischemia of cerebral tissue. This harmful effect on cerebral tissue is additionally reinforced by the reopening of vessels, finally resulting in reperfusion injury [45].

Initial interactions between platelets and leukocytes are mediated by P-selectin, an adhesive molecule that is stored in α-granules of platelets. In response to activating signals (e.g., oxidative stress), P-selectin is translocated to the surface of platelets and binds to P-selectin glycoprotein ligand-1 (PSGL-1), the main receptor for P-selectin on leukocytes [46]. In vitro studies have revealed that leukocytes can "roll" on and firmly "adhere" to a layer of immobilized (adherent) platelets [47]. The former results from the binding of PSGL-1 on leukocytes to P-selectin on activated platelets ("rolling"), the latter from an interaction of macrophage antigen-1 (Mac-1) on leukocytes with GPIb and/or fibrinogen on platelets ("adhesion"). An in vivo study corroborated these findings by revealing that platelets modulate leukocyte recruitment via P-selectin in a mouse tMCAo model [36]. Interactions between CD40 (on leukocytes) and the CD40 ligand (on platelets) may also contribute to the adhesion response [48]. Binding of CD40 ligand to CD40 induces the expression of adhesion molecules, such as intercellular adhesion molecule-1 (ICAM-1), vascular cell adhesion molecule-1 (VCAM-1), and E-selectin [48]. Regarding P-selectin, atherosclerotic lesions [49], as well as the number of leukocytes present in atherosclerotic lesions [50], are less frequent in P-selectin-deficient mice that lack apolipoprotein E. Interactions between platelets and leukocytes might also play a crucial role in patients with stroke as soluble P-selectin, surface P-selectin expression and/or platelet–leukocyte aggregates circulating in the blood were increased in these patients compared to controls [51]. These data from animal and human studies suggest that P-selectin, PSGL-1, and platelet–leukocyte aggregates modulate atherogenesis and may be important biomarkers in identifying patients who are at increased risk for stroke.

Activated platelets also release *r*egulated upon *a*ctivation, *n*ormal *T*-cells *e*xpressed and *s*ecreted (RANTES), which bind to the atherosclerotic endothelium and form a chemoattractant surface for monocytes [52]. In addition, platelets are involved in the generation of thromboxane A2, which promotes platelet and endothelial activation, and platelet-activating factor (PAF), a factor for signaling and adhesion of leukocytes to other cells [53]. Adherence of leukocytes and the subsequent release of pro-inflammatory factors are particular mechanisms that show how platelets induce inflammatory gene expression in target monocytes. For example, monocytes that are adherent to P-selectin-bearing platelets synthesize monocyte chemotactic protein-1 (MCP-1) and IL-8. MCP-1 and IL-8 are necessary for leukocytes to migrate into subendothelial layers [53]. In addition, activated platelets upregulate and induce monocytes to release COX-2 [54]. COX-2 is an enzyme that is responsible for the synthesis of pro-inflammatory eicosanoids. There is some evidence that the interaction of platelets with monocytes results in the expression of IL-1β, IL-6, and TNF-α [54]. IL-1β and IL-6, in turn, play several roles in inflammatory processes, e.g., activation of leukocytes and endothelial cells and induction of pro-inflammatory mediators [55, 56]. Activated platelets also increase the synthesis of matrix metalloproteinase-9 (MMP-9) in human monocytes in the presence of collagen [57]. Some of the aforementioned pro-inflammatory mediators are implicated in ischemic stroke. Patients with stroke have revealed a higher expression of MCP-1, IL-8, and IL-1β [58]. In addition, patients with stroke, in whom a high concentration of IL-6, TNF-α, or soluble VCAM-1 was detected, were at a higher risk of recurrent ischemic stroke [59]. Increased blood concentrations of MMP-9 and elevated urine concentrations of thromboxane A2, an indicator of platelet cyclooxygenase activity, have also been observed in patients with ischemic stroke [60]. Another study showed that leukocyte activation—assessed by decreased L-selectin expression—is prolonged in patients with stroke up to 3 months after the ischemic stroke [51]. These data suggest that platelet–leukocyte interactions regulate atherothrombosis and concomitant inflammatory events.

In the early phase in particular, T cells have been shown to contribute critically to stroke development: T cells were detectable in the postischemic brain as early as 24 h after reperfusion [61]. The adhesion of T cells to the activated endothelium and subsequent brain invasion after cerebral ischemia is a key factor in the induction of inflammatory cerebral damage. Invasion of T cells into cerebral parenchyma depends on the interaction of the leukocyte very late antigen-4 (VLA-4) with VCAM-1 on endothelial cells [62]. A recently published study revealed that VLA-4 blockade by monoclonal antibodies improved outcome in a mouse model of moderate stroke lesions by inhibiting cerebral leukocyte invasion and neurotoxic cytokine production, which finally resulted in an equally potent reduction of infarct volume and postischemic neuroinflammation [63]. However, a study by our group showed that VLA-4 blockade with the anti-CD49d antibody failed to improve stroke outcome irrespective of the model or the time point investigated [64]. A preclinical randomized controlled multicenter trial revealed that treatment with CD49d-specific antibodies significantly attenuated both leukocyte invasion and infarct volume after permanent distal occlusion of the middle cerebral artery, but did not reduce lesion size or affect leukocyte invasion after transient proximal occlusion of the middle cerebral artery, which induces large lesions [65]. Recently, we showed that recombination activating gene (Rag1)-deficient mice, which lack functional T cells, are largely resistant against ischemic neurodegeneration, whereas T cell-mediated brain damage was detectable 24 h after the induction of ischemic stroke [66]. There is also growing evidence that a T-cell subpopulation, i.e., the regulatory T cells (Tregs), contributes to neurodegeneration by interacting with cerebral endothelial cells via the lymphocyte function-associated antigen (LFA-1)/ICAM-1 pathway and platelets [67].

Thus, it has been documented that the leukocytes most frequently infiltrating the ischemic cerebral tissue include neutrophils, macrophages, and natural killer cells, as well as T-cell subpopulations [68]. Such cells, as detected within the infarcted and peri-infarcted areas of cerebral tissues, are involved in all stages of the ischemic cascade [69].

16.4 Role of the Plasma Contact System in Acute Ischemic Stroke

Thrombo-inflammation in stroke results in the interaction of platelets with inflammatory components and is also related to the plasma contact system. The plasma contact system encompasses five proteins that assemble when blood comes into contact with negatively charged surfaces such as phosphatidylserine [70] or inorganic polyphosphates [71]. The factors encompassing the plasma contact system are the serine proteases factor XII (FXII), factor XI (FXI), plasma prekallikrein (PPK), the nonenzymatic cofactor high-molecular-weight kininogen (HMWK), and the serpin C1 esterase inhibitor [72]. Activation of the plasma contact system triggers different pathways, such as the kallikrein–kinin system (KKS), the intrinsic pathway of coagulation, the classical complement cascade, and the fibrinolytic system. Recent studies have shown that the contact system plays a crucial part in thrombus formation [73]. However, the plasma contact system is not only involved in thrombus formation but also in inflammatory processes after cerebral ischemia, when it is involved in promoting endothelial leakage through bradykinin (Fig. 16.1) [74]. The contact–kinin system is initiated via activation of FXII: Activated platelets release negatively charged polyphosphates, which results in the shift of FXII in its activated form, FXIIa [71]. FXIIa initiates the intrinsic coagulation cascade via activation of FXI and is involved in the cleavage of PPK to plasma kallikrein (PK). In the next step, PK cleaves HMWK, which results in the release of bradykinin from HMWK [72].

Regarding activation of the intrinsic coagulation cascade, we recently showed that FXII is important in thrombus formation. Transgenic mice with a FXII deficiency showed impaired thrombus formation by generating unstable thrombi [75] and developed significantly smaller ischemic infarcts without an increase in ICH [76]. These findings are in line with another study by our group, revealing that inhibition of FXII through a complex of recombinant human albumin and infestin-4 is highly protective in a murine model of ischemic stroke [77]. The underlying mechanism of its neuroprotective efficacy has been investigated using high-field magnetic resonance imaging (17.6T) in a murine stroke model. In $FXII^{-/-}$ mice undergoing tMCAo, cortical cerebral blood flow (CBF) was restored by 36% within 24 h, whereas in wild-type mice, cortical CBF was further decreased by 30% [78]. Thus, an enhanced cortical reperfusion in $FXII^{-/-}$-mice might explain—at least, in part—the aforementioned effect after ischemic stroke. This finding is further corroborated by immunohistologic analyses, which revealed reduced fibrin formation in the infarct area of $FXII^{-/-}$ mice compared to wild-type controls, and indicates that FXII-dependent thrombin generation occurs to a considerable extent during cerebral infarction [76]. As already mentioned, FXIIa is related to inflammatory processes via activation of KKS. The significance of KKS in experimental stroke has been addressed using both a genetic and a pharmacologic approach to inhibit PK [79]. PK-deficient mice subjected to tMCAo developed significantly smaller brain infarcts and less severe neurologic deficits compared to controls without an increase in infarct-associated hemorrhage [79]. In addition, less cerebral inflammation was observed in PK-deficient mice subjected to tMCAo [79]. Similar results were found in our study on mice with HMWK deficiency undergoing tMCAo: These mice exhibited a reduction in thrombus formation, blood–brain barrier damage, and cerebral inflammation [80]. Thus, the plasma contact system seems to be a promising target for therapeutic intervention after ischemic stroke.

The relationship between ischemic stroke and FXII in humans remains uncertain. The *R*isk of *A*rterial *T*hrombosis *I*n Relation to *O*ral Contraceptives (RATIO) study showed that elevated levels of the FXIIa–C1–INH complex were associated with an increased risk of stroke [81]. In contrast, a case–control study of middle-aged men in the Second Northwick Park Study (NPHS-II) revealed the opposite—that is, lower levels of the FXIIa–C1–INH complex were a risk factor for ischemic stroke [82]. The *A*therosclerosis *R*isk *I*n *C*ommunities (ARIC) study did not identify any relationship between FXII levels and ischemic stroke [83]. Both a report on published cases of FXII deficiency and thrombosis [84], as well as a study on Swiss families with FXII deficiency [85], concluded that there was no association between FXII deficiency and stroke. Thus, the possibility that elevated FXII levels in humans represent a risk factor for ischemic stroke is, so far, not confirmed.

16.5 Thrombo-inflammation: Evidence in Human Stroke and Consecutive Therapeutic Options

Several drugs for the secondary prevention of ischemic stroke have been linked to a reduction in thrombus formation but also to an attenuation of inflammation in patients with stroke. Up to now, the vast majority of studies on secondary prevention of stroke in humans were mainly focused on thrombosis; studies addressing anti-inflammatory properties of these drugs are sparse. A recently published study [86] addressed this issue by investigating the efficacy of aspirin (150 mg daily) on platelet-related inflammatory factors, namely, CD62P and CD40L, in both patients with acute ischemic stroke and healthy volunteers. Although the study yielded evidence for the hyperactivation of platelets in the acute stage of cerebral ischemia, the platelet α-granule-derived inflammatory mediators and monocyte–platelet aggregation were only reduced in healthy individuals but not in patients with acute stroke after aspirin intake [86]. Additionally, Chronos and co-workers found that aspirin did not inhibit adenosine diphosphate- or thrombin-induced platelet α-granule secretion [87]. Another study revealed that aspirin treatment does not attenuate either resting P-selectin expression or the formation of leukocyte–platelet aggregates [88].

Clopidogrel, a second-generation thienopyridine, acts by binding to the platelet P2Y12 receptor, which results in a reduction in the release of pro-inflammatory mediators from platelet α-granules, such as soluble P-selectin and CD40L, and therefore in attenuated interactions between platelets and leukocytes [89]. Clopidogrel, in addition to aspirin, significantly decreases levels of TNF-α and CRP compared to aspirin alone in patients with acute coronary syndrome, whereas there are no data on this issue in patients with stroke.

In 2009, another P2Y12 receptor inhibitor, ticagrelor, was introduced as an antiplatelet agent for acute coronary syndrome. In the *Plat*elet Inhibition and Patient *O*utcomes (PLATO) trial, this drug reduced the incidence of adverse cardiovascular events compared to clopidogrel [90] and, unexpectedly, also reduced all-cause mortality to a greater degree than would be expected from other trials of P2Y12 receptor inhibitors. In the aforementioned trial, patients with a prior history of stroke or TIA had a relative risk reduction in the primary endpoint (i.e., composite of death from vascular causes, myocardial infarction, or stroke) similar to those who did not have a prior stroke. These findings might suggest that ticagrelor impacts immune signaling differently compared to other P2Y12 inhibitors, such as clopidogrel [91], which might be due to its effect on cellular uptake of adenosine by inhibiting the equilibrative nucleoside transporter ENT1 [92]. Furthermore, these findings have raised the question of whether ticagrelor might be used as a preventative treatment in stroke, an issue that is addressed currently by the Acute Stroke or Transient Ischaemic Attack Treated with Aspirin or Ticagrelor and Patient Outcomes (SOCRATES) trial (https://clinicaltrials.gov/ct2/show/NCT01994720?term=SOCRATES &rank=2).

As mentioned above, the glycoprotein GPIIb/IIIa receptor is also involved in platelet-mediated inflammatory processes [93], and thus GPIIb/IIIa receptor inhibitors such as abciximab, eptifibatide, and tirofiban probably also attenuate thrombo-inflammation [93]. However, a Cochrane meta-analysis [94] encompassing the three cohorts of the *Ab*ciximab in *E*mergency Treatment of *St*roke *T*rial (AbESTT-II) and the *S*tudy of *E*fficacy of *T*irofiban in Acute *I*schaemic *S*troke (SETIS) showed that these drugs are associated with a significant risk of ICH with no evidence of any reduction in death or neurologic symptoms in patients with stroke. Although the exact mechanism of this hazard is unclear, the excess fatality rate might be attributed to the release of pro-thrombotic and pro-inflammatory platelet CD40 ligands from suboptimal GPIIb/IIIa inhibition [95]. However, it is of note that the aforementioned analysis is mainly (89%) based on abciximab. A recently published study, the *Sa*fety of *T*irofiban in acute *I*schemic *S*troke (SaTIS) trial, showed that tirofiban did not result in more ICH compared to placebo but reduced mortality over the long term. There was no difference in neurologic/functional outcome between patients treated with tirofiban or with placebo [96].

Another therapeutic approach is targeting GPVI: In a Phase I study, revacept, a fusion protein consisting of an extracellular portion of GPVI receptor of humans for collagen (GPVI-Fc), efficiently inhibited collagen-induced platelet aggregation ex vivo, with no alteration of primary hemostasis in 30 healthy donors [97]. However, in another study, GPVI-Fc had only limited antithrombotic effects in an animal model, whereas the direct blockade of GPVI function was effective in preventing occlusive thrombus formation [98]. Currently, there is a revacept phase II trial, which aims to evaluate whether the incidence of microembolic signals can be reduced in patients with symptomatic carotid artery stenosis receiving revacept plus antiplatelet monotherapy (https://clinicaltrials.gov/ct2/show/NCT01645306?term=revacept&rank=1).

Thrombin-induced platelet activation is the most potent pathway in platelet aggregation [95]. Vorapaxar and atopaxar are two protease-activated receptor-1 (PAR-1) antagonists that selectively target thrombin-induced platelet activation. PAR-1 is also known to be involved in inflammatory processes during thrombus formation [99]. Vorapaxar was evaluated in two phase III trials—the *T*hrombin *R*eceptor *A*ntagonist for *C*linical *E*vent *R*eduction in Acute Coronary Syndrome (TRACER) [100] and the *T*hrombin *R*eceptor *A*ntagonist in Secondary Prevention of Atherothrombotic Ischemic Events (TRA 2°P TIMI 50) trial [101]. However, treatment of participants with prior ischemic stroke was terminated early in both trials due to a high ICH rate. Additionally, a subanalysis showed that vorapaxar did not reduce recurrent ischemic stroke among participants with prior ischemic stroke [102]. This finding has been explained by the different mechanisms of stroke type: In a cohort with a high prevalence of previous lacunar strokes, recurrent strokes were overwhelmingly likely to have lacunar mechanisms, and lacunar ischemic stroke may be less responsive to antiplatelet therapy [103].

Despite these rather sobering findings, there are some new promising studies on immune-modulatory agents in acute stroke therapy, such as fingolimod. Fingolimod is a sphingosine-1-phosphate (S1P) analog that attenuates the number of peripheral lymphocytes by blocking the egress of these cells from lymphoid organs [104]. Fingolimod has been clinically approved for the treatment of relapsing-remitting multiple sclerosis [105]. Recently, Fu and co-workers reported that oral fingolimod administered within 72 h of stroke onset was safe, limited secondary tissue injury from baseline to 7 days, decreased microvascular permeability, attenuated neurologic deficits, and promoted recovery [106]. Additionally, the combination therapy of fingolimod and alteplase has been shown to attenuate reperfusion injury and improve clinical outcomes in patients with stroke in a pilot study [107]. Another immunomodulator that is under evaluation for acute stroke treatment is natalizumab, a humanized monoclonal antibody that belongs to the selective adhesion molecule inhibitors. It binds to the α4-subunit of α4-integrins and blocks binding to adhesion molecules (i.e., VCAM-1 and mucosal addressin cell adhesion molecule-1), thereby attenuating inflammation [108]. A study on the effect of natalizumab in the acute phase of stroke was completed in April 2015 (Effect of natalizumab on infarct volume in acute ischemic stroke (ACTION); https://clinicaltrials.gov/ct2/show/NCT01955707?term=infarct+volume&rank=1) and publication of the first analyses is expected soon.

16.6 Conclusions

Both thrombotic and inflammatory mechanisms are highly intertwined in the pathophysiology of ischemic stroke. Thrombus formation and inflammation are therefore promising targets for the development of novel therapeutic strategies. The growing insights into thrombo-inflammation after cerebral ischemia might provide a platform for further exploration of the critical interface between inflammation and thrombosis after ischemic stroke. These interesting findings in the field of cerebral thrombo-inflammation should encourage stroke researchers to seek further treatments that target the reduction of thrombo-inflammation in stroke and other cardiovascular diseases.

Compliance with Ethical Standards

Conflict of Interest: Felix Fluri, Bernhard Nieswandt, Guido Stoll, and Christoph Kleinschnitz declares that they have no conflict of interest.

Ethical Approval: This article does not contain any studies with human participants or animals performed by any of the authors.

References

1. Lozano R, Naghavi M, Foreman K, et al. Global and regional mortality from 235 causes of death for 20 age groups in 1990 and 2010: a systematic analysis for the Global Burden of Disease Study 2010. Lancet. 2012;380:2095–128.
2. Murray CJL, Vos T, Lozano R, et al. Disability-adjusted life years (DALYs) for 291 diseases and injuries in 21 regions, 1990–2010: a systematic analysis for the Global Burden of Disease Study 2010. Lancet. 2012;380:2197–223.
3. Heuschmann PU, Wiedmann S, Wellwood I, Rudd A, Carlo AD, Bejot Y, Ryglewicz D, Rastenyte D, Wolfe CDA. Three-month stroke outcome The European Registers of Stroke (EROS) Investigators. Neurology. 2011;76:159–65.
4. Cardiogenic Brain Embolism. Cerebral Embolism Task Force. Arch Neurol. 1986;43:71–84.
5. Diener HC, Cunha L, Forbes C, Sivenius J, Smets P, Lowenthal A. European Stroke Prevention Study 2. Dipyridamole and acetylsalicylic acid in the secondary prevention of stroke1. J Neurol Sci. 1996;143:1–13.

6. Adams RJ, Albers G, Alberts MJ, Benavente O, Furie K, Goldstein LB, Gorelick P, Halperin J, Harbaugh R, Johnston SC, Katzan I, Kelly-Hayes M, Kenton EJ, Marks M, Sacco RL, Schwamm LH. Update to the AHA/ASA recommendations for the prevention of stroke in patients with stroke and transient ischemic attack. Stroke. 2008;39:1647–52.
7. Committee CS. A randomised, blinded, trial of clopidogrel versus aspirin in patients at risk of ischaemic events (CAPRIE). Lancet. 1996;348:1329–39.
8. Diener H-C, Bogousslavsky J, Brass LM, Cimminiello C, Csiba L, Kaste M, Leys D, Matias-Guiu J, Rupprecht H-J. Aspirin and clopidogrel compared with clopidogrel alone after recent ischaemic stroke or transient ischaemic attack in high-risk patients (MATCH): randomised, double-blind, placebo-controlled trial. Lancet. 2004;364:331–7.
9. Hankey GJ, Johnston SC, Easton JD, Hacke W, Mas J-L, Brennan D, Mak KH, Bhatt DL, KAA F, Topol EJ, on behalf of the CHARISMA trial investigators. Effect of clopidogrel plus ASA vs. ASA early after TIA and ischaemic stroke: a substudy of the CHARISMA trial. Int J Stroke. 2011;6:3–9.
10. Wang Y, Wang Y, Zhao X, Liu L, Wang D, Wang C, Wang C, Li H, Meng X, Cui L, Jia J, Dong Q, Xu A, Zeng J, Li Y, Wang Z, Xia H, Johnston SC. Clopidogrel with aspirin in acute minor stroke or transient ischemic attack. N Engl J Med. 2013;369:11–9.
11. Investigators TAWG on behalf of the A. Clopidogrel plus aspirin versus oral anticoagulation for atrial fibrillation in the Atrial fibrillation Clopidogrel Trial with Irbesartan for prevention of Vascular Events (ACTIVE W): a randomised controlled trial. Lancet. 2006;367:1903–12.
12. ACTIVE Investigators, Connolly SJ, Pogue J, Hart RG, Hohnloser SH, Pfeffer M, Chrolavicius S, Yusuf S. Effect of clopidogrel added to aspirin in patients with atrial fibrillation. N Engl J Med. 2009;360:2066–78.
13. Mateen FJ, Shuaib A. Progress in clinical neurosciences: the "antiplatelet" agents and the role of the endothelium. Can J Neurol Sci. 2007;34:270–9.
14. Jickling GC, Liu D, Ander BP, Stamova B, Zhan X, Sharp FR. Targeting neutrophils in ischemic stroke: translational insights from experimental studies. J Cereb Blood Flow Metab. 2015;35:888–901.
15. Löwenberg EC, Meijers JCM, Levi M. Platelet-vessel wall interaction in health and disease. Neth J Med. 2010;68(6):242–51.
16. Li R, Emsley J. The organizing principle of the platelet glycoprotein Ib–IX–V complex. J Thromb Haemost. 2013;11:605–14.
17. Kanaji S, Fahs SA, Shi Q, Haberichter SL, Montgomery RR. Contribution of platelet vs. endothelial VWF to platelet adhesion and hemostasis. J Thromb Haemost. 2012;10:1646–52.
18. Massberg S, Gawaz M, Grüner S, Schulte V, Konrad I, Zohlnhöfer D, Heinzmann U, Nieswandt B. A crucial role of glycoprotein VI for platelet recruitment to the injured arterial wall in vivo. J Exp Med. 2003;197:41–9.
19. Bergmeier W, Piffath CL, Goerge T, Cifuni SM, Ruggeri ZM, Ware J, Wagner DD. The role of platelet adhesion receptor GPIbalpha far exceeds that of its main ligand, von Willebrand factor, in arterial thrombosis. Proc Natl Acad Sci USA. 2006;103:16900–5.
20. Kleinschnitz C, Pozgajova M, Pham M, Bendszus M, Nieswandt B, Stoll G. Targeting platelets in acute experimental stroke impact of glycoprotein Ib, VI, and IIb/IIIa blockade on infarct size, functional outcome, and intracranial bleeding. Circulation. 2007;115:2323–30.
21. Kleinschnitz C, Meyer SFD, Schwarz T, Austinat M, Vanhoorelbeke K, Nieswandt B, Deckmyn H, Stoll G. Deficiency of von Willebrand factor protects mice from ischemic stroke. Blood. 2009;113:3600–3.
22. Seaman CD, Yabes J, Comer DM, Ragni MV. Does deficiency of von Willebrand factor protect against cardiovascular disease? Analysis of a national discharge register. J Thromb Haemost. 2015;13:1999–2003.
23. Nieswandt B, Watson SP. Platelet-collagen interaction: is GPVI the central receptor? Blood. 2003;102:449–61.

24. Nieswandt B, Brakebusch C, Bergmeier W, Schulte V, Bouvard D, Mokhtari-Nejad R, Lindhout T, Heemskerk JW, Zirngibl H, Fässler R. Glycoprotein VI but not alpha2beta1 integrin is essential for platelet interaction with collagen. EMBO J. 2001;20:2120–30.
25. Lecut C, Feijge MAH, Cosemans JMEM, Jandrot-Perrus M. Heemskerk1 JWM. Fibrillar type I collagens enhance platelet-dependent thrombin generation via glycoprotein VI with direct support of α2β 1 but not αIIbβ 3 integrin. Thromb Haemost. 2005;94:107–14.
26. Schulte V, Reusch HP, Pozgajová M, Varga-Szabó D, Gachet C, Nieswandt B. Two-phase antithrombotic protection after anti-glycoprotein VI treatment in mice. Arterioscler Thromb Vasc Biol. 2006;26:1640–7.
27. Bültmann A, Li Z, Wagner S, Peluso M, Schönberger T, Weis C, Konrad I, Stellos K, Massberg S, Nieswandt B, Gawaz M, Ungerer M, Münch G. Impact of glycoprotein VI and platelet adhesion on atherosclerosis—A possible role of fibronectin. J Mol Cell Cardiol. 2010;49:532–42.
28. Schönberger T, Ziegler M, Borst O, Konrad I, Nieswandt B, Massberg S, Ochmann C, Jürgens T, Seizer P, Langer H, Münch G, Ungerer M, Preissner KT, Elvers M, Gawaz M. The dimeric platelet collagen receptor GPVI-Fc reduces platelet adhesion to activated endothelium and preserves myocardial function after transient ischemia in mice. Am J Physiol Cell Physiol. 2012;303:C757–66.
29. Inoue O, Suzuki-Inoue K, McCarty OJT, Moroi M, Ruggeri ZM, Kunicki TJ, Ozaki Y, Watson SP. Laminin stimulates spreading of platelets through integrin α6β1–dependent activation of GPVI. Blood. 2006;107:1405–12.
30. Goebel S, Li Z, Vogelmann J, Holthoff H-P, Degen H, Hermann DM, Gawaz M, Ungerer M, Münch G. The GPVI-Fc fusion protein revacept improves cerebral infarct volume and functional outcome in stroke. PLoS ONE. 2013;8:e66960.
31. Bigalke B, Stellos K, Geisler T, Kremmer E, Seizer P, May AE, Lindemann S, Melms A, Luft A, Gawaz M. Expression of platelet glycoprotein VI is associated with transient ischemic attack and stroke. Eur J Neurol. 2010;17:111–7.
32. Wurster T, Poetz O, Stellos K, Kremmer E, Melms A, Schuster A, Nagel E, Joos T, Gawaz M, Bigalke B. Plasma levels of soluble glycoprotein VI (sGPVI) are associated with ischemic stroke. Platelets. 2013;24:560–5.
33. Savage B, Saldívar E, Ruggeri ZM. Initiation of platelet adhesion by arrest onto Fibrinogen or translocation on von Willebrand factor. Cell. 1996;84:289–97.
34. Choudhri TF, Hoh BL, Zerwes HG, Prestigiacomo CJ, Kim SC, Connolly ES, Kottirsch G, Pinsky DJ. Reduced microvascular thrombosis and improved outcome in acute murine stroke by inhibiting GP IIb/IIIa receptor-mediated platelet aggregation. J Clin Invest. 1998;102:1301–10.
35. Abumiya T, Fitridge R, Mazur C, Copeland BR, Koziol JA, Tschopp JF, Pierschbacher MD, del Zoppo GJ. Integrin alpha(IIb)beta(3) inhibitor preserves microvascular patency in experimental acute focal cerebral ischemia. Stroke J Cereb Circ. 2000;31:1402–9. discussion 1409–1410.
36. Ishikawa M, Cooper D, Arumugam TV, Zhang JH, Nanda A, Granger DN. Platelet-leukocyte-endothelial cell interactions after middle cerebral artery occlusion and reperfusion. J Cereb Blood Flow Metab. 2004;24:907–15.
37. Kraft P, Schuhmann MK, Fluri F, Lorenz K, Zernecke A, Stoll G, Nieswandt B, Kleinschnitz C. Efficacy and safety of platelet glycoprotein receptor blockade in aged and comorbid mice with acute experimental stroke. Stroke. 2015;46:3502–6.
38. Adams HP, Effron MB, Torner J, Dávalos A, Frayne J, Teal P, Leclerc J, Oemar B, Padgett L, Barnathan ES, Hacke W, Investigators for the A-I. Emergency administration of Abciximab for treatment of patients with acute ischemic stroke: results of an International Phase III Trial Abciximab in Emergency Treatment of Stroke Trial (AbESTT-II). Stroke. 2008;39:87–99.
39. Massberg S, Schürzinger K, Lorenz M, Konrad I, Schulz C, Plesnila N, Kennerknecht E, Rudelius M, Sauer S, Braun S, Kremmer E, Emambokus NR, Frampton J, Gawaz M. Platelet

adhesion via glycoprotein IIb integrin is critical for Atheroprogression and focal cerebral ischemia an in vivo study in mice lacking glycoprotein IIb. Circulation. 2005;112:1180–8.
40. Cox D, Smith R, Quinn M, Theroux P, Crean P, Fitzgerald DJ. Evidence of platelet activation during treatment with a GPIIb/IIIa antagonist in patients presenting with acute coronary syndromes. J Am Coll Cardiol. 2000;36:1514–9.
41. Klinkhardt U, Graff J, Harder S. Clopidogrel, but not abciximab, reduces platelet leukocyte conjugates and P-selectin expression in a human ex vivo in vitro model. Clin Pharmacol Ther. 2002;71:176–85.
42. Lincoff AM, Kereiakes DJ, Mascelli MA, Deckelbaum LI, Barnathan ES, Patel KK, Frederick B, Nakada MT, Topol EJ. Abciximab suppresses the rise in levels of circulating inflammatory markers after percutaneous coronary revascularization. Circulation. 2001;104:163–7.
43. Adderley SR, Fitzgerald DJ. Glycoprotein IIb/IIIa antagonists induce apoptosis in rat cardiomyocytes by caspase-3 activation. J Biol Chem. 2000;275:5760–6.
44. Okada Y, Copeland BR, Fitridge R, Koziol JA, del Zoppo GJ. Fibrin contributes to microvascular obstructions and parenchymal changes during early focal cerebral ischemia and reperfusion. Stroke. 1994;25:1847–53. discussion 1853–1854.
45. del Zoppo GJ, Mabuchi T. Cerebral microvessel responses to focal ischemia. J Cereb Blood Flow Metab. 2003;23:879–94.
46. Xia Y, Wang CZ, Liu J, Anastasio NC, Johnson KM. Brain-derived neurotrophic factor prevents phencyclidine-induced apoptosis in developing brain by parallel activation of both the ERK and PI-3K/Akt pathways. Neuropharmacology. 2010;58:330–6.
47. Hammer DA, Apte SM. Simulation of cell rolling and adhesion on surfaces in shear flow: general results and analysis of selectin-mediated neutrophil adhesion. Biophys J. 1992;63:35–57.
48. Smyth SS, Mcever RP, Weyrich AS, Morrell CN, Hoffman MR, Arepally GM, French PA, Dauerman HL, Becker RC. For the 2009 Platelet Colloquium Participants. Platelet functions beyond hemostasis. J Thromb Haemost. 2009;7:1759–66.
49. Dong ZM, Brown AA, Wagner DD. Prominent role of p-selectin in the development of advanced atherosclerosis in ApoE-deficient mice. Circulation. 2000;101:2290–5.
50. Manka D, Collins RG, Ley K, Beaudet AL, Sarembock IJ. Absence of p-selectin, but not intercellular adhesion molecule-1, attenuates neointimal growth after arterial injury in apolipoprotein e-deficient mice. Circulation. 2001;103:1000–5.
51. Htun P, Fateh-Moghadam S, Tomandl B, Handschu R, Klinger K, Stellos K, Garlichs C, Daniel W, Gawaz M. Course of platelet activation and platelet-leukocyte interaction in cerebrovascular ischemia. Stroke. 2006;37:2283–7.
52. von Hundelshausen P, Weber KSC, Huo Y, Proudfoot AEI, Nelson PJ, Ley K, Weber C. RANTES deposition by platelets triggers monocyte arrest on inflamed and atherosclerotic endothelium. Circulation. 2001;103:1772–7.
53. Weyrich AS, Zimmerman GA. Platelets: signaling cells in the immune continuum. Trends Immunol. 2004;25:489–95.
54. Dixon DA, Tolley ND, Bemis-Standoli K, Martinez ML, Weyrich AS, Morrow JD, Prescott SM, Zimmerman GA. Expression of COX-2 in platelet-monocyte interactions occurs via combinatorial regulation involving adhesion and cytokine signaling. J Clin Invest. 2006;116:2727–38.
55. Schuett H, Luchtefeld M, Grothusen C, Grote K, Schieffer B. How much is too much? Interleukin-6 and its signalling in atherosclerosis. Thromb Haemost. 2009;102:215–22.
56. Lindemann S, Tolley ND, Dixon DA, McIntyre TM, Prescott SM, Zimmerman GA, Weyrich AS. Activated platelets mediate inflammatory signaling by regulated interleukin 1β synthesis. J Cell Biol. 2001;154:485–90.
57. Galt SW, Lindemann S, Medd D, Allen LL, Kraiss LW, Harris ES, Prescott SM, McIntyre TM, Weyrich AS, Zimmerman GA. Differential regulation of matrix metalloproteinase-9 by monocytes adherent to collagen and platelets. Circ Res. 2001;89:509–16.

58. Kostulas N, Pelidou SH, Kivisäkk P, Kostulas V, Link H. Increased IL-1β, IL-8, and IL-17 mRNA expression in blood mononuclear cells observed in a prospective ischemic stroke study. Stroke. 1999;30:2174–9.
59. Welsh P, Lowe GDO, Chalmers J, Campbell DJ, Rumley A, Neal BC, MacMahon SW, Woodward M. Associations of proinflammatory cytokines with the risk of recurrent stroke. Stroke. 2008;39:2226–30.
60. Worthmann H, Tryc AB, Goldbecker A, Ma YT, Tountopoulou A, Hahn A, Dengler R, Lichtinghagen R, Weissenborn K. The temporal profile of inflammatory markers and mediators in blood after acute ischemic stroke differs depending on stroke outcome. Cerebrovasc Dis. 2010;30:85–92.
61. Jander S, Kraemer M, Schroeter M, Witte OW, Stoll G. Lymphocytic infiltration and expression of intercellular adhesion molecule-1 in photochemically induced ischemia of the rat cortex. J Cereb Blood Flow Metab. 1995;15:42–51.
62. Laschinger M, Engelhardt B. Interaction of α4-integrin with VCAM-1 is involved in adhesion of encephalitogenic T cell blasts to brain endothelium but not in their transendothelial migration in vitro. J Neuroimmunol. 2000;102:32–43.
63. Liesz A, Zhou W, Mracskó É, Karcher S, Bauer H, Schwarting S, Sun L, Bruder D, Stegemann S, Cerwenka A, Sommer C, Dalpke AH, Veltkamp R. Inhibition of lymphocyte trafficking shields the brain against deleterious neuroinflammation after stroke. Brain. 2011;134:704–20.
64. Langhauser F, Kraft P, Göb E, Leinweber J, Schuhmann MK, Lorenz K, Gelderblom M, Bittner S, Meuth SG, Wiendl H, Magnus T, Kleinschnitz C. Blocking of α4 integrin does not protect from acute ischemic stroke in mice. Stroke. 2014;45:1799–806.
65. Llovera G, Hofmann K, Roth S, Salas-Pérdomo A, Ferrer-Ferrer M, Perego C, Zanier ER, Mamrak U, Rex A, Party H, Agin V, Fauchon C, Orset C, Haelewyn B, De Simoni M-G, Dirnagl U, Grittner U, Planas AM, Plesnila N, Vivien D, Liesz A. Results of a preclinical randomized controlled multicenter trial (pRCT): anti-CD49d treatment for acute brain ischemia. Sci Transl Med. 2015;7:299ra121.
66. Kleinschnitz C, Schwab N, Kraft P, Hagedorn I, Dreykluft A, Schwarz T, Austinat M, Nieswandt B, Wiendl H, Stoll G. Early detrimental T-cell effects in experimental cerebral ischemia are neither related to adaptive immunity nor thrombus formation. Blood. 2010;115:3835–42.
67. Kleinschnitz C, Kraft P, Dreykluft A, Hagedorn I, Göbel K, Schuhmann MK, Langhauser F, Helluy X, Schwarz T, Bittner S, Mayer CT, Brede M, Varallyay C, Pham M, Bendszus M, Jakob P, Magnus T, Meuth SG, Iwakura Y, Zernecke A, Sparwasser T, Nieswandt B, Stoll G, Wiendl H. Regulatory T cells are strong promoters of acute ischemic stroke in mice by inducing dysfunction of the cerebral microvasculature. Blood. 2013;121:679–91.
68. Chamorro Á, Meisel A, Planas AM, Urra X, van de Beek D, Veltkamp R. The immunology of acute stroke. Nat Rev Neurol. 2012;8:401–10.
69. Iadecola C, Anrather J. The immunology of stroke: from mechanisms to translation. Nat Med. 2011;17:796–808.
70. Heemskerk JWM, Bevers EM, Lindhout T. Platelet activation and blood coagulation. Thromb Haemost. 2002;88:186–93.
71. Müller F, Mutch NJ, Schenk WA, Smith SA, Esterl L, Spronk HM, Schmidbauer S, Gahl WA, Morrissey JH, Renné T. Platelet polyphosphates are proinflammatory and procoagulant mediators in vivo. Cell. 2009;139:1143–56.
72. Maas C, Oschatz C, Renné T. The plasma contact system 2.0. Semin Thromb Hemost. 2011;37:375–81.
73. Nickel KF, Renné T. Crosstalk of the plasma contact system with bacteria. Thromb Res. 2012;130(Supplement 1):S78–83.
74. Albert-Weißenberger C, Sirén A-L, Kleinschnitz C. Ischemic stroke and traumatic brain injury: the role of the kallikrein–kinin system. Prog Neurobiol. 2013;101–102:65–82.

75. Renné T, Pozgajová M, Grüner S, Schuh K, Pauer H-U, Burfeind P, Gailani D, Nieswandt B. Defective thrombus formation in mice lacking coagulation factor XII. J Exp Med. 2005;202:271–81.
76. Kleinschnitz C, Stoll G, Bendszus M, Schuh K, Pauer H-U, Burfeind P, Renné C, Gailani D, Nieswandt B, Renné T. Targeting coagulation factor XII provides protection from pathological thrombosis in cerebral ischemia without interfering with hemostasis. J Exp Med. 2006;203:513–8.
77. Hagedorn I, Schmidbauer S, Pleines I, Kleinschnitz C, Kronthaler U, Stoll G, Dickneite G, Nieswandt B. Factor XIIa inhibitor recombinant human albumin infestin-4 abolishes occlusive arterial thrombus formation without affecting bleeding. Circulation. 2010;121:1510–7.
78. Pham M, Kleinschnitz C, Helluy X, Bartsch AJ, Austinat M, Behr VC, Renné T, Nieswandt B, Stoll G, Bendszus M. Enhanced cortical reperfusion protects coagulation factor XII-deficient mice from ischemic stroke as revealed by high-field MRI. NeuroImage. 2010;49:2907–14.
79. Göb E, Reymann S, Langhauser F, Schuhmann MK, Kraft P, Thielmann I, Göbel K, Brede M, Homola G, Solymosi L, Stoll G, Geis C, Meuth SG, Nieswandt B, Kleinschnitz C. Blocking of plasma kallikrein ameliorates stroke by reducing thromboinflammation. Ann Neurol. 2015;77(5):784–803.
80. Langhauser F, Göb E, Kraft P, Geis C, Schmitt J, Brede M, Göbel K, Helluy X, Pham M, Bendszus M, Jakob P, Stoll G, Meuth SG, Nieswandt B, McCrae KR, Kleinschnitz C. Kininogen deficiency protects from ischemic neurodegeneration in mice by reducing thrombosis, blood-brain barrier damage, and inflammation. Blood. 2012;120:4082–92.
81. Siegerink B, Govers-Riemslag JWP, Rosendaal FR, Ten Cate H, Algra A. Intrinsic coagulation activation and the risk of arterial thrombosis in young women: results from the Risk of Arterial Thrombosis in relation to Oral contraceptives (RATIO) case-control study. Circulation. 2010;122:1854–61.
82. Govers-Riemslag JWP, Smid M, Cooper JA, Bauer KA, Rosenberg RD, Hack CE, Hamulyak K, Spronk HMH, Miller GJ, ten Cate H. The plasma kallikrein-kinin system and risk of cardiovascular disease in men. J Thromb Haemost. 2007;5:1896–903.
83. Suri MFK, Yamagishi K, Aleksic N, Hannan PJ, Folsom AR. Novel hemostatic factor levels and risk of ischemic stroke: the Atherosclerosis Risk in Communities (ARIC) Study. Cerebrovasc Dis. 2010;29:497–502.
84. Girolami A, Candeo N, De Marinis GB, Bonamigo E, Girolami B. Comparative incidence of thrombosis in reported cases of deficiencies of factors of the contact phase of blood coagulation. J Thromb Thrombolysis. 2011;31:57–63.
85. Zeerleder S, Schloesser M, Redondo M, Wuillemin WA, Engel W, Furlan M, Lämmle B. Reevaluation of the incidence of thromboembolic complications in congenital factor XII deficiency—a study on 73 subjects from 14 Swiss families. Thromb Haemost. 1999;82:1240–6.
86. Lukasik M, Dworacki G, Michalak S, Kufel-Grabowska J, Golanski J, Watala C, Kozubski W. Aspirin treatment influences platelet-related inflammatory biomarkers in healthy individuals but not in acute stroke patients. Thromb Res. 2011;128:e73–80.
87. Chronos NA, Wilson DJ, Janes SL, Hutton RA, Buller NP, Goodall AH. Aspirin does not affect the flow cytometric detection of fibrinogen binding to, or release of alpha-granules or lysosomes from, human platelets. Clin Sci Lond. 1994;87:575–80.
88. Li N, Hu H, Hjemdahl P. Aspirin treatment does not attenuate platelet or leukocyte activation as monitored by whole blood flow cytometry. Thromb Res. 2003;111:165–70.
89. Xiao Z, Théroux P. Clopidogrel inhibits platelet-leukocyte interactions and thrombin receptor agonist peptide-induced platelet activation in patients with an acute coronary syndrome. J Am Coll Cardiol. 2004;43:1982–8.
90. Wallentin L, Becker RC, Budaj A, Cannon CP, Emanuelsson H, Held C, Horrow J, Husted S, James S, Katus H, Mahaffey KW, Scirica BM, Skene A, Steg PG, Storey RF, Harrington

RA. Ticagrelor versus clopidogrel in patients with acute coronary syndromes. N Engl J Med. 2009;361:1045–57.

91. Storey RF, James SK, Siegbahn A, Varenhorst C, Held C, Ycas J, Husted SE, Cannon CP, Becker RC, Steg PG, Åsenblad N, Wallentin L. Lower mortality following pulmonary adverse events and sepsis with ticagrelor compared to clopidogrel in the PLATO study. Platelets. 2014;25:517–25.
92. Armstrong D, Summers C, Ewart L, Nylander S, Sidaway JE, van Giezen JJ, Simkhovich B. Characterization of the adenosine pharmacology of ticagrelor reveals therapeutically relevant inhibition of equilibrative nucleoside transporter 1. J Cardiovasc Pharmacol Ther. 2014;19:209–19.
93. Salanova B, Choi M, Rolle S, Wellner M, Luft FC, Kettritz R. β2-integrins and acquired glycoprotein IIb/IIIa (GPIIb/IIIa) receptors cooperate in NF-κB activation of human neutrophils. J Biol Chem. 2007;282:27960–9.
94. Ciccone A, Motto C, Abraha I, Cozzolino F, Santilli I. Glycoprotein IIb-IIIa inhibitors for acute ischaemic stroke. In: Cochrane database of systematic reviews. Wiley; 2014. http://onlinelibrary.wiley.com/doi/10.1002/14651858.CD005208.pub3/abstract. Cited 29 Nov 2015.
95. Yousuf O, Bhatt DL. The evolution of antiplatelet therapy in cardiovascular disease. Nat Rev Cardiol. 2011;8:547–59.
96. Siebler M, Hennerici MG, Schneider D, von Reutern GM, Seitz RJ, Röther J, Witte OW, Hamann G, Junghans U, Villringer A, Fiebach JB. Safety of Tirofiban in acute Ischemic Stroke: the SaTIS trial. Stroke. 2011;42:2388–92.
97. Ungerer M, Rosport K, Bültmann A, Piechatzek R, Uhland K, Schlieper P, Gawaz M, Münch G. Novel antiplatelet drug revacept (Dimeric Glycoprotein VI-Fc) specifically and efficiently inhibited collagen-induced platelet aggregation without affecting general hemostasis in humans. Circulation. 2011;123:1891–9.
98. Grüner S, Prostredna M, Koch M, Miura Y, Schulte V, Jung SM, Moroi M, Nieswandt B. Relative antithrombotic effect of soluble GPVI dimer compared with anti-GPVI antibodies in mice. Blood. 2005;105:1492–9.
99. Rohatgi T, Henrich-Noack P, Sedehizade F, Goertler M, Wallesch CW, Reymann KG, Reiser G. Transient focal ischemia in rat brain differentially regulates mRNA expression of protease-activated receptors 1 to 4. J Neurosci Res. 2004;75:273–9.
100. Tricoci P, Huang Z, Held C, Moliterno DJ, Armstrong PW, Van de Werf F, White HD, Aylward PE, Wallentin L, Chen E, Lokhnygina Y, Pei J, Leonardi S, Rorick TL, Kilian AM, LHK J, Ambrosio G, Bode C, Cequier A, Cornel JH, Diaz R, Erkan A, Huber K, Hudson MP, Jiang L, Jukema JW, Lewis BS, Lincoff AM, Montalescot G, Nicolau JC, Ogawa H, Pfisterer M, Prieto JC, Ruzyllo W, Sinnaeve PR, Storey RF, Valgimigli M, Whellan DJ, Widimsky P, Strony J, Harrington RA, Mahaffey KW. Thrombin-receptor antagonist vorapaxar in acute coronary syndromes. N Engl J Med. 2012;366:20–33.
101. Morrow DA, Braunwald E, Bonaca MP, Ameriso SF, Dalby AJ, Fish MP, Fox KAA, Lipka LJ, Liu X, Nicolau JC, Ophuis AJO, Paolasso E, Scirica BM, Spinar J, Theroux P, Wiviott SD, Strony J, Murphy SA. Vorapaxar in the secondary prevention of atherothrombotic events. N Engl J Med. 2012;366:1404–13.
102. Morrow DA, Alberts MJ, Mohr JP, Ameriso SF, Bonaca MP, Goto S, Hankey GJ, Murphy SA, Scirica BM, Braunwald E. Efficacy and safety of vorapaxar in patients with prior ischemic stroke. Stroke. 2013;44:691–8.
103. SPS3 Investigators, Benavente OR, Hart RG, LA MC, Szychowski JM, Coffey CS, Pearce LA. Effects of clopidogrel added to aspirin in patients with recent lacunar stroke. N Engl J Med. 2012;367:817–25.
104. Mandala S, Hajdu R, Bergstrom J, Quackenbush E, Xie J, Milligan J, Thornton R, Shei G-J, Card D, Keohane C, Rosenbach M, Hale J, Lynch CL, Rupprecht K, Parsons W, Rosen H. Alteration of lymphocyte trafficking by sphingosine-1-phosphate receptor agonists. Science. 2002;296:346–9.

105. Kappos L, Radue E-W, O'Connor P, Polman C, Hohlfeld R, Calabresi P, Selmaj K, Agoropoulou C, Leyk M, Zhang-Auberson L, Burtin P. A placebo-controlled trial of oral fingolimod in relapsing multiple sclerosis. N Engl J Med. 2010;362:387–401.
106. Fu Y, Zhang N, Ren L, Yan Y, Sun N, Li Y-J, Han W, Xue R, Liu Q, Hao J, Yu C, Shi F-D. Impact of an immune modulator fingolimod on acute ischemic stroke. Proc Natl Acad Sci. 2014;111:18315–20.
107. Zhu Z, Fu Y, Tian D, Sun N, Han W, Chang G, Dong Y, Xu X, Liu Q, Huang D, Shi F-D. Combination of the immune modulator fingolimod with alteplase in acute ischemic stroke a pilot trial. Circulation. 2015;132:1104–12.
108. Yednock TA, Cannon C, Fritz LC, Sanchez-Madrid F, Steinman L, Karin N. Prevention of experimental autoimmune encephalomyelitis by antibodies against alpha 4 beta 1 integrin. Nature. 1992;356:63–6.

Platelets and Polymorphisms

17

Tobias Geisler, Elke Schaeffeler, and Matthias Schwab

Abstract

Functional genome-wide association studies and pharmacogenomics of antiplatelet therapy are of high clinical interest to explain the interindividual variability of drug response. Here, we review recent progress in knowledge of platelet genomics with focus on clinical implications. While there are numerous genetic variants associated with platelet biology with focus on platelet-specific phenotypes and cardiovascular outcome, the clinical evidence is missing so far. However, platelet pharmacogenomics concentrating on antiplatelet drug metabolism and transport is already used in personalized medicine.

Supported by the Deutsche Forschungsgemeinschaft KFO 274 (SCHW 858/1-2) and in part by the Robert-Bosch-Stiftung (Stuttgart, Germany).

T. Geisler • E. Schaeffeler
Dr. Margarete Fischer-Bosch Institute of Clinical Pharmacology, Auerbachstraße 112, 70376 Stuttgart, Germany

University of Tübingen, Tübingen, Germany

M. Schwab (✉)
Dr. Margarete Fischer-Bosch Institute of Clinical Pharmacology, Auerbachstraße 112, 70376 Stuttgart, Germany

University of Tübingen, Tübingen, Germany

Department of Clinical Pharmacology, University Hospital Tübingen, Auf der Morgenstelle 8, 72076 Tübingen, Germany

Department of Pharmacy and Biochemistry, University of Tübingen, Auf der Morgenstelle 8, 72076 Tübingen, Germany
e-mail: matthias.schwab@ikp-stuttgart.de

A. Zirlik et al. (eds.), *Platelets, Haemostasis and Inflammation*,
Cardiac and Vascular Biology 5, https://doi.org/10.1007/978-3-319-66224-4_17

Contents

17.1 Introduction

Traditionally, research in platelet genetics was focused on the investigation of the inheritance of bleeding disorders. In recent years, advances and many technical innovations including deep sequencing and functional genome-wide association studies stimulated the field of platelet genomics. Thus this field of research aims to elucidate in more broader and complex context the role of genomic mechanisms for thrombotic and bleeding diseases but also to apply genetic profiles to cardiovascular risk assessment. Platelet mRNA expression significantly contributes to total mRNA expression of the human genome up to one third. Since the armamentarium of antiplatelet drugs constantly increased in the last decade, platelet pharmacogenomics became of high clinical interest. This research area not only comprises issues related to alteration of platelet proteins by genetic reasons but also covers the genetic background of the absorption, disposition, metabolism, and excretion of antiplatelet agents. A mandatory prerequisite for all these investigations is the definition of the precise platelet phenotype which involves platelet protein expression and function, quantitative or qualitative platelet characteristics (e.g., platelet volume, platelet count, ratio of immature/reticulated platelets), platelet-associated thrombotic events (e.g., stent thrombosis), and/or bleeding events. This chapter will give an overview about the current knowledge in platelet genomics with focus on clinical implications.

17.2 Genetic Background of Platelet Biology, Insights from Association Studies

Platelets are critically involved in a number of signaling pathways. Although being anucleate cells, platelet-related genetics is of important relevance. There is innumerous pathophysiological and clinical evidence that platelets play a crucial role in the development of occlusive thrombotic coronary artery disease, e.g., in the setting of acute coronary syndromes (ACS) at the site of ruptured plaques [1] but also in promoting atherosclerosis and chronic vascular inflammation [2]. Therefore, it seems likely that genetic variants of platelet proteins are associated with platelet-mediated cardiovascular events including myocardial infarction (MI), stenotic coronary artery disease (CAD), and ischemic stroke. The challenge of platelet genetic association studies is to detect a quantitative trait locus. This relies on the number of enrolled individuals and a low number of protein-protein interactions associated with the phenotype. In contrast to other genetic risk markers, for instance, in lipid research, the effect size of platelet polymorphic variants is quite low lying in the range of an odds ratio up to 1.5, thereby requiring large patient cohorts. Intermediate phenotypes have been suggested to be used in platelet gene association studies as more suitable effect parameter to increase the statistical power and to reduce sample size. There is a large variety of possible intermediate phenotypes comprising platelet activation markers measured by flow cytometry (i.e., activation of integrin alphaIIbbeta3), markers of platelet degranulation (e.g., P-selectin, platelet factor 4 release, adenosine triphosphate, serotonin), platelet miRNA expression, platelet aggregation, mean platelet volume (MPV), and platelet turnover (e.g., number of immature/reticulated platelets). Light transmission aggregometry (LTA) turned out as a robust parameter. It possesses a large interindividual variability and has a sufficient hereditary background [3–5].

Classical approaches to investigate geno-phenotype relationships are linkage and association studies. *Association studies* aim to discover genetic variation associated with a specific phenotype or outcome by analyzing population-based cohorts. *Linkage studies* seek to identify genetic variants that differentiate the phenotype among pedigrees. A challenge of association studies is the requirement of a strong linkage disequilibrium (usual $R > 0.8$) between the allele of risk and the genetic alteration such as single nucleotide polymorphisms (SNPs) or copy number variation (CNV).

17.3 Platelet Genome-Wide Association Studies

Genome-wide association studies (GWAS) have demonstrated a high number of single nucleotide polymorphisms (SNPs) associated with cardiovascular diseases. GWAS evaluate the statistical association of genetic variants with a specified disease or trait. This approach has been proved effective in exploring novel loci; however the interpretation of GWAS findings faces several challenges in particular in the field of platelet genomics. A large proportion of identified SNPs correspond to noncoding regions of the genome, hindering assignment of a functional property of the genetic variant. In addition, many genetic variants are linked with the phenotype at any given genomic locus due to linkage disequilibrium (LD), making it hard to determine the specific variant that causes the effect (causal variant). As with any GWAS, it is essential to identify the target genes through which identified variants influence traits or phenotypes. The first GWAS aimed to investigate genomic loci associated with platelet reactivity and covered 2.5×10^6 SNPs related to platelet aggregation response to ADP, collagen, and epinephrine. European ancestry populations from two large cohorts (FHS, $n = 2753$, and the GeneSTAR cohort, $n = 1238$) were included. Genetic variants at seven loci (*GPVI*, *PEAR1*, *SHH*, *ADRA2A*, *MRVI1*, *PIK3CG*, *JMJD1C*) showed significant genome-wide association (Table 17.1). Results were replicated in an independent cohort of African origin ($n = 840$) [5]. Of note, this genome-wide functional approach did not confirm results of early classical candidate association studies including SNPs associated with the human platelet antigen (HPA) complex comprising HPA-1 (*ITGB3*), HPA-2 (*GP1BA*), and HPA-3 (*ITGA2B*) [14, 15].

A number of other intermediate platelet phenotypes have been investigated by genome-wide approaches. Thus, several GWAS and meta-analyses reported SNP associations with the platelet count and the MPV [12, 13, 16, 17]. A consortium analyzing 66,867 subjects reported GWAS data indicating (1) that loci encoding for proteins are functionally responsible for megakaryocytopoiesis, platelet survival, and thrombopoiesis (i.e., *ITGA2B*, *GP1BA*, *and F2R*), (2) that proteins encoding for platelet membrane proteins are transcription factors (i.e., *NFE2*, *MEF2C*, and *MYB*), and (3) that cytoplasmatic proteins (*CBL*, *PIK3CG*, *PTPN11*, *SH2B3*, and *TUBB1*) account for around 5% of the phenotypic variability in platelet count and around 10% in MPV [13]. In combination, the results of the GWAS by Johnson et al. and those by the Bloodomics and HaemGen consortia together with large meta-analyses suggest nine loci (*ARHGEF3*, *CDKN2A*, *GPIBA*, *ITGA2B*, *JMJD1C*, *PEAR1*, *PIK3CG*, *TAOK1*, and *WDR66*) associated with platelet count/MPV [17] or platelet function [5, 8] at a genome-wide significance threshold (Table 17.1). Of note, most of these loci have an impact on platelet function. While about 30 loci have been associated with the clinical manifestation of stroke or MI [18, 19], so far there is lack of genome-wide approaches elucidating an association with arterial thrombosis.

Only two loci (*CDKN2A* at Chr9p.21 and SH2B3 at Chr12q23.3) related to platelet count/MPV and platelet function and the occurrence of CAD and MI have been discovered by GWAS approaches. Possible reasons for the mismatch of platelet gene variants, their effects on platelet functional parameters, and atherothrombotic events are manifold and might be explained by several reasons: (1) platelet and CAD/MI loci have not yet been identified and/or are too rare to be detected by recent genotyping approaches such as GWAS, (2) markers of platelet function that have been previously used are weak surrogates and do not sufficiently mirror the more complex platelet activation signaling pathways in vivo and in case of thrombotic events, and (3) highly specific platelet-associated outcome definitions are required (e.g., platelet-dependent thrombus formation) rather than atherogenic events in general that are a consequence of underlying diverse pathophysiological processes including chronic inflammation.

Table 17.1 Genetic variation and impact of diverse markers on platelet biology identified by genome-wide association studies and functional genomic approaches with focus on the platelet function phenotype and cardiovascular outcome (modified according to [6] + [7])

Gene	Genetic variation	Functional phenotypes							References	Clinical outcome	References
		PSADP	FADP	PCRPXL	FCRPXL	PA	MPV	Plt Count		Cardiovascular events/thrombotic events	
GPVI	rs1613662	+		+	+				[8]	+	[9]
	rs1671152					Coll ↑			[5]	+	[10]
PEAR1	rs11264579			+	+				[8, 11]	?	
	rs3737224	+	+						[8]	?	
	rs12566888					ADP↓, EPI ↓			[5]	?	
JAK2	rs10429491	+	+						[8]	?	
P2RY12	rs1472122	+	+	+						?	
RAF1	rs3729931		+	+	+					?	
FCER1G	rs3557			+	+					?	
GNAZ	rs3788337	+								?	
VAV3	rs17229705	+	+	+						?	
CD36	rs1049654			+	+					?	
MAP2K2	rs350916		+							?	
MAPK14	rs851007	+	+	+						?	
MAP2K4	rs41307923			+	+					′?	
ITGA2	rs41305896	+	+							?	
AKT2	rs41275750			+	+					?	
ITPR1	rs17786144	+	+							?	

(continued)

Table 17.1 (continued)

Gene	Genetic variation	Functional phenotypes							References	Clinical outcome	References
		PSADP	FADP	PCRPXL	FCRPXL	PA	MPV	Plt Count		Cardiovascular events/thrombotic events	
ADRA2A	rs4311994					EPI ↓			[5]	?	
MRVI1	rs7940646					ADP ↑, EPI↑				?	
SHH	rs2363910									?	
PIK3CG	rs342286					EPI ↓				?	
JMJD1C	rs10761741					EPI↑				?	
DNM3	rs10914144						+		[12, 13]	?	
TPM1	rs11071720						+			?	
BET1L	rs11602954						+			?	
ARHGEF3	rs12485738						+			?	
TMCC2	rs1668873						+			?	
TAOK1	rs2138852						+			?	
JMJD1C	rs2393967						+			?	
PIK3CG	rs342293						+			?	
SIRPA	rs6136489						+			?	
EHD3	rs647316						+			?	
WDR66	rs7961894						+			?	
CD226	rs893001						+			?	
ATXN2	rs11065987							+		?	
SH2B3	rs3184504							+		?	
PTPN11	rs11066301							+		?	
BAK1	rs210135							+		?	
AK3	rs385893							+		?	

PSADP adenosine diphosphate (ADP)-induced P-selectin expression, *FADP* ADP-induced fibrinogen binding, *PCRPXL* CRP-XL-induced P-selectin expression, *FCRPXL* CRP-XL-induced fibrinogen binding, *PA* agonist-induced platelet aggregation, *SNP* single nucleotide polymorphism, *MPV* mean platelet volume, *Plt count* platelet count

17.4 MicroRNA

Human platelets express around 300 miRNAs. Besides their potential as biomarkers, miRNAs have a role as regulators of interindividual variation in platelet function via their effects on mRNA and protein levels. For some miRNAs effects on platelet function have been described [20–22]. Analysis of platelet microRNA-mRNA coexpression profiles and their effects on platelet phenotype revealed identification of genes previously not being described to have a function in platelets. Network analyses including integrated system biological approaches are needed to better understand the complex regulatory mechanisms on the platelet miRNA/mRNA level including demographic information like age and gender [23].

17.5 Platelet Pharmacogenetics/Pharmacogenomics

Pharmacogenetics or pharmacogenomics deal with underlying genetic mechanisms to explain interindividual variability in drug response including adverse drug reactions (ADR) [24, 25]. In recent years both terms pharmacogenetics and pharmacogenomics are used synonymously covering the concept that drug response is a more complex scenario. Based on novel omics technologies and big data resources, the consideration of multiple genes (germline and somatic genome in case of cancer studies) and gene-gene and gene-environment interactions offers a high potential to identify better predictive biomarkers [26].

Platelet pharmacogenomics have become of particular interest in the emerging era of personalized medicine. Antiplatelet agents are the most frequently prescribed drugs worldwide and are being used for primary and secondary prevention in a broad spectrum of cerebral, peripheral, and cardiovascular disease and thrombotic disorders. Despite the development and utilization of newer antiplatelet agents in the recent decade, there is still a substantial proportion of patients who develop atherothrombotic and platelet-associated events ranging up to 10% per year depending on individual risk profiles. Thus, platelet pharmacogenomics is an attractive and promising research area to identify patients who might benefit from more specific or intensified antiplatelet strategies by genetic prediction of drug response and/or ADR (e.g., bleeding events). Antiplatelet drugs currently used in clinical practice include cyclooxygenase (COX) inhibitors, $P2Y_{12}$ ADP receptor antagonists, glycoprotein IIb-IIIa inhibitors, and protease-activated receptor 1 (PAR1) antagonists. Table 17.2 provides a short overview about genetic variants in candidate genes associated with clinical and/or pharmacological endpoints in patients treated with the $P2Y_{12}$ ADP receptor antagonists clopidogrel, prasugrel, and ticagrelor. Data are derived from large-scale candidate association studies, GWAS, and meta-analyses. Of note, the impact on interindividual variability of drug response depends strongly on the specific in vitro and/or in vivo (patient) setting, i.e., phenotypically characteristics are critical determinants in clinical trials. Therefore replication studies and prospectively designed pharmacogenomic trials are warranted before pharmacogenomic information could be implemented into

Table 17.2 Variation in selected candidate genes and consequences on pharmacokinetic and/or pharmacodynamic parameters related to antiplatelet drug therapy (modified according to [7])

Antiplatelet drug	Gene	Genetic variant	Effects on pharmacokinetic and/or pharmacodynamics parameters	References
Clopidogrel	*CYP2C19*	rs4244285 (*CYP2C19*2*) rs4986893 (*CYP2C19*3*)	Reduced active metabolite levels, reduced inhibition of platelet aggregation, and worse cardiovascular outcome in ACS/PCI patients	Case-control studies: [27–34] Meta-analyses: [35–44]
		rs12248560 (*CYP2C19*17*)	Enhanced platelet aggregation inhibition, increased bleeding rates, controversial data on outcome and effect sizes	[40, 41, 45–47]
	PON1	rs662	Missing replication of effects on clopidogrel metabolite levels and clopidogrel-dependent aggregation inhibition by multiple trials	[48]
	CES1	rs71647871	Association with increased systemic exposure of clopidogrel and its active metabolite, enhanced inhibition of platelet aggregation	[49, 50]
	ABCB1/ *MDR1*	rs1045642 (3435C>T) rs1128503 (1236C>T) rs2032582 (2677G>T)	Patients (ACS/PCI) are at increased risk for cardiovascular events; inconsistencies of findings	[28, 51] *Meta-Analyses:* [52, 53]
Prasugrel	*CYP2C19*	rs4244285 (*CYP2C19*2*) rs12248560 (*CYP2C19*17*)	No association with prasugrel response in multiple trials Significant impact on platelet reactivity index VASP in ACS/PCI patients	[51, 54–57]
	PEAR1	rs3737224 rs822442 rs1214331 rs12566888	Extended platelet inhibition in Chinese healthy volunteers	[58]
Ticagrelor	*SLCO1B1*[a]	rs113681054	Association with higher drug levels of ticagrelor and its active metabolite No association with primary clinical outcome parameters (i.e., cardiovascular death, myocardial infarction, stroke) in ACS patients	[59]
	UGT2B7	rs61361928		
	CYP3A4	rs62471956 rs56324128		

[a]In close LD with the functional *SLCO1B1* rs4149056 variant (c.521T>C; *SLCO1B1*5*) associated with statin-related myopathy [60, 61]

clinical practice. Currently there are only limited examples based on robust data from in vitro/in vivo studies that genetic variation may significantly alter drug efficacy of antiplatelet therapy.

Clopidogrel a thienopyridine prodrug is extensively metabolized by several cytochrome P450 enzymes (CYP2C19, CYP3A4, CYP2B6, CYP1A2, CYP2C9) for which functional consequences related to genetic variation are very well established [62]. The active *cis* 5-thiol clopidogrel metabolite irreversibly binds to $P2Y_{12}$ receptor, thereby potently inhibiting platelet aggregation for the life span of platelets for about 10 days. Several in part prospective clinical studies provide abundant evidence that the bioactivation of clopidogrel via CYP2C19 is strongly influenced by loss-of-function variants (e.g., *CYP2C19*2*) resulting in lower plasma levels of the active metabolite and subsequent pharmacodynamic consequences for selected cardiovascular risk groups (Table 17.2). In particular the *CYP2C19*2* (c.681G>A; rs4244285) allele majorly interferes with clopidogrel-dependent platelet inhibition by altering the mRNA reading frame and resulting in a truncated nonfunctional protein. However rare variants (e.g., *CYP2C19*3* to **8*) beyond *CYP2C19*2* depending on the patient's ethnic background additionally result in missing or decreased CYP2C19 enzyme activity with consequences on clopidogrel response. Of note the opposite effect of an increased catalytic activity of the CYP2C19 enzyme in carriers of the gain-of-function *CYP2C19*17* allele (c.-806C>T; rs12248560) leading to an enhanced transcription was recently reported [63]. The *CYP2C19*17* allele has been associated with clopidogrel extensive metabolism, thus resulting in enhanced clopidogrel-associated platelet inhibition and reduced cardiovascular events but higher bleeding risk [64]. The effect related to *CYP2C19*17* and clopidogrel use, however, is inconsistent between different studies. This may be explained by the limitation that mostly only an isolated analysis of the *CYP2C19*17* allele status without simultaneous consideration of the loss-of-function variant *CYP2C19*2* in each patient has been performed. This assumption is supported by recent data suggesting that both alleles are genetically linked through a high LD, the nonrandom association between various genetic variants localized on the same chromosome in close proximity. Thus the adjustment of a positive association of the *CYP2C19*17* allele with clinical outcome for the individual *CYP2C19*2* genotype seems to be crucial to avoid misinterpretation of *CYP2C19* pharmacogenetics and clopidogrel response [45].

In 2013 the international Clinical Pharmacogenetics Implementation Consortium (CPIC) published an update version summarizing currently existing evidence on *CYP2C19* genotype-directed clopidogrel treatment based on a systematic literature and expert review [65]. Several prospective studies as well as large meta-analyses support the evidence for an *indication-specific CYP2C19* pharmacogenetic approach of clopidogrel treatment demonstrating that predominantly ACS patients receiving percutaneous coronary intervention (PCI) who are *CYP2C19* poor metabolizers (i.e., carrier of two loss-of-function alleles) are at a significantly increased risk for major (recurrent) adverse cardiovascular events (e.g., stent thrombosis). In contrast *CYP2C19* pharmacogenetics appears to be of limited impact regarding to clopidogrel-treated patients without PCI or without coronary disease or non-cardiovascular outcomes. Notably, the frequency distribution of subjects carrying at least one *CYP2C19*2* allele is up to 30% in Caucasians and Africans and approximately 60% in Asians [62]. Combining genotyping results for the loss-of-function alleles *CYP2C19*2* and **3* and the gain-of-function *CYP2C19*17* allele, the CPIC guideline suggests four different CYP2C19 phenotypes that should be considered in ACS/PCI patients to select appropriate prescribing of antiplatelet agents: poor metabolizer (**2/*2*, **2/*3*, **3/*3*), intermediate metabolizer (**1/*2*, **1/*3*, **2/*17*), extensive metabolizer (**1/*1*), and ultrarapid metabolizer (**1/*17*, **17/*17*) [65]. Alternate treatment options for ACS/PCI patients with a CYP2C19 poor metabolizer phenotype have been recommended such as prasugrel or ticagrelor. Based on these facts, the FDA included already a boxed warning in the clopidogrel label indicating that a minor benefit of clopidogrel response in ACS/PCI patients with a CYP2C19 poor metabolizer phenotype could be expected.

Beyond *CYP2C19* genetics other nongenetic confounders such as diabetes mellitus, body mass index, and co-medication (e.g., selected proton pump inhibitors or tricyclic antidepressants) contributing to residual platelet aggregation and/or clopidogrel nonresponse may have clinical relevance [63]. However so far a well-established algorithm like the warfarin-dosing algorithm [66] covering clopidogrel-relevant genetic and nongenetic clinical factors [67] is missing to guide a personalized treatment approach. Nevertheless a first proof-of-concept (POC) trial elucidating prospectively point-of-care genetic testing for *CYP2C19*2* in PCI patients and subsequent genotype-guided antiplatelet therapy demonstrated that none of the patients who carried the *CYP2C19*2* allele receiving prasugrel treatment showed high on-treatment platelet reactivity compared to 30% of control patients carrying the *CYP2C19*2* allele and were treated with clopidogrel. This trial provided first evidence that POC genetic testing of *CYP2C19*2* can be implemented effectively into clinical practice [68]. These data has been confirmed by a smaller study investigating *CYP2C19*2* POC genotyping in patients with ACS/PCI and dual antiplatelet therapy in the emergency setting. Platelet inhibition in patients treated with prasugrel and ticagrelor compared to clopidogrel was more pronounced in *CYP2C19*2* carriers demonstrating again that POC genotyping might be helpful for the identification of clopidogrel poor responder [69]. It is remarkable to note that at the University of Maryland Medical Center and the Baltimore Veterans Administration Medical Center, a *Personalized Anti-Platelet Pharmacogenetics Program* has been already implemented in clinical routine since 2013. To ensure a turnaround time of approximately 5 h after blood sampling, a clinical decision support system is used including interpretation of the genetic test result and respective prescribing recommendations [70].

Independent from *CYP2C19* other candidate genes have been considered to alter clopidogrel pharmacokinetics and pharmacodynamics. While genetic variation of the paroxonase-1 enzyme (*PON-1*, rs662 (192Q>R)), which is involved in clopidogrel activation, could not be replicated in multiple clinical trials [48] indicating no clinical relevance for clopidogrel responsiveness, the enzyme carboxylesterase 1 (CES1) appears to be of particular interest. CES1 hydrolyses both 2-oxo-clopidogrel, the methyl ester of clopidogrel, and the active metabolite

5-thiol clopidogrel resulting in carboxylic acid derivatives which are pharmacologically inactive. There is clear evidence that the *CES1* variant c.428G>A (rs71647871) strongly reducing the CES1 enzyme activity in vitro and in vivo [71] is significantly associated with an increased systemic exposure of clopidogrel and its active metabolite. This indicates an opposite direction compared with the effects of *CYP2C19* loss-of-function variants. Moreover the *CES1* c.428G>A variant was associated with an enhanced inhibition of platelet aggregation in independent studies [49, 50]. Currently, there is no evidence that *CES1* pharmacogenetics may impact on cardiovascular events in clopidogrel-treated ACS/PCI patients, but larger trials are required since the allele frequency of the *CES1* c.428G>A variant is rather low (about 2 to 4% in Caucasians, 4% in African Americans, 0% in Asians).

Although currently the impact of heritability on the interindividual variability of response to the newer $P2Y_{12}$ receptor inhibitors prasugrel or ticagrelor has not been investigated in larger clinical trials, recent gene association studies suggest some promising candidates (Table 17.2). For instance novel findings regarding the probe-drug prasugrel suggest that two regions of the *PEAR1* gene locus may explain extended platelet inhibition in Han Chinese healthy volunteers [58]. Regarding the active antiplatelet agent ticagrelor, recently a GWAS approach of the clinical PLATO trial considering a discovery ($n = 1812$) and replication cohort ($n = 1941$) provides the first evidence that genetic variation of the membrane transporter SLCO1B1 as well as of the phase I and II drug metabolizing enzymes CYP3A4 and UGT2B7, respectively, is significantly associated with higher drug levels of the parent drug ticagrelor and its active metabolite. However no association with primary clinical outcome parameters (i.e., cardiovascular death, myocardial infarction, stroke) in ticagrelor-treated patients has been observed [59].

17.6 Summary and Perspectives

Pharmacogenomics of antiplatelet drugs appears to be a promising area to elucidate underlying mechanisms for the better understanding of the so far unexplained interindividual variability of drug response. The challenge of such research activities depends on several aspects. Very well-characterized clinical phenotypes of study participants as well as comprehensive and standardized follow-up approaches on drug response and/or on the development of ADR are required. Confounding factors like disease status, co-medication, and environmental aspects are critical and must be considered carefully. Particularly better definition and implementation of platelet-specific endpoints and in vivo models to quantify platelet-dependent thrombus formation and hemorrhage are needed. In this context novel high-throughput technologies to systematically assess the patient's phenotype in a standardized manner are mandatory like computer-based, direct interview methods [72].

GWAS approaches in platelet genomics have opened a door to assess an individual risk and provide biomarkers for a personalized antiplatelet therapy. However the majority of previously identified SNPs in platelet genetic association studies are noncoding, and their functional relevance is so far poorly understood. The effect sizes of any given platelet SNPs are mostly low, thus making it hard to adequately determine the risk for disease and major cardiovascular-/platelet-driven endpoints from the results of GWAS. Furthermore, rare variants have escaped current GWAS approaches due to technical limitations of previous generation genotyping arrays that have been designed to capture only common variants. Novel genotyping platforms like next-generation sequencing approaches are now capable to detect rare variants with larger effect sizes [24]. However, associations between variants with low minor allele frequencies and a moderate effect size will be still challenging and require global meta-analysis approaches performed in large-scale pooled, well-phenotyped, and harmonized consortia cohorts. In this context an initiative of the international clopidogrel consortium (http://www.pharmgkb.org/page/icpc) has been started to establish large patient cohorts including different ethnic backgrounds to identify novel genetic variants with rare allele frequencies for better prediction of clopidogrel response. In addition to SNPs not only located in coding but also in regulatory noncoding gene regions, other types of sequence variation including inversions, aberrations, copy number variants, etc. as well as epigenetic posttranslational modifications like DNA methylation and/or miRNAs have been proposed to be risk-/disease- and drug response-modifying factors. These approaches and the consideration of additional omics technologies (e.g., pharmacometabolomics) will likely enable us to link the platelet genome with platelet biology and atherothrombotic processes in the context of cardiovascular disease and, by this, to personalize platelet-targeted therapy in the future.

Compliance with Ethical Standards

Conflict of Interest: Tobias Geisler, Elke Schaeffeler, and Matthias Schwab declares that they have no conflict of interest.

Ethical Approval: This article contains studies with human participants or animals performed by the authors with ethical approval.

References

1. Davì G, Patrono C. Platelet activation and atherothrombosis. N Engl J Med. 2007 Dec 13;357 (24):2482–94.
2. Müller KA, Chatterjee M, Rath D, Geisler T. Platelets, inflammation and anti-inflammatory effects of antiplatelet drugs in ACS and CAD. Thromb Haemost. 2015;114(3):498–518.
3. O'Donnell CJ, Larson MG, Feng D, Sutherland PA, Lindpaintner K, Myers RH, D'Agostino RA, Levy D, Tofler GH. Framingham heart study genetic and environmental contributions to platelet aggregation: the Framingham heart study. Circulation. 2001;103(25):3051–6.

4. Bray PF, Mathias RA, Faraday N, Yanek LR, Fallin MD, Herrera-Galeano JE, Wilson AF, Becker LC, Becker DM. Heritability of platelet function in families with premature coronary artery disease. J Thromb Haemost. 2007;5(8):1617–23.
5. Johnson AD, Yanek LR, Chen M-H, et al. Genome-wide meta-analyses identifies seven loci associated with platelet aggregation in response to agonists. Nat Genet. 2010;42:608–13.
6. Kunicki TJ, Nugent DJ. The genetics of normal platelet reactivity. Blood. 2010;116 (15):2627–34.
7. Geisler T, Schaeffeler E, Gawaz M, Schwab M. Genetic variation of platelet function and pharmacology: an update of current knowledge. Thromb Haemost. 2013;110(5):876–87.
8. Jones CI, Bray S, Garner SF, Stephens J, de Bono B, Angenent WG, Bentley D, Burns P, Coffey A, Deloukas P, Earthrowl M, Farndale RW, Hoylaerts MF, Koch K, Rankin A, Rice CM, Rogers J, Samani NJ, Steward M, Walker A, Watkins NA, Akkerman JW, Dudbridge F, Goodall AH, Ouwehand WH, Bloodomics Consortium. A functional genomics approach reveals novel quantitative trait loci associated with platelet signaling pathways. Blood. 2009;114(7):1405–16.
9. Bezemer ID, Bare LA, Doggen CJ, Arellano AR, Tong C, Rowland CM, Catanese J, Young BA, Reitsma PH, Devlin JJ, Rosendaal FR. Gene variants associated with deep vein thrombosis. JAMA. 2008 Mar 19;299(11):1306–14. https://doi.org/10.1001/jama.299.11.1306
10. Snoep JD, Gaussem P, Eikenboom JC, Emmerich J, Zwaginga JJ, Holmes CE, Vos HL, de Groot PG, Herrington DM, Bray PF, Rosendaal FR, van der Bom JG. The minor allele of GP6 T13254C is associated with decreased platelet activation and a reduced risk of recurrent cardiovascular events and mortality: results from the SMILE-Platelets project. J Thromb Haemost. 2010;8(11):2377–84.
11. Faraday N, Yanek LR, Yang XP, Mathias R, Herrera-Galeano JE, Suktitipat B, Qayyum R, Johnson AD, Chen MH, Tofler GH, Ruczinski I, Friedman AD, Gylfason A, Thorsteinsdottir U, Bray PF, O'Donnell CJ, Becker DM, Becker LC. Identification of a specific intronic PEAR1 gene variant associated with greater platelet aggregability and protein expression. Blood. 2011 Sep 22;118(12):3367–75. https://doi.org/10.1182/blood-2010-11-320788
12. Meisinger C, Prokisch H, Gieger C, et al. A genome-wide association study identifies three loci associated with mean platelet volume. Am J Hum Genet. 2009;84(1):66–71.
13. Soranzo N, Spector TD, Mangino M, et al. A genome-wide meta-analysis identifies 22 loci associated with eight hematological parameters in the HaemGen consortium. Nat Genet. 2009;41(11):1182–90.
14. Sajid M, Vijayan KV, Souza S, Bray PF. PlA polymorphism of integrin beta 3 differentially modulates cellular migration on extra-cellular matrix proteins. Arterioscler Thromb Vasc Biol. 2002;22:1984–9.
15. Michelson AD, Furman MI, Goldschmidt-Clermont P, Mascelli MA, Hendrix C, Coleman L, Hamlington J, Barnard MR, Kickler T, Christie DJ, Kundu S, Bray PF. Platelet GPIIIa Pl (A) polymorphisms display different sensitivities to agonists. Circulation. 2000;101:1013–8.
16. Weiss EJ, Bray PF, Tayback M, Schulman SP, Kickler TS, Becker LC, Weiss JL, Gerstenblith G, Goldschmidt-Clermont PJ. A polymorphism of a platelet glycoprotein receptor as an inherited risk factor for coronary thrombosis. N Engl J Med. 1996;334:1090–4.
17. Gieger C, Radhakrishnan A, Cvejic A, Tang W, Porcu E, Pistis G, Serbanovic-Canic J, Elling U, Goodall AH, Labrune Y, Lopez LM, Mägi R, Meacham S, Okada Y, Pirastu N, Sorice R, Teumer A, Voss K, Zhang W, Ramirez-Solis R, Bis JC, Ellinghaus D, Gögele M, Hottenga JJ, Langenberg C, Kovacs P, O'Reilly PF, Shin SY, Esko T, Hartiala J, Kanoni S, Murgia F, Parsa A, Stephens J, van der Harst P, Ellen van der Schoot C, Allayee H, Attwood A, Balkau B, Bastardot F, Basu S, Baumeister SE, Biino G, Bomba L, Bonnefond A, Cambien F, Chambers JC, Cucca F, D'Adamo P, Davies G, de Boer RA, de Geus EJ, Döring A, Elliott P, Erdmann J, Evans DM, Falchi M, Feng W, Folsom AR, Frazer IH, Gibson QD, Glazer NL, Hammond C, Hartikainen AL, Heckbert SR, Hengstenberg C, Hersch M, Illig T, Loos RJ, Jolley J, Khaw KT, Kühnel B, Kyrtsonis MC, Lagou V, Lloyd-Jones H, Lumley T,

Mangino M, Maschio A, Mateo Leach I, McKnight B, Memari Y, Mitchell BD, Montgomery GW, Nakamura Y, Nauck M, Navis G, Nöthlings U, Nolte IM, Porteous DJ, Pouta A, Pramstaller PP, Pullat J, Ring SM, Rotter JI, Ruggiero D, Ruokonen A, Sala C, Samani NJ, Sambrook J, Schlessinger D, Schreiber S, Schunkert H, Scott J, Smith NL, Snieder H, Starr JM, Stumvoll M, Takahashi A, Tang WH, Taylor K, Tenesa A, Lay Thein S, Tönjes A, Uda M, Ulivi S, van Veldhuisen DJ, Visscher PM, Völker U, Wichmann HE, Wiggins KL, Willemsen G, Yang TP, Hua Zhao J, Zitting P, Bradley JR, Dedoussis GV, Gasparini P, Hazen SL, Metspalu A, Pirastu M, Shuldiner AR, Joost van Pelt L, Zwaginga JJ, Boomsma DI, Deary IJ, Franke A, Froguel P, Ganesh SK, Jarvelin MR, Martin NG, Meisinger C, Psaty BM, Spector TD, Wareham NJ, Akkerman JW, Ciullo M, Deloukas P, Greinacher A, Jupe S, Kamatani N, Khadake J, Kooner JS, Penninger J, Prokopenko I, Stemple D, Toniolo D, Wernisch L, Sanna S, Hicks AA, Rendon A, Ferreira MA, Ouwehand WH, Soranzo N. New gene functions in megakaryopoiesis and platelet formation. Nature. 2011;480(7376):201–8. https://doi.org/10.1038/nature10659

18. Coronary Artery Disease (C4D) Genetics Consortium. A genome-wide association study in Europeans and South Asians identifies five new loci for coronary artery disease. Nat Genet. 2011;43(4):339–44.
19. Schunkert H, König IR, Kathiresan S, Reilly MP, Assimes TL, Holm H, Preuss M, Stewart AF, Barbalic M, Gieger C, Absher D, Aherrahrou Z, Allayee H, Altshuler D, Anand SS, Andersen K, Anderson JL, Ardissino D, Ball SG, Balmforth AJ, Barnes TA, Becker DM, Becker LC, Berger K, Bis JC, Boekholdt SM, Boerwinkle E, Braund PS, Brown MJ, Burnett MS, Buysschaert I, Cardiogenics, Carlquist JF, Chen L, Cichon S, Codd V, Davies RW, Dedoussis G, Dehghan A, Demissie S, Devaney JM, Diemert P, Do R, Doering A, Eifert S, Mokhtari NE, Ellis SG, Elosua R, Engert JC, Epstein SE, de Faire U, Fischer M, Folsom AR, Freyer J, Gigante B, Girelli D, Gretarsdottir S, Gudnason V, Gulcher JR, Halperin E, Hammond N, Hazen SL, Hofman A, Horne BD, Illig T, Iribarren C, Jones GT, Jukema JW, Kaiser MA, Kaplan LM, Kastelein JJ, Khaw KT, Knowles JW, Kolovou G, Kong A, Laaksonen R, Lambrechts D, Leander K, Lettre G, Li M, Lieb W, Loley C, Lotery AJ, Mannucci PM, Maouche S, Martinelli N, PP MK, Meisinger C, Meitinger T, Melander O, Merlini PA, Mooser V, Morgan T, Mühleisen TW, Muhlestein JB, Münzel T, Musunuru K, Nahrstaedt J, Nelson CP, Nöthen MM, Olivieri O, Patel RS, Patterson CC, Peters A, Peyvandi F, Qu L, Quyyumi AA, Rader DJ, Rallidis LS, Rice C, Rosendaal FR, Rubin D, Salomaa V, Sampietro ML, Sandhu MS, Schadt E, Schäfer A, Schillert A, Schreiber S, Schrezenmeir J, Schwartz SM, Siscovick DS, Sivananthan M, Sivapalaratnam S, Smith A, Smith TB, Snoep JD, Soranzo N, Spertus JA, Stark K, Stirrups K, Stoll M, Tang WH, Tennstedt S, Thorgeirsson G, Thorleifsson G, Tomaszewski M, Uitterlinden AG, van Rij AM, Voight BF, Wareham NJ, Wells GA, Wichmann HE, Wild PS, Willenborg C, Witteman JC, Wright BJ, Ye S, Zeller T, Ziegler A, Cambien F, Goodall AH, Cupples LA, Quertermous T, März W, Hengstenberg C, Blankenberg S, Ouwehand WH, Hall AS, Deloukas P, Thompson JR, Stefansson K, Roberts R, Thorsteinsdottir U, O'Donnell CJ, McPherson R, Erdmann J, CARDIoGRAM Consortium, Samani NJ. Large-scale association analysis identifies 13 new susceptibility loci for coronary artery disease. Nat Genet. 2011;43 (4):333–8. https://doi.org/10.1038/ng.784
20. Sl N, Shaw C, Kong X, Kondkar AA, Edelstein LC, Ma L, Chen J, McKnight GS, López JA, Yang L, Jin Y, Bray MS, Leal SM, Dong JF, Bray PF. Platelet microRNA-mRNA coexpression profiles correlate with platelet reactivity. Blood. 2011;117(19):5189–97.
21. Kaudewitz D, Skroblin P, Bender LH, Barwari T, Willeit P, Pechlaner R, Sunderland NP, Willeit K, Morton A, Armstrong PC, Chan MV, Lu R, Yin X, Gracio F, Dudek K, Langley S, Zampetaki A, de Rinaldis E, Ye S, Warner TD, Saxena A, Kiechl S, Storey R, Mayr M. Association of microRNAs and YRNAs with platelet function. Circ Res. 2015 Dec 8. https://doi.org/10.1161/CIRCRESAHA.114.305663
22. McManus DD, Freedman JE. MicroRNAs in platelet function and cardiovascular disease. Nat Rev Cardiol. 2015;12(12):711–7.

23. Simon LM, Edelstein LC, Nagalla S, Woodley AB, Chen ES, Kong X, Ma L, Fortina P, Kunapuli S, Holinstat M, McKenzie SE, Dong JF, Shaw CA, Bray PF. Human platelet microRNA-mRNA networks associated with age and gender revealed by integrated plateletomics. Blood. 2014;123(16):e37–45.
24. Meyer UA, Zanger UM, Schwab M. Omics and drug response. Annu Rev Pharmacol Toxicol. 2013;53:475–502.
25. Relling MV, Evans WE. Pharmacogenomics in the clinic. Nature. 2015 Oct 15;526 (7573):343–50.
26. Schwab M, Schaeffeler E. Pharmacogenomics: a key component of personalized therapy. Genome Med. 2012;4(11):93.
27. Geisler T, Schaeffeler E, Dippon J, Winter S, Buse V, Bischofs C, Zuern C, Moerike K, Gawaz M, Schwab M. CYP2C19 and nongenetic factors predict poor responsiveness to clopidogrel loading dose after coronary stent implantation. Pharmacogenomics. 2008 Sep;9 (9):1251–9.
28. Simon T, Verstuyft C, Mary-Krause M, et al. French Registry of Acute ST-Elevation and Non-ST-Elevation Myocardial Infarction (FAST-MI) Investigators. Genetic determinants of response to clopidogrel and cardiovascular events. N Engl J Med. 2009;360:363–75.
29. Mega JL, Close SL, Wiviott SD, et al. Cytochrome p-450 genetic polymorphisms and the response to clopidogrel. N Engl J Med. 2009;360:354–62.
30. Collet JP, Hulot JS, Pena A, Villard E, Esteve JB, Silvain J, Payot L, Brugier D, Cayla G, Beygui F, Bensimon G, Funck-Brentano C, Montalescot G. Cytochrome P450 2C19 polymorphism in young patients treated with clopidogrel after myocardial infarction: a cohort study. Lancet. 2009;373(9660):309–17.
31. Sibbing D, Stegherr J, Latz W, Koch W, Mehilli J, Dörrler K, Morath T, Schömig A, Kastrati A, von Beckerath N. Cytochrome P450 2C19 loss-of-function polymorphism and stent thrombosis following percutaneous coronary intervention. Eur Heart J. 2009;30 (8):916–22.
32. Giusti B, Gori AM, Marcucci R, Saracini C, Sestini I, Paniccia R, Buonamici P, Antoniucci D, Abbate R, Gensini GF. Relation of cytochrome P450 2C19 loss-of-function polymorphism to occurrence of drug-eluting coronary stent thrombosis. Am J Cardiol. 2009;103(6):806–11.
33. Shuldiner AR, O'Connell JR, Bliden KP, Gandhi A, Ryan K, Horenstein RB, Damcott CM, Pakyz R, Tantry US, Gibson Q, Pollin TI, Post W, Parsa A, Mitchell BD, Faraday N, Herzog W, Gurbel PA. Association of cytochrome P450 2C19 genotype with the antiplatelet effect and clinical efficacy of clopidogrel therapy. JAMA. 2009;302(8):849–57.
34. Cayla G, Hulot JS, O'Connor SA, Pathak A, Scott SA, Gruel Y, Silvain J, Vignalou JB, Huerre Y, de la Briolle A, Allanic F, Beygui F, Barthélémy O, Montalescot G, Collet JP. Clinical, angiographic, and genetic factors associated with early coronary stent thrombosis. JAMA. 2011;306(16):1765–74.
35. Hulot JS, Collet JP, Silvain J, Pena A, Bellemain-Appaix A, Barthélémy O, Cayla G, Beygui F, Montalescot G. Cardiovascular risk in clopidogrel-treated patients according to cytochrome P450 2C19*2 loss-of-function allele or proton pump inhibitor coadministration: a systematic meta-analysis. J Am Coll Cardiol. 2010;56(2):134–43.
36. Mega JL, Simon T, Collet JP, Anderson JL, Antman EM, Bliden K, Cannon CP, Danchin N, Giusti B, Gurbel P, Horne BD, Hulot JS, Kastrati A, Montalescot G, Neumann FJ, Shen L, Sibbing D, Steg PG, Trenk D, Wiviott SD, Sabatine MS. Reduced-function CYP2C19 genotype and risk of adverse clinical outcomes among patients treated with clopidogrel predominantly for PCI: a meta-analysis. JAMA. 2010 Oct 27;304(16):1821–30.
37. Sofi F, Giusti B, Marcucci R, Gori AM, Abbate R, Gensini GF. Cytochrome P450 2C19*2 polymorphism and cardiovascular recurrences in patients taking clopidogrel: a meta-analysis. Pharmacogenomics J. 2011 Jun;11(3):199–206.
38. Holmes MV, Perel P, Shah T, Hingorani AD, Casas JP. CYP2C19 genotype, clopidogrel metabolism, platelet function, and cardiovascular events: a systematic review and meta-analysis. JAMA. 2011;306(24):2704–14.

39. Jin B, Ni HC, Shen W, Li J, Shi HM, Li Y. Cytochrome P450 2C19 polymorphism is associated with poor clinical outcomes in coronary artery disease patients treated with clopidogrel. Mol Biol Rep. 2011;38(3):1697–702.
40. Bauer T, Bouman HJ, van Werkum JW, Ford NF, ten Berg JM, Taubert D. Impact of CYP2C19 variant genotypes on clinical efficacy of antiplatelet treatment with clopidogrel: systematic review and meta-analysis. BMJ. 2011;343:d4588.
41. Zabalza M, Subirana I, Sala J, Lluis-Ganella C, Lucas G, Tomás M, Masiá R, Marrugat J, Brugada R, Elosua R. Meta-analyses of the association between cytochrome CYP2C19 loss- and gain-of-function polymorphisms and cardiovascular outcomes in patients with coronary artery disease treated with clopidogrel. Heart. 2012;98(2):100–8.
42. Jang JS, Cho KI, Jin HY, Seo JS, Yang TH, Kim DK, Kim DS, Seol SH, Kim DI, Kim BH, Park YH, Je HG, Jeong YH, Lee SW. Meta-analysis of cytochrome P450 2C19 polymorphism and risk of adverse clinical outcomes among coronary artery disease patients of different ethnic groups treated with clopidogrel. Am J Cardiol. 2012;110(4):502–8.
43. Yamaguchi Y, Abe T, Sato Y, Matsubara Y, Moriki T, Murata M. Effects of VerifyNow P2Y12 test and CYP2C19*2 testing on clinical outcomes of patients with cardiovascular disease: a systematic review and meta-analysis. Platelets. 2013;24(5):352–61.
44. Singh M, Shah T, Adigopula S, Molnar J, Ahmed A, Khosla S, Arora R. CYP2C19*2/ABCB1-C3435T polymorphism and risk of cardiovascular events in coronary artery disease patients on clopidogrel: is clinical testing helpful? Indian Heart J. 2012;64(4):341–52.
45. Lewis JP, Stephens SH, Horenstein RB, O'Connell JR, Ryan K, Peer CJ, Figg WD, Spencer SD, Pacanowski MA, Mitchell BD, Shuldiner AR. The CYP2C19*17 variant is not independently associated with clopidogrel response. J Thromb Haemost. 2013;11(9):1640–6.
46. Frére C, Cuisset T, Gaborit B, Alessi MC, Hulot JS. The CYP2C19*17 allele is associated with better platelet response to clopidogrel in patients admitted for non-ST acute coronary syndrome. J Thromb Haemost. 2009;7(8):1409–11.
47. Li Y, Tang HL, Hu YF, Xie HG. The gain-of-function variant allele CYP2C19*17: a double-edged sword between thrombosis and bleeding in clopidogrel-treated patients. J Thromb Haemost. 2012;10(2):199–206.
48. Reny JL, Combescure C, Daali Y, Fontana P, PON1 Meta-Analysis Group. Influence of the paraoxonase-1 Q192R genetic variant on clopidogrel responsiveness and recurrent cardiovascular events: a systematic review and meta-analysis. J Thromb Haemost. 2012 July;10(7):1242–51.
49. Tarkiainen EK, Holmberg MT, Tornio A, Neuvonen M, Neuvonen PJ, Backman JT, Niemi M. Carboxylesterase 1 c.428G>A single nucleotide variation increases the antiplatelet effects of clopidogrel by reducing its hydrolysis in humans. Clin Pharmacol Ther. 2015;97(6):650–8.
50. Lewis JP, Horenstein RB, Ryan K, O'Connell JR, Gibson Q, Mitchell BD, Tanner K, Chai S, Bliden KP, Tantry US, Peer CJ, Figg WD, Spencer SD, Pacanowski MA, Gurbel PA, Shuldiner AR. The functional G143E variant of carboxylesterase 1 is associated with increased clopidogrel active metabolite levels and greater clopidogrel response. Pharmacogenet Genomics. 2013;23(1):1–8.
51. Mega JL, Close SL, Wiviott SD, Shen L, Walker JR, Simon T, Antman EM, Braunwald E, Sabatine MS. Genetic variants in ABCB1 and CYP2C19 and cardiovascular outcomes after treatment with clopidogrel and prasugrel in the TRITON-TIMI 38 trial: a pharmacogenetic analysis. Lancet. 2010 Oct 16;376(9749):1312–9.
52. Luo M, Li J, Xu X, Sun X, Sheng W. ABCB1 C3435T polymorphism and risk of adverse clinical events in clopidogrel treated patients: a meta-analysis. Thromb Res. 2012;129(6):754–9.
53. Su J, Xu J, Li X, Zhang H, Hu J, Fang R, Chen X. ABCB1 C3435T polymorphism and response to clopidogrel treatment in coronary artery disease (CAD) patients: a meta-analysis. PLoS One. 2012;7(10):e46366.
54. Brandt JT, Close SL, Iturria SJ, Payne CD, Farid NA, Ernest CS 2nd, Lachno DR, Salazar D, Winters KJ. Common polymorphisms of CYP2C19 and CYP2C9 affect the pharmacokinetic and pharmacodynamic response to clopidogrel but not prasugrel. J Thromb Haemost. 2007 Dec;5(12):2429–36.

55. Mega JL, Close SL, Wiviott SD, Shen L, Hockett RD, Brandt JT, Walker JR, Antman EM, Macias WL, Braunwald E, Sabatine MS. Cytochrome P450 genetic polymorphisms and the response to prasugrel: relationship to pharmacokinetic, pharmacodynamic, and clinical outcomes. Circulation. 2009 May 19;119(19):2553–60.
56. Franken CC, Kaiser AF, Krüger JC, Overbeck K, Mügge A, Neubauer H. Cytochrome P450 2B6 and 2C9 genotype polymorphism—a possible cause of prasugrel low responsiveness. Thromb Haemost. 2013;110(1):131–40.
57. Cuisset T, Loosveld M, Morange PE, Quilici J, Moro PJ, Saut N, Gaborit B, Castelli C, Beguin S, Grosdidier C, Fourcade L, Bonnet JL, Alessi MC. CYP2C19*2 and *17 alleles have a significant impact on platelet response and bleeding risk in patients treated with prasugrel after acute coronary syndrome. JACC Cardiovasc Interv. 2012;5(12):1280–7.
58. Xiang Q, Cui Y, Zhao X, Zhao N. Identification of PEAR1 SNPs and their influences on the variation in prasugrel pharmacodynamics. Pharmacogenomics. 2013;14(10):1179–89.
59. Varenhorst C, Eriksson N, Johansson Å, Barratt BJ, Hagström E, Åkerblom A, Syvänen AC, Becker RC, James SK, Katus HA, Husted S, Steg PG, Siegbahn A, Voora D, Teng R, Storey RF, Wallentin L, PLATO Investigators. Effect of genetic variations on ticagrelor plasma levels and clinical outcomes. Eur Heart J. 2015;36(29):1901–12.
60. Moßhammer D, Schaeffeler E, Schwab M, Mörike K. Mechanisms and assessment of statin-related muscular adverse effects. Br J Clin Pharmacol. 2014 Sep;78(3):454–66.
61. Nies AT, Niemi M, Burk O, Winter S, Zanger UM, Stieger B, Schwab M, Schaeffeler E. Genetics is a major determinant of expression of the human hepatic uptake transporter OATP1B1, but not of OATP1B3 and OATP2B1. Genome Med. 2013;5(1):1.
62. Zanger UM, Schwab M. Cytochrome P450 enzymes in drug metabolism: regulation of gene expression, enzyme activities, and impact of genetic variation. Pharmacol Ther. 2013;138 (1):103–41.
63. Scott SA, Sangkuhl K, Shuldiner AR, Hulot JS, Thorn CF, Altman RB, Klein TE. PharmGKB summary: very important pharmacogene information for cytochrome P450, family 2, subfamily C, polypeptide 19. Pharmacogenet Genomics. 2012;22(2):159–65.
64. Zabalza M, Subirana I, Sala J, Lluis-Ganella C, Lucas G, Tomás M, Masiá R, Marrugat J, Brugada R, Elosua R. Meta-analyses of the association between cytochrome CYP2C19 loss- and gain-of-function polymorphisms and cardiovascular outcomes in patients with coronary artery disease treated with clopidogrel. Heart. 2012 Jan;98(2):100–8.
65. Scott SA, Sangkuhl K, Stein CM, Hulot JS, Mega JL, Roden DM, Klein TE, Sabatine MS, Johnson JA, Shuldiner AR, Clinical Pharmacogenetics Implementation Consortium. Clinical Pharmacogenetics Implementation Consortium guidelines for CYP2C19 genotype and clopidogrel therapy: 2013 update. Clin Pharmacol Ther. 2013;94(3):317–23.
66. Schwab M, Schaeffeler E. Warfarin pharmacogenetics meets clinical use. Blood. 2011 Sep 15;118(11):2938–9.
67. Droppa M, Tschernow D, Müller KA, Tavlaki E, Karathanos A, Stimpfle F, Schaeffeler E, Schwab M, Tolios A, Siller-Matula JM, Gawaz M, Geisler T. Evaluation of clinical risk factors to predict high on-treatment platelet reactivity and outcome in patients with stable coronary artery disease (PREDICT-STABLE). PLoS One. 2015 Mar 23;10(3):e0121620.
68. Roberts JD, Wells GA, Le May MR, Labinaz M, Glover C, Froeschl M, Dick A, Marquis JF, O'Brien E, Goncalves S, Druce I, Stewart A, Gollob MH, So DY. Point-of-care genetic testing for personalisation of antiplatelet treatment (RAPID GENE): a prospective, randomised, proof-of-concept trial. Lancet. 2012 May 5;379(9827):1705–11.
69. Stimpfle F, Karathanos A, Droppa M, Metzger J, Rath D, Müller K, Tavlaki E, Schäffeler E, Winter S, Schwab M, Gawaz M, Geisler T. Impact of point-of-care testing for CYP2C19 on platelet inhibition in patients with acute coronary syndrome and early dual antiplatelet therapy in the emergency setting. Thromb Res. 2014;134(1):105–10.
70. Shuldiner AR, Palmer K, Pakyz RE, Alestock TD, Maloney KA, O'Neill C, Bhatty S, Schub J, Overby CL, Horenstein RB, Pollin TI, Kelemen MD, Beitelshees AL, Robinson SW, Blitzer MG, McArdle PF, Brown L, Jeng LJ, Zhao RY, Ambulos N, Vesely MR. Implementation of

pharmacogenetics: the University of Maryland Personalized Anti-platelet Pharmacogenetics Program. Am J Med Genet C Semin Med Genet. 2014 Mar;166C(1):76–84.
71. Tarkiainen EK, Backman JT, Neuvonen M, Neuvonen PJ, Schwab M, Niemi M. Carboxylesterase 1 polymorphism impairs oseltamivir bioactivation in humans. Clin Pharmacol Ther. 2012 Jul;92(1):68–71.
72. Zakim D, Schwab M. Data collection as a barrier to personalized medicine. Trends Pharmacol Sci. 2015;36(2):68–71.

Correction to: Platelets, Haemostasis and Inflammation

Andreas Zirlik, Christoph Bode, and Meinrad Gawaz

Correction to:
A. Zirlik et al. (eds.), *Platelets, Haemostasis and Inflammation*, Cardiac and Vascular Biology 5, https://doi.org/10.1007/978-3-319-66224-4

The original version of this volume was revised as it was originally published unnumbered. The revised version now has been numbered; numbering is done following the order of appearance in the book series Cardiac and Vascular Biology.

The updated online version of the book can be found at https://doi.org/10.1007/978-3-319-66224-4

A. Zirlik et al. (eds.), *Platelets, Haemostasis and Inflammation*,
Cardiac and Vascular Biology 5, https://doi.org/10.1007/978-3-319-66224-4_18

The manufacturer's authorised representative in the EU is Springer
Nature Customer Service Centre GmbH, Europaplatz 3, 69115 Heidelberg,
Germany. If you have any concerns regarding our products, please
contact ProductSafety@springernature.com

Printed and bound by CPI Group (UK) Ltd, Croydon, CR0 4YY
01/10/2026
02225866-0007